Springer Series on Environmental Management

Robert S. DeSanto, Series Editor

Karin E. Limburg   Mary Ann Moran
William H. McDowell

The Hudson River Ecosystem

With Contributions by
Janet M. Buckley, Edward H. Buckley, Dooley S. Kiefer,
and Peter S. Walczak

With a Foreword by Simon A. Levin

Springer-Verlag
New York Berlin Heidelberg Tokyo

Karin E. Limburg
Ecosystems Research Center
Cornell University
Ithaca, New York 14853
U.S.A.

Mary Ann Moran
Department of Microbiology
University of Georgia
Athens, Georgia 30602
U.S.A.

William H. McDowell
SURCO
State University of New York
Oswego, New York 13126
U.S.A.

Library of Congress Cataloging-in-Publication Data
Limburg, K.E. (Karin E.)
The Hudson River ecosystem.
(Springer series on environmental management)
Bibliography: p.
Includes index.
1. Stream ecology—Hudson River (N.Y. and N.J.)
2. Water—Pollution—Environmental aspects—Hudson River (N.Y. and N.J.) I. Moran, M. A. (Mary Ann)
II. McDowell, W. H. (William H.) III. Title.
IV. Series.
QH104.5.H83L56 1985   574.5'26323'097473   85-22167

© 1986 by Springer-Verlag New York Inc.
All rights reserved. No part of this book may be translated or reproduced in any form without written permission from Springer-Verlag, 175 Fifth Avenue, New York, New York 10010, U.S.A.
The use of general descriptive names, trade names, trademarks, etc. in this publication, even if the former are not especially identified, is not to be taken as a sign that such names, as understood by the Trade Marks and Merchandise Marks Act, may accordingly be used freely by anyone.

Media conversion by Bi-Comp, Incorporated, York, Pennsylvania.
Printed and bound by R.R. Donnelley & Sons, Harrisonburg, Virginia.
Printed in the United States of America.

9 8 7 6 5 4 3 2 1

ISBN 0-387-96220-4 Springer-Verlag New York Berlin Heidelberg Tokyo
ISBN 3-540-96220-4 Springer-Verlag Berlin Heidelberg New York Tokyo

Foreword

The Ecosystems Research Center (ERC) was established at Cornell University in October 1980 by the Environmental Protection Agency (EPA) with the goals of:

1. Identifying fundamental principles and concepts of ecosystems science and the determination of their importance in understanding and predicting the responses of ecosystems to stress, the description of the basic mechanisms that operate within ecosystems, and an examination of the stability of ecosystem structure and function in the face of stress.

2. Testing the applicability of those theoretical concepts to problems of concern to the EPA through a consideration of retrospective and other case studies.

In line with these goals, the Hudson River ecosystem provided the basis for the first major retrospective study undertaken by the ERC. The goal of the project was to develop recommendations concerning how ecosystem monitoring can and should be carried out in support of EPA's regulatory responsibilities. Our hope was and is that the experience gained from this study will be broadly applicable to a range of management problems involving estuarine ecosystems, and will lead to more effective regulation.

As an initial effort, a report (Baslow and Logan, 1982) summarizing much of the utility literature on the Hudson was prepared as a resource document for the ERC. Furthermore, the present document is complemented by the report "Principles for Estuarine Impact Assessment: Lessons Learned from the Hudson River and Other Estuarine Experiences" by Karin E. Limburg, Christine C. Harwell, and Simon A. Levin (Ecosystems Research Center, Cornell University, September 1984). The focus of that document is upon the regulatory and legislative issues surrounding estuarine management, with emphasis upon the role of the scientist, the problems associated with measuring ecosystem health, and the evaluative tools available for assessing ecological effects. The present document is more specifically directed to the Hudson River. It provides a physical, chemical, and biological description of the river, including a synopsis of our present understanding of ecosystem processes and a survey of studies that have been carried out. The study summarizes the present extent of the Hudson River data base, drawing from the scientific literature and utility company studies.

We hope that these reports will serve as a guide to those interested in the Hudson River ecosystem and the anthropogenic stresses it receives. The Hudson is a prototype of a large, multiple-use estuarine ecosystem subject to multiple stresses. It is also one of the great natural treasures of the United States, and the subject of a rich, diverse, and colorful literature and tradition.

We dedicate our efforts to the memory of our friend and colleague Tibor Polgar, First Chairman of the Panel of the Hudson River Foundation. Tibor died suddenly on January 21, 1985 at the age of 41. Through his efforts with the Hudson River Foundation, he became one of the major spokesmen for the preservation of the resources of the Hudson River. He will be sorely missed.

Simon A. Levin
Cornell University
Ithaca, New York

Series Preface

This series is dedicated to serving the growing community of scholars and practitioners concerned with the principles and applications of environmental management. Each volume is a thorough treatment of a specific topic of importance for proper management practices. A fundamental objective of these books is to help the reader discern and implement man's stewardship of our environment and the world's renewable resources. For we must strive to understand the relationship between man and nature, act to bring harmony to it, and nurture an environment that is both stable and productive.

These objectives have often eluded us because the pursuit of other individual and societal goals has diverted us from a course of living in balance with the environment. At times, therefore, the environmental manager may have to exert restrictive control, which is usually best applied to man, not nature. Attempts to alter or harness nature have often failed or backfired, as exemplified by the results of imprudent use of herbicides, fertilizers, water, and other agents.

Each book in this series will shed light on the fundamental and applied aspects of environmental management. It is hoped that each will help solve a practical and serious environmental problem.

Robert S. DeSanto
East Lyme, Connecticut

Acknowledgments

The task of reviewing the scientific assessments of the Hudson River has been an awesome one. It was necessary—in fact essential—to seek the advice of highly experienced individuals, all of whom very willingly obliged us and provided a great deal of help. We wish to extend wholehearted thanks to these people: Lawrence Barnthouse, Robert Boyle, Mark Brown, Sigurd Christensesn, John Sanders, Ross Sandler, Douglas Sheppard, and Ronald Sloan. Special thanks are due to Edward Ted and Janet Buckley, who, in addition to providing a great deal of literature and advice, developed the extensive PCB bibliography included in Section B of the Bibliography.

The support, comments, and help of the staff of the Ecosystems Research Center is also gratefully acknowledged, particularly that of Simon Levin as coordinator of the Hudson project; Dooley S. Kiefer, who coauthored Chapter 5 on the Westway; Peter S. Walczak, who painstakingly assembled much of the bibliography; John Beatty, who did background research on estuarine impact assessment models; Jack Kelly for extensive review of Chapter 2; and Carin Rundle, Hanna Barker, and Bertie Sardo, who processed and reprocessed our words ad nauseam.

Finally, we wish to acknowledge the community of ecosystems ecologists and resource managers we have learned from and who have taught us to "think big," but not to ignore the details.

Contents

1. **Introduction**
   Mary Ann Moran and Karin E. Limburg   1

2. **The Hudson River Ecosystem**
   Mary Ann Moran and Karin E. Limburg   6

3. **Power Plant Operation on the Hudson River**
   William H. McDowell   40

4. **PCBs in the Hudson**
   Karin E. Limburg   83

5. **The Westway**
   Mary Ann Moran and Dooley S. Kiefer   131

6. **Synthesis and Evaluation**
   Karin E. Limburg and Mary Ann Moran   155

7. **Hudson River Data Base**
   Mary Ann Moran, Peter S. Walczak, and Karin E. Limburg   173

   Hudson River Bibliography
   Peter S. Walczak, Mary Ann Moran, Janet M. Buckley, Edward H. Buckley, and Karin E. Limburg   241

   Index   327

Contents of Bibliography

Bibliography A: Utility-Sponsored Literature	241
Bibliography B: Open Literature	262
I. Organisms	262
A. Phytoplankton	262
B. Macrophytes	265
C. Invertebrates	266
D. Birds, Reptiles, and Amphibians	269
E. Fish	269
1. Striped Bass	269
2. White Perch	273
3. Atlantic Tomcod	274
4. Shad and Blueback Herring	274
5. Sturgeon	276
6. Fish—General	277
F. Diseases and Parasites	281
G. Entrainment and Impingement	282
II. Physical and Chemical Characteristics	285
A. Geology, Hydrology	285
B. Temperature	290
C. Water Chemistry	290
D. Salinity	292
E. Water Quality	292
III. Pollutants	295
A. Sewage	295
B. Polychlorinated Biphenyls	296

	C. Radionuclides	309
	D. Other Toxic Substances	312
IV.	General Ecological Studies and Surveys	317
V.	Regulations and Impact Statements	318
VI.	Bibliographies	321
VII.	General Papers	322
VIII.	Non-Hudson	324

1
Introduction

MARY ANN MORAN and KARIN E. LIMBURG

Objectives

A major objective of this book is the documentation and retrospective examination of the most recent, major, human-induced impacts on the Hudson River, with special attention given to the procedure known as environmental impact assessment (EIA). Specifically, we shall focus on the role that science and scientists have played in the EIA process.

Our examination covers three conflicts: 1) the uptake and discharge of river water for cooling power generating stations located along the middle reach of the estuary; 2) polychlorinated biphenyls' (PCBs) release and fate in the ecosystem, and PCB cleanup procedures; and 3) the proposed construction of an interstate highway (the Westway) on a landfill that would claim 200 acres of the estuary at the lower end of Manhattan. Of interest in our review of the EIA process are the points at which science entered into the assessment of Hudson River alterations and the role scientists played during litigation that followed on the heels of impact assessment.

The Hudson River issues serve as a valuable forum for another objective of this report: an evaluation of the utility of EIA from an ecosystem perspective. We wish to affirm the importance of this perspective with regard to estuarine management, and in particular to impact assessments of estuaries. The ecosystem perspective is defined here as an ecological problem-solving approach that structurally and functionally connects biotic and abiotic components of a system. It includes population- and community-level approaches as well.

The ecosystem approach is essentially a framework within which to organize knowledge of the habitat, or system, its abiotic and biological components, and their interactions both internally and with other systems. The approach is useful for several reasons: first, it allows the problem at hand to define the system to be studied; second, it delineates interconnections that might otherwise go unnoticed, with potentially unfortunate consequences; third, it provides a context in which to study the interactions of individual organisms, populations, and communities; and fourth, it provides methods for evaluations of large-scale, integrative properties of the system that would not necessarily be evident by summing properties of the individuals.

A final objective of our Hudson River case study is the compilation of Hudson data bases into a "baseline" resource document for future river studies. Data collection efforts on the Hudson have been monumental in recent years, due mainly to utility company investigations into power plant impacts and, to a lesser extent, to tracing the dispersal of PCBs throughout the upper river and estuary. Our acquired familiarity with utility company data bases is discussed in the closing chapter of this report. A review of Hudson River open literature is also included, as well as an extensive Hudson bibliography of scientific research.

In our review of the Hudson cases, we examine the impact assessment process and ultimately attempt to answer the following questions:

Have the right questions been asked? Have appropriate aspects of the river/estuarine ecosystem been emphasized, and if so, have the data collected been proven adequate for the estimation of the impacts under consideration?

If not, what could have been done differently to provide an adequate estimation?

Was the EIA work subjected to continual peer review, rather than solely after the fact for publication purposes? (Or, was the work ever reviewed at all?)

Did the scientific work carry any regulatory clout? If adverse impacts were predicted, was the regulatory agency of concern able to alter the design of the proposed project to minimize effects?

What is to be done now with the collected data, and how can they best be complemented in future monitoring/assessment studies?

The Hudson, Human Impacts, and the Development of Environmental Policy

The Hudson ecosystem has been heavily influenced by humans long before environmental impacts were ever perceived, let alone understood. The dumping of refuse from the streets and houses of New York was a recognized health problem in the nineteenth century, and doomed the once prosperous oyster fishery (Franz, 1982). Changing patterns of land use exerted more subtle but perhaps more significant changes to the ecosystem; for example, the construction of railways along both shores of the river diminished the amount of wetland available to organisms and at the same time brought more people and more development to the river's edge. Both agriculture and industry radically altered the nature of the landscape and, in turn, the inputs delivered to the Hudson and its tributaries.

With the exception of its pristine upper reaches, the Hudson River today still serves in part as a major waste disposal system for the eastern region of New York State. Along with the river's service as a transport and food source, this waste-receiving capacity was instrumental in settlement and, more importantly later on, development. Cumulative uses and abuses have had their effects over the years. The once luxuriant estuarine flora and fauna have been adversely affected by long-term pollution. Recreational activities on the lower Hudson had all but ceased by the early 1960s.

Over the past 20 years, new and growing awareness of the Hudson's value as an environmental resource has sparked concern over continual human abuse. Locally based citizen groups, born during the days of ambitious expansion of powershed development, have been the single most important force in bringing environmental issues to the forefront; today they form a solid base of public opinion that exerts strong influence in determining policy for the river's future.

Not only the environmental organizations, but many other parties as well, have participated in the Hudson's convoluted history of management. Environmental issues began as face-offs between conservationists, industry, and regulatory bodies. In short time, lawyers and scientists became involved, and as they did so, existing rules and regulations for development came under question. Similar events were occurring in other parts of the nation, and altogether these snowballed into Congress as a general mandate for some sort of national environmental approach.

National Environmental Policy Act and Impact Assessment: Brief History of Inception and Success

Two pieces of federal legislation have been featured in Hudson River environmental conflicts. The first was the 1899 Rivers and Harbors Act, which was resurrected by Hudson valley conservationists to fight industrial dumping into the river. This was essentially superseded by passage of the Federal Water Pollution Control Act as amended in 1972 (also known as the Clean Water Act or CWA). The CWA replaced sections of the Rivers and Harbors Act pertaining to discharge of pollutants and dredge permits.

The CWA was developed in the same spirit of national environmental concern that gave rise to the National Environmental Policy Act (NEPA) in 1969. NEPA formalized the federal government's stance on protection of the natural environment against the rising tide of development. As a part of this comprehensive statement was the requirement that a "detailed statement" regarding "major federal actions significantly affecting the quality of the human environment (Section 102(2)(c))" be included. Thus was born the notion of environmental impact assessment or EIA.

Nearly 15 years after NEPA's passage and thousands of EIAs later, the environmental review process has come under close examination more than a few times (e.g., Anderson, 1973; Kennedy and Hanshaw, 1974; Kibby and Glass, 1980; Rees and Davis, 1978; Rosenberg et al., 1981). The main conclusions have been that EIAs, while useful in identifying areas susceptible to damage from proposed activities as well as bringing in outside parties into the evaluation process, are not terribly effective at halting projects with known adverse consequences. Lack of success in impact assessment has been attributed to undervaluation of environmental resources in economic-based approaches and to the inability of EIAs to address cumulative ecological effects on a regional basis (Rees and Davis, 1978).

Other problems have afflicted the evaluation process. Consulting firms, kept in business by preparing EIAs, may be troubled by client review prior to release of their reports and therefore pushed into a biased position (Hanson, 1976); lack of scientific credibility and outdated scientific tools and understanding may further hamper their effectiveness (Schindler, 1976). The short time frames usually given for impact assessments limit their success in evaluating all but the gross and obvious problems (Lewis, 1980). Finally, although the EIA process itself is required, protection of a threatened resource (except endangered species) is not. As a result, EIAs ultimately have no teeth.

Why, then, does this report deal so extensively with impact assessment? The answer is that the EIA has become institutionalized as the primary paradigm for addressing questions of major environmental alteration. By reviewing the experiences with EIA work in the Hudson region, we hope that specific strengths and weaknesses will be highlighted. Furthermore, such a review can lead to suggestions for improvement in the way to conduct assessments and, perhaps, can avoid adding to past mistakes.

Organization of this Book

This book is divided into seven chapters. Chapter 1 is followed by a presentation of the physical, chemical, and biological characteristics of the Hudson ecosystem. The three succeeding chapters recount the histories of the impacts chosen for study: Chapter 3 describes the power plant controversies, Chapter 4 relates the events and assessments of PCB pollution, and Chapter 5 discusses the issue of dredging and filling about lower Manhattan for the Westway project. A summary and critical synthesis is presented in Chapter 6. Finally, Chapter 7 and the remaining Appendices summarize the Hudson data bases and contain a review and bibliography of scientific literature pertaining to the Hudson.

One last point to be mentioned is that, like old soldiers, environmental issues do not seem to die. Of the three conflicts described in this report, only one, that of the power plants, has been decisively settled. Even that issue will be reborn, however, when the formal settlement between the utility companies and the U.S. Environmental Protection Agency expires in 1990. The process of making decisions is ultimately a political one. Good science can provide an important base of knowledge from which to make those decisions, and is thus valuable as a regulatory tool. We hope the reader will bear this in mind as he or she follows the science—good and bad—that documented the impacts on the Hudson.

References

Anderson, F.R. 1973. NEPA in the courts. Resources for the Future, Inc. Johns Hopkins University Press, Baltimore, MD. 324 pp.

Franz, D.R. 1982. An historical perspective on molluscs in Lower New York Harbor, with emphasis on oysters, pp. 181–197 In Ecological Stress and the New York Bight: Science and Management (G.F. Mayer, ed.). Estuarine Research Federation, Columbia, SC.

Hanson, C.H. 1976. Commentary—ethics in the business of science. Ecology 57:627–628.

Kennedy, W.V. and B.B. Hanshaw. 1974. The effectiveness of impact statements: The U.S. Environmental Policy Act of 1969. Ekistics 218:19–22.

Kibby, H. and N. Glass. 1980. Evaluating the evaluations: A review perspective on environmental impact assessment, pp. 40–48 In Biological Evaluation of Environmental Impacts: The Proceedings of a Symposium. Council on Environmental Quality and U.S. Dept. of Interior, Fish and Wildlife Service. FWS/OBS-80/26.

Lewis, J.R. 1980. Options and problems in environmental management and evaluation. Helgol. Meeresunters. 33:452–466.

Rees, W.E. and H.C. Davis. 1978. Coastal ecosystem planning and impact evaluation. University of British Columbia School of Community and Regional Planning. (Draft mss.)

Rosenberg, D.M., V.H. Resh, S.S. Balling, M.A. Barnaby, J.N. Collins, D.V. Durbin, T.S. Flynn, D.D. Hart, G.A. Lamberti, E.P. McElravy, J.R. Wood, T.E. Blank, D.M. Schultz, D.L. Marrin, and D.G. Price. 1981. Recent trends in environmental impact assessment. Can. J. Fish. Aquat. Sci. 38:591–624.

Schindler, D.W. 1976. The impact statement boondoggle. Science 192:509.

2
The Hudson River Ecosystem

MARY ANN MORAN and KARIN E. LIMBURG

Geographic Setting

The Hudson River springs from Lake Tear of the Clouds in the Adirondack Mountains of northern New York State. From its source, the river flows in a southerly direction for about 315 river miles (RM) (507 km) to the Battery, New York City, where it discharges into upper New York Bay (Figure 2.1). The Mohawk River, the largest tributary of the Hudson River, flows in a generally east-southeast direction to its junction with the Hudson River just north of Albany.

The Hudson drains a total of 13,390 square miles (34,680 km^2) in eastern and northern New York State and parts of Vermont, Massachusetts, Connecticut, and New Jersey (LMS, 1976). The basin can be divided into three principal drainage areas: the upper Hudson from Mt. Marcy to Troy, the Mohawk from Rome to Troy, and the lower Hudson from Troy to New York Bay. The upper Hudson and Mohawk drainage basins contain fresh water; the lower Hudson is estuarine.

Physical and Chemical Aspects of the Ecosystem

Morphometric and Hydrologic Characteristics

The gradient of the upper Hudson channel averages 1 m/mi (0.62 m/km) between Troy and Fort Edward; north of Fort Edward to the Sacandaga River, the gradient steepens to approximately 5 m/mi (3.1 m/km)

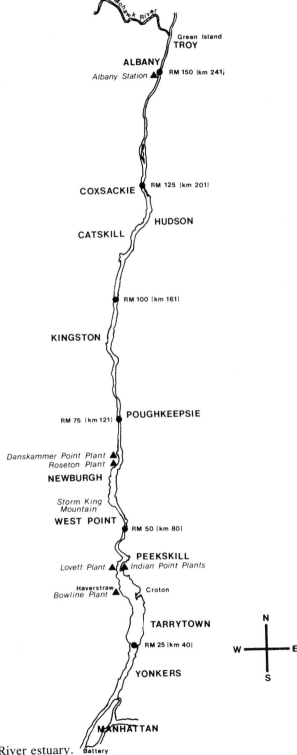

Figure 2.1. The Hudson River estuary.

(Sanders, 1982). In contrast, the lower Hudson is a drowned river valley rising only 1.5 m in the 150 miles between New York City and Troy (0.006 m/km) (Helsinger and Friedman, 1982). The channel of the lower Hudson runs straight north and south with few meanders. Three wide bays are found in the lower reaches: Tappan Zee (4 km wide), Haverstraw (4.8 km wide), and Newburgh (1.6 km wide) (Schureman, 1934). Elsewhere, the river is generally less than 1.5 km in width (Hudson River Policy Committee, 1969).

In order to maintain a shipping route to Albany, the Hudson channel is kept at a minimum depth of 9 to 11 m (McFadden *et al.*, 1978) by the U.S. Army Corps of Engineers. Many portions of the river are much deeper, particularly around the area of the Hudson Highlands from Newburgh-Beacon down past West Point. A maximum sounding of 66 m (216 ft) was reported at West Point in 1934 by Schureman (1934). The volume of the Hudson is greatest in the Haverstraw Bay area (RM 25 to 40; km 40 to 64) and decreases toward the head of the estuary at Troy (Figure 2.2). Approximately 20 km^2 of wetlands are contiguous with the Hudson. A little more than half of this is marshes and wooded swamps; the remainder consists of mud flats inundated during high tides. Wetlands are in greatest abundance in the upper third of the estuary (Figure 2.2).

The Hudson River is tidally influenced from the Battery at New York City to the Federal Dam at Troy, N.Y. (153.4 river miles). The estuary receives salt water from upper New York Bay during the flood phase of a tidal cycle, and there is a discharge of less saline water from the estuary into the bay during the ebb phase. The mean tidal flow varies from 425,000 cfs (12,040 m^3/s) at the Battery to zero at the Federal Dam at Troy. This substantial tidal flow can be 10 to 100 times greater than freshwater flow (Texas Instruments, 1979a). The mean ebb current velocity is 0.4 m/s, and the mean tidal flood current velocity is 0.36 m/s (LMS, 1978a). During each tidal cycle of 24 hours and 50 minutes (Stewart, 1958), two high tides and two low tides occur in the Hudson River, producing an average tidal range of 1.4 m at the Battery, 0.8 m at West Point, and 1.4 m at the Troy Dam (Darmer, 1969). Tidal flow can be temporarily suppressed when extreme north or south winds persist (Busby and Darmer, 1970).

The flow of fresh water in the Hudson River follows a typical seasonal pattern for temperate climates, with the highest flow during the spring and lowest flow during late summer/early fall (Figure 2.3). Approximately 80% of the fresh water enters the river above Troy. Most of the remainder joins the Hudson from tributaries in the upper reaches of the estuary (McFadden *et al.*, 1977). Regulated releases from Sacandaga Reservoir supplement the freshwater flow at Green Island, just above Troy. The annual average (1946–1980) freshwater flow at Green Island is 13,820 cfs (392 m^3/s) (USGS, 1981), which corresponds to an estimated lower Hudson freshwater flow of approximately 19,000 to 20,000 cfs (538 to 567 m^3/s)

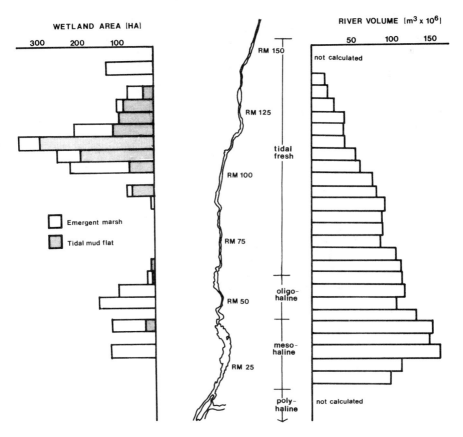

Figure 2.2. Hudson estuary river volume (Texas Instruments, 1977), salinity zones (Ristich et al., 1977), and wetland areas.

(Central Hudson Gas and Electric, 1977). Below Manhattan, additional fresh water enters with New York City sewage and the Hackensack, Passaic, and Raritan Rivers (Deck, 1981). Severe droughts occurred in the Hudson in the early 1960s, in 1980, and again in 1985; and high flow conditions prevailed during the early 1970s. Between 1971 and 1976, above-normal rainfall resulted in a number of record-setting freshwater flows (Central Hudson Gas and Electric, 1977) (Figure 2.3).

The ratio of water volume to mean annual freshwater flow, a rough estimate of flushing time, is 0.35 years (126 days) in the Hudson estuary. Thus the Hudson is flushed, on the average, faster than many other large east coast estuaries (e.g., Delaware, Potomac, Chesapeake) (Simpson et al., 1974). Flushing time varies substantially from month to month as well as by reach of the estuary. In the portion of the river between the George Washington Bridge (RM 11) and Poughkeepsie (RM 76), the ratio of vol-

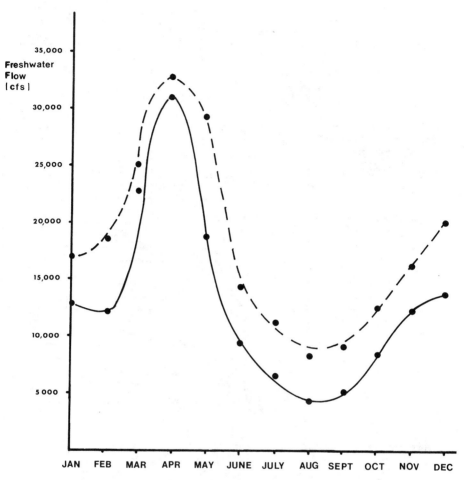

Figure 2.3. Mean monthly fresh water flow in the Hudson River, gauged at Green Island. The solid line represents the longterm average, 1918–1980. The broken line shows the above average freshwater flow during the period from 1971–1976 (Central Hudson Gas and Electric), 1977; USGS 1977–1981).

ume to mean freshwater flow is approximately one month. During May and June, however, the ratio can drop to less than 10 days (Simpson *et al.*, 1974).

Salinity

The Hudson estuary can be divided into four salinity zones based on salt concentration: polyhaline (18 to 30%), mesohaline (5 to 18%), oligohaline (0.5 to 5%), and tidal fresh (0.5%). The location of these zones will vary both daily and seasonally, depending on the tidal surge of saline water from the ocean and the magnitude of freshwater flow into the upper estuary. Under an average runoff regime, the salinity intrusion (0.5%) reaches West Point or Newburgh, 50 to 60 miles above the Battery. During conditions of high freshwater runoff, typically during spring months, the salt water in the Hudson can be pushed as far south as the Bronx, at RM 15. Summer drought conditions in 1965 and 1966 allowed salt water to reach the Kingston area, approximately 100 miles upriver (Buckley, 1971). In general, then, seasonal patterns in freshwater flow cause saline water to proceed further upriver in the summer and early fall than in the winter and spring (Baslow and Logan, 1982). Figure 2.4 shows longitudinal conductivity patterns from May to November, from which it is possible to see the movement of the salt front. Approximate late summer locations of the salinity zones are shown in Figure 2.2 (Ristich et al., 1977).

Vertical gradients of salinity (measured as chloride concentration) show that salt in the Hudson is generally well mixed with fresh water during low flow conditions. Only a 10% increase is found, on the average, from the top of the water column to the bottom layers. Under high flow conditions, fresh water overrides the salt water layer, and salinity differences of up to 20% can be established (Busby and Darmer, 1970).

River depth is also important to vertical salinity gradients. A critical region for freshwater and salt water mixing occurs near the southern boundary of the Hudson Highlands (RM 40 to 44), where the channel suddenly deepens from approximately 10 to 12 m to an average depth of nearly 30 m. Following high flow periods and the establishment of salinity gradients, such channel irregularities and holes create water turbulence and act to promote vertical mixing. A sharp salinity gradient (or halocline) exists at RM 25 (Piermont), indicating distinct salt and flow layers; the absence of a strong gradient near RM 41 (Tomkins Cove) (Figure 2.5),

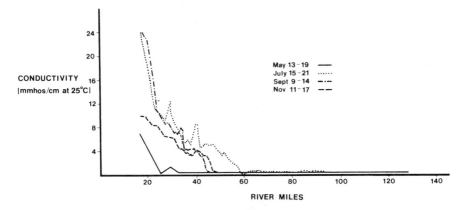

Figure 2.4. Longitudinal Hudson River conductivity measurements for late spring through fall, 1974 (data from Texas Instruments, 1977).

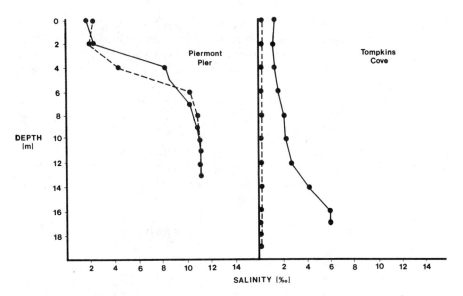

Figure 2.5. July 1974 vertical salinity profiles at Piermont Pier (RM 25) and Tompkins Cove (RM 41) during high-water slack (solid line) and low-water slack (dashed line) (data from Simpson et al., 1974).

however, indicates relatively complete mixing in this deeper zone. Depth further influences salinity in that salt water tends to move at higher velocities in the deeper areas and along the channel bottom during intrusion. The parallel shallow areas thus receive less salt water, have delayed mixing, and experience reduced ranges in salinity (Texas Instruments, 1976a).

Temperature

Five years of data at the Green Island gauging station show that the temperature of fresh water entering the estuary ranged between a January minimum of 0°C (32°F) and a July maximum of 27°C (81°F) (USGS, 1977–1981). In the spring and summer, temperature decreases toward New York City, as colder ocean water enters the river with the tides. By the late fall and winter, the horizontal temperature gradient has reversed and the temperature increases from the source to the mouth, since oceanic waters cool to a lesser extent than shallow sources of fresh water (McFadden et al., 1977). The magnitude of temperature differences at any one time between the upper and lower reaches of the channel can approach 11°C (20°F) (Abood et al., 1976).

Average temperatures in the estuary generally follow the mean air temperature. Lowest average temperatures are 1°C (34°F) to 2°C (36°F), and occur in January. Highest average temperatures may reach 25°C (77°F) during July or August in the main channel or up to 30°C (86°F) in areas along the shore (Texas Instruments, 1979a).

Dissolved Oxygen

During summer, dissolved oxygen (DO) values show undersaturation throughout much of the estuary; there is a particularly sharp reduction as water flows through the New York City area, and this has been related to biological oxygen demand (BOD) associated with sewage input. McFadden *et al.* (1977) reported a general drop in DO levels below Albany, a recovery and peak near Saugerties (RM 100), relatively high levels south to Croton Point (RM 35), and a decline again in the New York City region. Figure 2.6 shows dissolved oxygen levels expressed as percent of O_2 values in saturated water. There is considerable fluctuation with the seasons (Figure 2.6a). In general, the highest DO levels occur during the late winter/early spring months when the river is coolest and least saline. As the growing season progresses, oxygen is depleted in some areas (Storm and Heffner, 1976). However, as can be seen in Figure 2.6b, DO levels have improved somewhat since the implementation of the Clean Water Act (CWA), particularly north of Kingston (RM 90).

pH, Alkalinity, and Hardness

In an early survey by Faigenbaum (1937), the pH of the Hudson ranged between 7.0 and 8.0. The only deviations from this range were found in a few productive coves and bays, where pH values above 8.0 were recorded. Forty years later, channel-long water quality surveys conducted by Texas Instruments (1979b) revealed a similar river profile. Most measurements were above 7.0, and some exceptions of slightly lower pH occurred during the late spring and summer. A New York State water quality survey performed in the mid-1960s for the entire river showed average pH values to range from 6.8 to 7.8 (NYSDEC, 1967). This constancy is most likely a result of large inputs of calcareous and magnesium carbonates in runoff from adjacent limestone bedrock (Texas Instruments, 1976a).

Alkalinity is used as a measure of the acid-neutralizing capacity of natural waters; in recent years, it has been an important indicator of the relative susceptibility of water bodies to acid deposition effects. Alkalinity data collected monthly at the Green Island gauging station (1976–1980) show that fresh water entering the estuary is well buffered, with values ranging between 29 and 71 mg/l $CaCO_3$. Hardness ranges between 49 and 93 mg/l. Both alkalinity and hardness show a slight decline during the early spring months when runoff is high (USGS, 1977–1980). River-long profiles compiled by McFadden *et al.* (1978) show that hardness and alkalinity increase towards the lower, saline end of the estuary. Thus, the

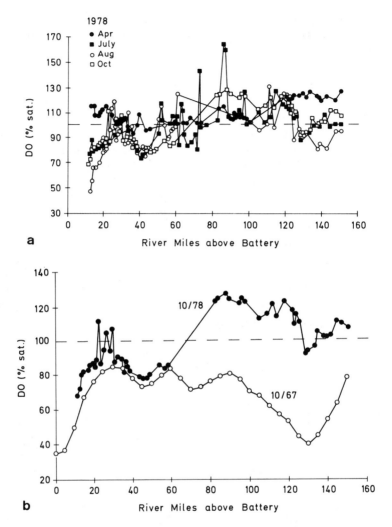

Figure 2.6. Hudson River-dissolved oxygen (DO) profiles, expressed as percentages of DO in saturated water. (a) Profiles for April, July, August, and October 1978. (b) Profiles comparing October 1967 with October 1978. Note in particular the severe "oxygen sag" downstream of Albany (RM 150) in 1967 (calculated with data from Texas Instruments 1979a; Abood *et al.*, 1976; Fleming *et al.*, 1976; and USGS 1969).

main stem of the Hudson, for its entire length, is relatively protected from adverse effects of acid deposition.

Major Pollutants

Since 1952, there has been a steady increase in the annual wastewater flow into the Hudson from municipal sources. In 1974, total municipal

Table 2.1. BOD inputs (metric tons/day) into the lower Hudson River.

	Industrial	Municipal
Lower Hudson (RM 14–152)	22.8[a]	55.5[a]
Manhattan area (RM 0–14)	7.6[b]	131.5[a]
Total	30.4	187.0

[a]From Hetling (1976).
[b]From McFadden et al. (1978).

wastewater input excluding the New York-New Jersey metropolitan area (RM 14 to 152) was 300 million gallons daily (MGD), including BOD loading of 56 metric tons/day (mt/day). Inputs from the metropolitan area totaled over 666 MGD and 132 mt/day BOD (Hetling, 1976) (Table 2.1).

Despite the greater volumes of wastewater flow, a general increase in Hudson River water quality has been evident in recent years. Peak BOD loading from municipal sources above New York City (86 mt/day) occurred in 1965 and has declined since then (Hetling, 1976) (Figure 2.7). The New York State Department of Environmental Conservation water quality monitoring program similarly indicates a general rise in quality of Hudson River water (1966–1974) (Mt. Pleasant and Bagley, 1975). Hetling (1976) suggests that the New York State Pure Waters Program, which subsidized construction of wastewater treatment facilities, is responsible

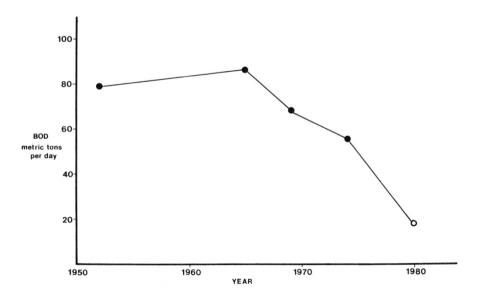

Figure 2.7. Trends in BOD loading to the lower Hudson River (above New York City) from municipal wastewater sources (Hetling, 1976). The 1980 value (open circle) was estimated in 1976 based on planned improvements in sewage treatment. No recent compilation of BOD loading has been undertaken.

for the improved water quality. The BOD load to the Hudson-Raritan estuary area (the Hudson River south of Poughkeepsie, plus the East River, New York Bay, and New Jersey rivers) has decreased 27% between 1974 and 1980, and is expected to decline an additional 50% if all planned treatment facilities are constructed (Mueller et al., 1982).

Industrial wastes are discharged to the Hudson either through the municipal wastewater systems or directly into the river through separate outflows. Although the 3134 MGD of direct industrial discharges amounted to over 10 times the volume of municipal discharges in 1974, over 95% of the former entered the estuary as cooling water (Hetling, 1976) and contributed little to the BOD input to the lower Hudson (McFadden et al., 1977) (Table 2.1). Industrial and agricultural wastes, however, can serve as a major source of pesticides, radionuclides, heavy metals, and other toxic chemicals, and many of these have not been well studied. Pollutants of known importance in the Hudson River ecosystem include polychlorinated biphenyls (PCBs), cadmium, and nickel. The release of PCBs, their distribution, and what is known of their effects on the biotic components of the Hudson are detailed in Chapter 4.

Municipal sewage sources above New York City release 9 mt/day of phosphorus (3.3×10^6 kg P/yr) and 20 mt/day of nitrogen (7.3×10^6 kg N/yr) into the Hudson (Hetling, 1976). In the Hudson-Raritan estuary, which covers a large area around the New York-New Jersey metropolitan area including the lower Hudson River, nitrogen loading has remained steady during the past decade at approximately 200 mt/day total N. Sewage treatment plants have been installed along the Hudson, but were not designed for nitrogen removal. Phosphorus loads, however, have decreased by 40% (from 45 to 27 mt/day total P in the Hudson-Raritan complex) due to better phosphorus removal (Mueller et al., 1982).

Nutrients

Phosphorus

Phosphorus sources to the Hudson estuary include natural components (primarily organic detritus), non-point source runoff, and the above-mentioned sewage point sources. Deck (1981) attributes approximately 17% (1.8 moles/s or 1.8×10^6 kg P/yr) of the phosphorus of the lower estuary to upstream (above RM 25) sources. Sewage and urban flows between RM 24 and the narrows (RM −6) account for the remaining inputs: 73% of the total (7.7×10^6 kg P/yr) from direct sewage inputs and 10% (1.1×10^6 kg P/yr) from indirect urban sources. These large inputs near the mouth of the Hudson are reflected in the steep increase in phosphorus levels near Yonkers (RM 19), compared to only about a 2-fold increase in phosphorus concentration from RM 200 to RM 50 (Figure 2.8).

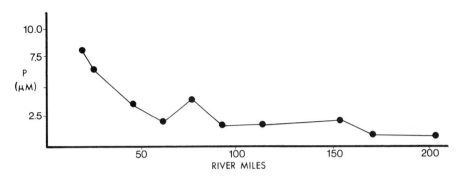

Figure 2.8. Total phosphorus concentration in Hudson River during August 1974. Each point represents a single water sample (USGS, 1975).

Phosphorus as phosphate is distributed in the lower estuary through a balance between sewage inputs and water movement (Simpson et al., 1975; Malone, 1977). This contrasts with other polluted estuarine systems in which biological uptake is a major controlling factor (Simpson et al., 1975). In the lower Hudson, algal uptake is small with respect to the enormous amount of available nutrients, and therefore biological activity plays a very minor role in determining phosphate concentrations (Deck, 1981). Conversely, macronutrients are not major constraints on biological productivity.

Because the major phosphate source (sewage) occurs near the mouth of the Hudson estuary, phosphate is introduced into the saline water and spread throughout the region of freshwater mixing. The minor effect of biological uptake results in a strong correlation between phosphate and salinity, which indicates near-conservative mixing behavior (Deck, 1981). Phosphate mixes upstream by tidal inflow and downstream by surface layer outflow (Figure 2.9) (Deck, 1981; Hammond, 1975). Low-flow phosphate concentrations are greater than high-flow concentrations; thus, summer is a critical period when the concentration of many introduced elements or compounds are likely to reach highest levels.

Nitrogen

Approximately 55 million kg of nitrogen enters the Hudson annually in the form of ammonia, nitrate, nitrite, and organic particulates (Table 2.2). The major source of ammonia comes from sewage inputs, particularly near the Albany and New York City urban areas (Figure 2.10; Table 2.2). Approximately one-half the total inorganic nitrogen added to the Hudson estuary can be traced to ammonia additions from sewage and urban runoff (Deck, 1981). Ammonium profiles of the lower estuary show both a peak in the area of greatest sewage effluent release and relatively conservative

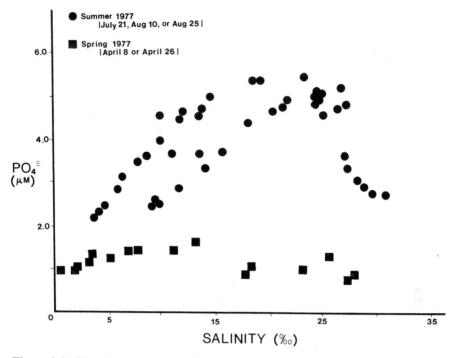

Figure 2.9. Phosphate concentrations as a function of salinity in the lower Hudson estuary under low flow (summer) and high flow (spring) conditions (redrawn from Deck, 1981).

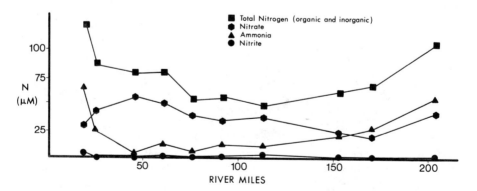

Figure 2.10. Ammonia, nitrate, nitrite and total nitrogen (unfiltered) concentrations in Hudson River water during August, 1974. Each point represents a single sample (USGS, 1975).

Table 2.2. Dissolved inorganic nitrogen inputs to the lower Hudson estuary (RM 20 to RM −6).[a]

	NH_4-N	NO_3-N	NO_2-N
Fresh water	0.58 (spring)	11.1	0.1–0.2
	0.33 (summer)	—	—
Sewage			
Direct	23.7	1.1	0.4
Indirect	2.4	0.2	—
Total[b]	26.9	12.4	0.6

[a]Units are 10^6 kg N/yr. (recalculated values from Deck, 1981).
[b]1 Dissolved inorganic nitrogen accounts for 72% of total nitrogen inputs (Deck, 1981); hence, total nitrogenous inputs (dissolved and particulate organic) are 55.4×10^6 kg/yr.

mixing upstream and downstream from this point (Deck, 1981) (Figure 2.11).

Except for the estuary endpoints where ammonia from sewage dominates, nitrate is the predominant nitrogen form in the river. The largest source of nitrate enters the top of the estuary with fresh water (Table 2.2) and mixes conservatively with salt water. Since nitrate concentrations in

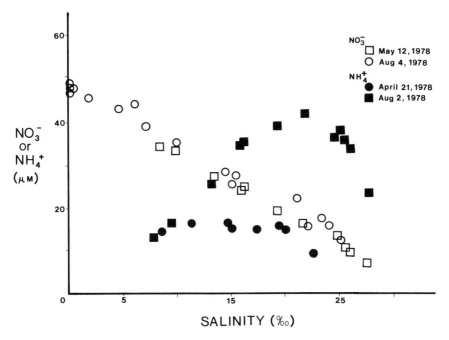

Figure 2.11. Ammonium and nitrate concentrations as a function of salinity in the lower Hudson estuary conditions (redrawn from Deck, 1981).

sewage are probably comparable to ambient lower estuary concentrations, a sewage input peak does not occur (Deck, 1981) (Figure 2.11). Sewage does provide the major source of nitrate to the Hudson estuary, but nitrite makes up only a small fraction of the total nitrogen budget (Table 2.2).

The Food Web

Primary production in the Hudson River ecosystem is dominated by planktonic algae. Gross phytoplankton production is estimated to range between 100 and 250 g $C/m^2 \cdot$ yr (Sirois and Fredrick, 1978), placing the system well within the range of northern temperate estuaries (Boynton *et al.*, 1983). Substantial variations are found along the length of the river, with a range in maximum gross productivity of 0.7 to 4.8 g $C/m^2 \cdot$ day between New York City and Poughkeepsie (Weinstein, 1977). Semiannual productivity measurements during 1972 were highest for the Hudson in the Tappan Zee and Haverstraw Bays. Phytoplankton tend to concentrate in bays and shallow regions of the river where low flushing rates allow populations to build up (Gladden *et al.*, 1984) (Figure 2.12).

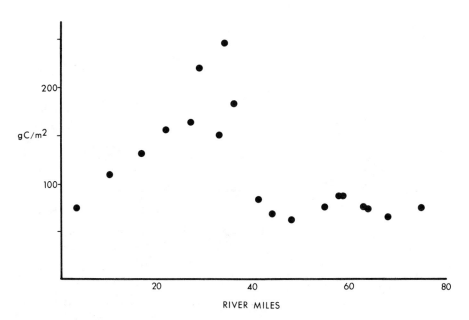

Figure 2.12. Estimated 1972 semiannual primary productivity in the lower Hudson for six-month period from May through October (Sirois and Fredrick, 1978). Semiannual productivity for this period is approximately 90% of annual productivity.

From a low during the winter months, phytoplankton production increases with the spring diatom blooms (McFadden et al., 1978), reaches its peak by midsummer, and declines in late summer (Sirois and Fredrick, 1978). During most of the year, flushing rates are high enough to prohibit the build-up of phytoplankton in the estuary, and most phytoplankton are washed out to New York Bay; summer low-flow periods may be the exception, since growth rates at these times can exceed the rate of dilution (Malone, 1977). In the lower estuary during high freshwater flow, Neale et al. (1981) found that the main source of chlorophyll a actually travelled up the mouth of the estuary in the saline water. During low flow, the major chlorophyll a source was from upstream water containing endemic estuarine phytoplankton. Nanoplankton of the lower Hudson dominate the summer biomass and productivity within the river. Larger net plankton, originating for the most part outside the Hudson and carried into the estuary with ocean waters, account for most productivity between November and June (Malone, 1977).

Light penetration is a major determinant of Hudson River primary productivity, especially in the lower parts of the estuary (Howells and Weaver, 1969). Malone (1977) measured insolation in the lower estuary; the average for 1973–1974 was about 610 kcal $\cdot$ m^{-2} $\cdot$ d^{-1} or 2.2×10^5 kcal $\cdot$ m^2 $\cdot$ yr^{-1}. The euphotic zone can be fairly shallow, with a range of sunlight penetration from 1 to 5 m below the river surface (Weinstein, 1977; Malone, 1977). In the lower estuary, most of the turbidity is due to suspended inorganic solids (Malone, 1977).

Because of light limitations, the large and steady nutrient inputs from sewage effluent do not substantially affect phytoplankton growth. In the lower Hudson and New York Bay, nitrogen inputs from sewage discharge usually greatly exceed the assimilative capabilities of the phytoplankton (Garside et al., 1976). Conversely, phytoplankton growth has little impact on dissolved nitrogen distribution in the Hudson estuary. If the nitrogen discharged in sewage from New York City is assumed to equal 160 mt/day, primary productivity results in assimilation within the estuary of 1.6% of the nitrogen during winter months and 26.9% during the summer (Garside et al., 1976). Most of the dissolved nitrogen is transported into the New York Bight (Garside et al. 1976) where it is gradually assimilated by phytoplankton in the plume of the estuary (Malone, 1977, 1984). Sewage nitrogen shows seasonal influences on new phytoplankton production, and may be responsible for 30% of production above non-sewage water quality conditions (Malone, 1984).

Phytoplankton species are distributed along the estuary in a spatial gradient corresponding to salinity conditions. Marine phytoplankton dominate near the mouth of the river; moving northward, river waters soon contain algae typical of brackish water and then fresh water (Weinstein, 1977). Movement of fresh water through the estuary, inputs from tributary streams, and tidal action blur the boundaries between marine, brackish, and freshwater phytoplankton. Algae normally occurring in terrestrial habitats have also been found in the Hudson, presumably entering with overland runoff (Fredrick *et al.,* 1976).

Diatoms constitute the group of algae appearing with greatest constancy in the river. *Melosira, Asterionella,* and *Cyclotella* are the dominant genera (Howells and Weaver, 1969; Fredrick *et al.,* 1976; Storm and Heffner, 1976; Weinstein, 1977). Although there is substantial longitudinal variation in species, typical green algae representatives include *Pediastrum, Scenedesmus,* and *Ankistrodesmus.* Dinoflagellaes are represented by *Ceratium* and *Porocentrum,* and blue-greens by *Anacystis* and *Anabaena* (Weinstein, 1977). During warm summer months, Hudson waters are dominated by green and blue-green algae. Throughout the late fall and through the winter to spring, diatoms are the major algal species (McFadden *et al.,* 1978).

The second group of primary producers, rooted macrophytes, are limited in distribution to shallow bays and shoal areas at the mouths of tributaries wherever water is less than 3 m deep (Muenscher, 1937); vascular plants seldom extend offshore more than 90 m (McFadden *et al.,* 1978). Muenscher (1937) classified the macrophyte communities of the Hudson according to their interaction with water level. The submerged community (plants that always remain below water) are represented by *Elodea, Vallisneria,* and *Potamogeton.* Mud flat plants, which are submerged during high tides but exposed at low tides, include *Sagittaria, Isoetes,* and *Eleocharis.* The emergent plant group, those species that rise above water level even at high tide, includes *Typha, Spartina, Zizania, Pontederia, Scirpus,* and *Sparganium,* among others. Species richness within the Hudson River macrophyte communities was reported as fairly low by Muenscher in 1937. Today, three introduced plants of European or Asian origin (*Trapa natans, Myriophyllum spicatum,* and *Lythrum salicaria*) form monocultures at the expense of native flora and further reduce species richness of the Hudson macrophyte communities.

Salinity is a major factor affecting distribution of macrophytes in the estuary. Salt-tolerant plants common in the lower estuary disappear rapidly between Peekskill and Beacon (Muenscher, 1937). North of Peekskill, dominant emergents in the freshwater marshes are *Lythrum, Peltandra, Nuphar,* and *Typha,* while dominant submergents include *Vallisneria, Myriophyllum,* and *Potamogeton.* South of Croton, the marshes are dominated by species of cordgrass (*Spartina spp.*) (McFadden *et al.,* 1978). Since most of the Hudson River tidal marshes are influenced by terrestrial drainage (Croton Point and Con Hook marshes are the exceptions), the reduced salinity allows some freshwater species to survive as far south as Piermont marsh (RM 25; km 40) (Buckley and Ristich, 1976).

Human transportation needs also have influenced the distribution of the Hudson's tidal wetlands (Houston, 1985). To the south, in the vicinity of New York City and up to Yonkers (RM 25), the estuary's shoreline has been artificially straightened and riprapped in order to facilitate shipping and prevent sediment buildup. Farther north, railroad tracks often have been laid as near to the shoreline as possible, at times cutting across wetlands on raised beams. This has probably significantly affected the rate of sediment accretion in these enclosed wetlands, and hence transition to terrestrial ecological communities (Houston, 1985).

Consumers in the Hudson estuary include microzooplankton, macrozooplankton, benthic invertebrates, and fish. Microzooplankton are distributed with water movements, feeding on phytoplankton as well as the bacteria and protozoa associated with detritus (Figure 2.13). When favorable conditions are encountered, microzooplankton are able to reproduce rapidly with short generation times. Macrozooplankton, generalist feeders that consume bacteria, algae, particulate matter, and other zooplankton, have a longer generation time and usually produce only one or two generations each year (Gladden et al., 1984). Zooplankton densities vary considerably with season and location in the Hudson estuary. Macrozooplankton densities are reported to vary 10-fold throughout the year in the Indian Point vicinity (NYU, 1978). Copepod larvae in upper New York Bay ranged in density from 10 mg/m^3 during winter and spring to 300 mg/m^3 during August (Malone, 1977).

Zooplankton characteristic of the Hudson River include *Bosmina longirostris, Diaphanosoma spp., Moina spp., Acartia tonsa, Eurytemora affinis, Temora longicornis,* larval forms of *Balanus spp.* (Crustacea), and *Valvata sincera* (Gastropoda) (Weinstein, 1977). The microcrustaceans are the most consistently dominant zooplankton group, and are important in the ecosystem both as grazers and as fish food (Howells et al., 1969). This group includes the copepod *Eurytemora affinis*, a dominant food for larvae of striped bass, herring, white perch, menhaden, and smelt, and the copepod *Acartia tonsa*, which replaces *Eurytemora affinis* as a major food source in more saline areas and during summer months (Weinstein, 1977).

Benthic invertebrate communities, inhabiting the surface and subsurface of the river bottom, are generally dominated by a small number of species (McFadden et al., 1978). Important benthos of the Hudson include *Scolecolepides* and *Limnodrilus* (Annelida), *Crangon, Palaemonetes, Gammarus, Cyathura, Leptocheiris,* and *Edotea* (Crustacea), *Hydrobia* (Gastropoda), chironomids, and oligochaetes (Ristich et al., 1977; McFadden et al., 1978). Benthic organisms as a group consume an assortment of living and non-living food items (some are filter feeders, some deposit feeders, some feed on other benthos) (Figure 2.13) and are tolerant of a wide range of environmental conditions.

Benthic species vary in distribution along the estuary. Haverstraw Bay (RM 37; km 59) supports a typically marine benthos. Marine worms and Crustacea dominate, along with a few barnacles, crabs, and oysters. Insects are absent in this area (Townes, 1937). At Beacon (RM 60; km 97), a

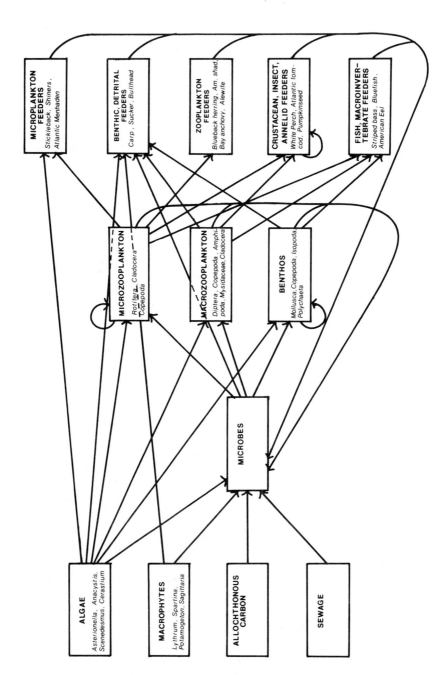

Figure 2.13. Representation of the Hudson River food web (adapted from Weinstein, 1977).

mixture of freshwater and marine organisms inhabit the river bottom, with freshwater fauna being common in the coves. Above Poughkeepsie (RM 76; km 122), there are freshwater snails, clams, *Limnodrilus, Gammarus,* and chironomids. Occasional caddisflies, dragonflies, and mayflies are found (Townes, 1937). Changes in dominant species and heterogeneity in faunal composition are most pronounced near the river mouth, where physical and chemical factors are most spatially variable. Upstream in the freshwater reaches, a homogeneous fauna is found (Ristich *et al.,* 1977) (Figure 2.14). Benthic invertebrates have been found to have a fairly constant density throughout the year. For example, at Indian Point, only a 3-fold variation in population size was found, with peak densities in October and November (Texas Instruments, 1976b).

Hudson River fish that feed on invertebrates can be characterized by food habits (Figure 2.13; Table 2.3). Benthic invertebrates are steady and dependable in abundance, and many resident fish rely on this year-round resource. Examples of resident fish species feeding on benthic and epibenthic invertebrates, as well as on detritus associated with sediments, are carp, golden and common shiner, white and yellow perch, and pumpkinseed (Gladden *et al.,* 1984). By contrast, 10 of the 11 migrant fish species studied by Gladden *et al.* (1984) were found to feed within the water column rather than on the benthos; blueback herring, American shad, rainbow smelt, striped bass, and bluefish, among others, are migrant carnivores that feed on zooplankton, macroinvertebrates, or fish. This suggests that migrant and resident fish in the Hudson may compete very little for resources as a result of trophic segregation. Spatial segrega-

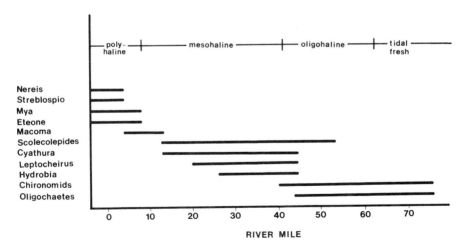

Figure 2.14. Progression of dominant benthic organisms in the lower Hudson River in 1972. Dominance is measured as frequency of occurrence (redrawn from Ristich *et al.,* 1977).

Table 2.3. Trophic strategies and salinity preferences of Hudson River fish.[a]

Salinity regime	Food preference			
	Planktonic	Benthic, detrital	Epibenthic crustacean, insect, and/or annelid	Fish and/or macroinvertebrate
Freshwater	Emerald shiner Spottail shiner Tesselated darter[b]	Goldfish Carp Silvery minnow Golden shiner White catfish Brown bullhead Bluegill White sucker	Yellow perch Pumpkinseed Redbreast sunfish	Largemouth bass
Euryhaline	Fourspine stickleback Banded killifish[b] Tidewater silverside[b] Alewife[b,c] American shad[b] Bay anchovy[b,c] Blueback herring[b,c] Rainbow smelt[b,c]	Mummichog	——————— White perch ——————— ——————— Shortnose sturgeon ——————— Hogchoker Striped bass[c] Atlantic tomcod[c] American eel[c]	
Marine	Atlantic menhaden[c] Atlantic silverside[b,c] Northern pipefish[b,c]	——————— Winter flounder[b,c] ———————	——————— Longhorn sculpin ——————— Fourbeard rockling[c]	Atlantic sturgeon[c] Bluefish[c] Weakfish[c]

[a] From Gladden et al. (1984); Texas Instruments (1981); and Weinstein (1977).
[b] Includes zooplankton, eggs, or larvae.
[c] Migratory species.

tion by salinity tolerance and river depth also helps reduce competition (Gladden et al., 1984).

Far from being "dead," as many thought it to be, the Hudson sports a wealth of species (even considering only fish), a complex food web, and a variety of trophic strategies. Yet only one species, striped bass (*Morone saxatilis*), has been extensively examined with respect to interaction with abiotic and biotic components of the river system. Those factors that influence the population levels of striped bass and determine year-class strength have been important questions because of the commercial and recreational value of these fish, and because the operation of several power generation plants in the estuary was thought to pose yet another substantial threat to their survival. Several hypotheses have been proposed to explain the relative success of a given year-class of striped bass (Texas Instruments, 1981). These include the occasional direct influences of water temperature, dissolved oxygen, and freshwater flow, as well as indirect, synergistic effects of temperature, food availability, and feeding behavior in young age classes (Texas Instruments, 1981). It has been suggested (Klauda et al., 1980) that high freshwater flows in spring carry carbon and nutrients into the estuary, stimulating lower food chain production at a time that is critical in larval bass development. Additionally, high flows may shift spawning grounds to river areas more suitable for larval feeding and survival (Klauda et al., 1980). The importance of specific abiotic factors varies from year to year, making prediction of population levels in the Hudson striped bass year-classes difficult.

Organic Carbon Inputs

The organic carbon inputs to the Hudson estuary include autochthonous sources (originating within the river) and allochthonous sources (originating outside the river). Although budget data are scarce, a synthesis of available information (shown in Table 2.4) produces an estimate of total inputs equaling roughly 246×10^6 kg C/yr.

Autochthonous inputs include both phytoplanktonic and macrophytic primary productivity. On an annual basis, primary productivity is estimated to account for 15% (36.4×10^6 kg C/yr) of the organic carbon available in the river (Table 2.4). Macrophyte productivity is only about 15% of the magnitude of phytoplankton productivity, because rooted aquatic vegetation has a restricted distribution along the Hudson shoreline.

Sources of allochthonous carbon include upper watershed inputs, lower watershed inputs, and sewage effluent released directly into the river. The upper watershed allochthonous input includes the organic carbon carried into the estuary from the non-tidal river above Troy Dam. These inputs are the largest source of organic carbon to the Hudson; but because they

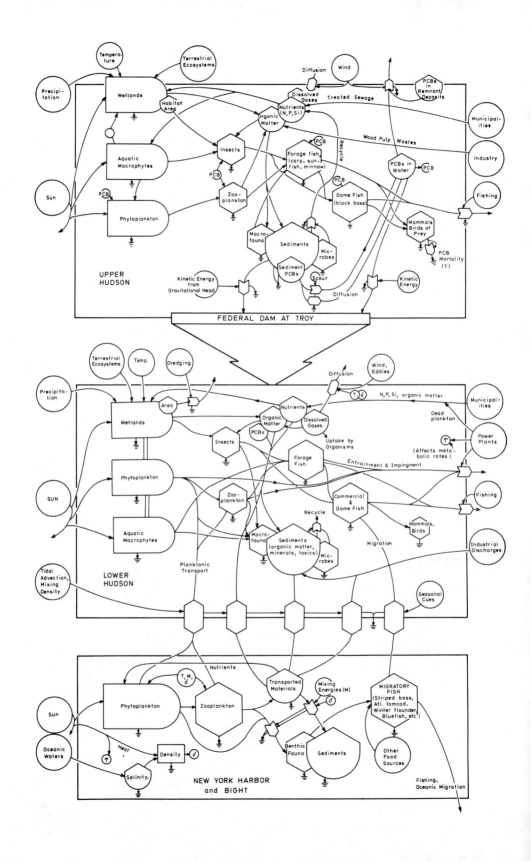

Table 2.4. Organic carbon inputs to the Hudson River estuary.

Source	Metric tons (mt) carbon/yr	Percent of total
Phytoplankton	36,364[a]	14.8
Macrophytes		
Emergent	3,936[b]	1.6
Tidal Flats	1,428[c]	0.6
Upper Watershed (above RM 152)	66,024[d]	26.8
Lower Watershed (RM 0-152)	43,254[e]	17.6
Sewage (RM 0-152)	57,649[f]	23.5
Marine	36,898[g]	15.0
Total	245,553	100

[a] Estimated semiannual (May through October) primary productivity (Sirois and Fredrick, 1978) increased by 5%.
[b] Emergent marsh area estimated at 9.6 km^2; net annual production assumed to be 1000 g/m^2/yr; energy content of biomass assumed to be 4.1 kcal/g dry weight (McFadden et al., 1978); carbon equivalents assumed to be 1 g C/10 kcal (Odum, 1971).
[c] Tidal flat area estimated at 10.2 km^2; annual production assumed to be 140 g C/m^2/yr (Marshall et al., 1971; Cadee and Hegeman, 1974).
[d] From Gladden et al. (1984) (8.633 kg C/km^2/day; upper watershed area equals 20,953 km2).
[e] Lower watershed area equals 13,727 km^2.
[f] Municipal and industrial BOD discharges 217.4 mt/day (Hetling, 1976; McFadden et al., 1978); 0.7265 kg C/kg BOD.
[g] Phytoplankton from marine sources (winter through summer) estimated at 6,200 mt C/yr (Malone et al., 1980). Sewage input from the New York bight assumed to be 30,680 mt C/yr (Malone et al., 1980). Fish eggs from anadromous fish estimated to contribute 18 mt C/yr (McFadden et al., 1978).

are transported into the estuary following spring snowmelt, when freshwater flow rates are highest, the materials tend to be flushed through the system rather than remain in the river (Q. Ross, pers. comm.). The estimated allochthonous organic carbon from both upper and lower watersheds totals about 109×10^6 kg/yr, nearly three times the contribution from phytoplankton and macrophytes. During the summer, when river flows have decreased, these inputs decline, and primary productivity and sewage become the major organic carbon sources. It is difficult to say at

◄ **Figure 2.15.** Diagram of the Hudson River ecosystem, using the symbols from Odum (1972) (circles denote mostly exogenous forces, "bullets" are producers, hexagons are consumers, tanks are storages, and wide arrows denote interactions). The top box illustrates the upper Hudson, which flows into the estuary. Connection with the New York Bight is shown in the bottom box. Note that materials generally flow downstream, unless brought upriver by tidal advection or migratory pattern. Flows and storages represent energy and matter.

present which input exerts greatest control over total food chain productivity.

Since the Hudson is an estuarine system, saline water moving into the river transports carbon into the lower estuary. Malone (1977) and Neale *et al.* (1981) showed that oceanic water is an important source of the phytoplankton in the lower Hudson. One would also expect zooplankton, fish, and detritus to enter the Hudson ecosystem with the tidal excursion. Lack of information, however, has made marine sources of carbon the most difficult to estimate. Movement of sewage and organisms (phytoplankton, fish eggs) into the Hudson with tidal flow is roughly quantified as 37×10^6 kg C/y to indicate the potential importance of marine sources (Table 2.4).

Other Human Influences

A description of the Hudson ecosystem would be incomplete without mention of the influences that society exerts. Besides the extensive use of the river as the bearer of wastes, the Hudson estuary and tributaries are also exploited for power generation (Chapter 3). The average annual steam generation of about 22×10^3 GWh (1.66×10^{17} kcal/yr) requires withdrawals of 4.66×10^9 m^3/yr of river water for once-through cooling (Horne, 1979), which represents approximately 13% of the total annual freshwater flow in the estuary. Less than 1% of the withdrawals are volatilized, so most of the water is returned to the estuary in a heated state.

Dredging of sediments for channel maintenance by the U.S. Army Corps of Engineers and New York State Department of Transportation amounts to 8.5×10^5 m^3/yr of sediments removed from the estuary and 4.0×10^5 m^3/yr removed above the Troy dam (Horne, 1979). Private dredging activities are estimated at 3.4×10^5 m^3/yr. In the 1970s, concern arose over high concentrations of toxic materials in the sediments, particularly cadmium in the estuary and PCBs in the upper river. Proposed dredging of 1.1×10^6 m^3 of PCB-contaminated river bottom has aroused much debate (Chapter 4). Intentional filling-in of the river channel is another ongoing process; the most massive fill project yet proposed is the Westway Highway Project, which would have added 200 acres to Manhattan at the expense of the estuary (Chapter 5).

Finally, man uses the river for fishing and such recreational uses as boating, swimming, and camping. The average total commercial catch between 1965 and 1974 was 137.1 mt/yr, with American shad comprising the bulk of the catch (73%); recreational fisheries are not evaluated in terms of catch, but probably fall within the range of commercial efforts (Sheppard, 1976). The estuary has been officially closed to commercial fishing for some species (including striped bass) since 1976 because of PCB contamination. Humans, as the largest biological influence on Hud-

Table 2.5. Annual flows and storages in the Hudson River estuary.

Parameter	Lower[a] New York Bay	Upper[b] New York Bay	Tappan Zee	Bowline[c] Point	Indian[d] Point	Roseton-Danskammer[e]
Primary productivity (gC/m^2·yr^{-1})	800	200	250[f]	170	290	200
Phytoplankton biomass (gC/m^2)	n.a.	n.a.	n.a.	0.01	n.a.	0.5–2.4
Chlorophyll-a (μg/l)	10.4	2.62	26.4[f]	9.92	2.1	5.6
Nutrients:						
N (mg/l)	0.43	0.71	n.a.	0.88	n.a.	0.77
PO$_4$-P (mg/l)	0.05	0.09	n.a.	0.05	n.a.	0.04
Si (mg/l)	0.32	n.a.	n.a.	2.4	n.a.	
Zooplankton (g wet weight/m^3)	n.a.	0.01–0.3	n.a.	0.014	0.031	n.a.
Benthic faunal biomass (g wet weight/m^2)	4.07–804.0[i]	n.a.	n.a.	2.91 (1975)	16.2 (1973)	6.89–19.8
		n.a.	n.a.	24.8 (1974)	31.0 (1974)	
Benthic respiration (gC/m^2·yr^{-1})	126–1018[g]	n.a.	407–1347[g,h]	n.a.	n.a.	n.a.

[a]From O'Reilly et al. (1976) unless otherwise specified.
[b]From Malone (1977) unless otherwise specified.
[c]From EA (1981), LMS (1977a, 1980).
[d]From NYU (1974), Texas Instruments (1974, 1975).
[e]From LMS (1977b, 1981).
[f]From Sirois and Frederick (1978).
[g]From Thomas et al. (1976).
[h]Spuyten Duyvil to Narrows.
[i]Pearce et al. (1981).
n.a., not available.

Table 2.6. Physical and environmental characteristics of the upper and lower Hudson River.

Parameter	Upper HR	Lower HR	Reference
River Dimensions			
Length (km)	257	250	—[a]
Avg. width (m)	—	1280	
(min. to max.)		(260–5520)	
Avg. depth (m)	—	10	—[a]
(min. to max.)		(4–34)	
Cross-sectional area (m)		10,420	—[a]
(min. to max.)		(1770–21,400)	
Elevation gradient (m/km)	2.6	0.006	Sanders (1982); Helsinger and Freidman (1982); USGS (1981); Busby (1966)
River Flow (m^3/s)	392	623	
Tidal flow	n.a.	5670–8500	
Environment			
Insolation (kcal m^2/yr)		9.6×10^5	SERI (1981)
Avg. temperature (°C)			—[b]
Air	15.3	15.9	NOAA (1982)
Water	—	12.3	
Precipitation (mm)	980	1140	NOAA (1982)
Drainage area (km^2)	20,917	13,740	Sanders (1982); Helsinger and Friedman (1982)
Allochthonous carbon loading (mt/yr)	66,024	43,431	Gladden et al. (1982)
Anthropogenic influence			
Power plants			
Combined water withdrawal (m^3/yr^1)	n.a.	4.66×10^9	Horne (1979)

Other Human Influences

Avg. within-plant temperature increase (ΔT, °C)	n.a.	8.3	Prorated from Chapter 3, Table 3
PCBs			
Estimated load in sediments (mt)	153.2 (including remnant deposits)	78.8	Sanders (1984)
Flow over Troy Dam (mt/yr)	0.4 (in 1981)	n.a.	Schroeder and Barnes (1983);
Flow to New York Harbor	n.a.	2.5	Bopp et al. (1981);
Loss to atmosphere (mt/yr)	2.9	—	Tofflemire and Quinn (1979)
Amount in biota (mt)	—	—	
Sewage			
BOD (mt/yr)	19000	6.8×10^4	Hetling (1976)
Nitrogen (mt/yr)	398	4.0×10^4	Deck (1981)
Phosphorus (mt/yr)	183	0.9×10^4	McFadden et al. (1978)
Dredging: amount of sediment removed (does not include Westway) (m³/yr)	4×10^5	12×10^5	Horne (1979)
Fishing			
Commercial catch			
mt/yr	n.a.	2600 (mid-1970s)	
Value ($1000)	500	—[c]	
Recreational catch			
Angler-days/yr	8–10,000	70,000	
Estimated value ($1000)	88–110	770	—[d]

[a] Data sources for estuarine dimensions: Albany to New Hamburg; Stedfast (1982); New Hamburg to the Battery, H. Darmer (1969); and measurements from 7.5-min topographic maps.
[b] Calculated with data for ambient monthly water temperatures: Roseton (LMS, 1978); Bowline-Lovett (LMS, 1977a); upper New York Harbor (Malone, 1977).
[c] Sheppard (1983) developed rough estimates of the contribution of the Hudson River to various fisheries, including marine fisheries.
[d] Assuming a conservative estimate for an angler-day value is $11.00 (New York State Economic Development Board, 1975).
HR, Hudson River; n.a., not applicable; T, temperature; C, celsius; PCBs, polychlorinated biphenyls.

son River ecology, will continue to face decisions on "best use" while protecting the river's ability to function as an ecosystem.

Synthesis

Figure 2.15 and Tables 2.5 and 2.6 present the information on the Hudson ecosystem in an integrated manner. Figure 2.15 is a diagram showing how processes and components of the ecosystem are linked together. In this way, it is possible to see many of the indirect relationships that characterize this and other complex systems. An example of an indirect relationship is that of fishing activity to primary production; another is the stimulation of secondary (animal) production by the release of organic sewage from municipalities or agricultural runoff. The accompanying tables provide characteristic environmental data that can be used to compare the Hudson with other estuarine ecosystems, for making various calculations, or for general information about the river.

References

Abood, K.A., E.A. Maikish, and R.R. Kimmel. 1976. Field and analytical investigations of ambient temperature distribution in the Hudson River. Paper 6 *In* Hudson River Ecology, 4th Symp. Hudson River Environ. Soc., Bronx, NY.

Baslow, M.H. and D.T. Logan. 1982. The Hudson River ecosystem. A case study. Report to the Ecosystems Research Center. Cornell University, Ithaca, NY. 460 pp.

Bopp, R.I., H.J. Simpson, C.R. Olsen, and N. Kostyk. 1981. Polychlorinated biphenyls in sediments of the tidal Hudson River, New York. Environ. Sci. Tech. 15(2):210–216.

Boynton, W.R., C.A. Hall, P.G. Falkowski, C.W. Keefe, and W.M. Kemp. 1983. Phytoplankton productivity in aquatic ecosystems, pp. 305–327 *In* Physiological Plant Ecology IV (O.L. Lange, P.S. Nobel, C.B. Osmond, and H. Ziegler, eds.). Springer-Verlag, Berlin.

Buckley, E.H. 1971. Maintenance of a functional environment in the lower portion of the Hudson River estuary: An appraisal of the estuary. Contrib. Boyce Thompson Institute 24:387–396.

Buckley, E.H. and S.S. Ristich. 1976. Distribution of rooted vegetation in the estuarine marshes of the Hudson River. Paper 20 *In* Hudson River Ecology, 4th Symp. Hudson River Environ. Soc., Bronx, NY.

Busby, M.W. 1966. Flow quality and salinity profiles in the Hudson River estuary, pp. 135–146 *In* Hudson River Ecology, 1st Symp. Hudson River Valley Comm., NY.

Busby, M.W. and K.I. Darmer. 1970. A look at the Hudson estuary. Water Resources Bull. 6:802–812.

Cadee, G.C. and J. Hegeman. 1974. Primary production of the benthic microflora living on tidal flats in the Dutch Wadden Sea. Neth. J. Sea Res. 8:260–291.

Central Hudson Gas and Electric. 1977. Roseton Generation Station. Near-field effects of once-through cooling system operation on Hudson River biota. Central Hudson Gas and Electric Corp., Poughkeepsie, NY.

Darmer, K.I. 1969. Hydrologic characteristics of the Hudson River estuary, pp. 40–55 *In* 2nd Hudson River Ecology Symposium, Tuxedo, NY (G.P. Howells and G.J. Lauer, eds.). N.Y.S. Dept. of Environmental Conservation, Albany, NY.

Deck, B.L. 1981. Nutrient-element distributions in the Hudson estuary. Ph.D. Dissertation. Columbia University, New York, 396 pp.

EA. 1978. Final report: Hudson River thermal effects studies for representative species. Prepared by Ecological Analysts, Inc. for Central Hudson Gas and Electric Corp., Consolidated Edison Co. of NY, Inc., and Orange and Rockland Utilities, Inc.

EA. 1981. Bowline Point Generating Station entrainment abundance and survival studies. 1979 Annual Report with Overview of 1975–1979 Studies. Prepared by Ecological Analysts, Inc. for Orange and Rockland Utilities, Inc.

Faigenbaum, H.M. 1937. Chemical investigation of the lower Hudson area., pp. 146–216 *In* A Biological Survey of the Lower Hudson Watershed. Suppl. to 26th Ann. Rep. N.Y.S. Conservation Dept. Albany, NY.

Fleming, A., K.A. Abood, H.F. Mulligan, C.B. Dew, C.A. Menzie, W. Sydor, and W. Su. 1976. The environmental impact of PL 92-500 on the Hudson River estuary. Paper 16 *In* Hudson River Ecology, 4th Symp. Hudson River Environ. Soc., Bronx, NY.

Fredrick, S.W., R.L. Heffner, A.T. Packard, P.M. Eldridge, J.C. Eldridge, G.J. Schumacher, K.L. Eichorn, J.H. Currie, J.N. Richards, and O.C. Boody, IV. 1976. Notes on phytoplankton distribution in the Hudson River estuary. Paper 34 *In* Hudson River Ecology, 4th Symp. Hudson River Environ. Soc., Bronx, NY.

Garside, C., T.C. Malone, O.A. Roels, and B.A. Shartstein. 1976. An evaluation of sewage derived nutrients and their influence on the Hudson estuary and New York bight. Est. Coastal Mar. Sci. 4:281–289.

Gladden, J.B., F.C. Cantelmo, J.M. Croom, and R. Shapot. 1984. An evaluation of the Hudson River ecosystem in relation to the dynamics of fish populations. Trans. Am. Fish. Soc. (in press).

Hammond, D.E. 1975. Dissolved gases and kinetic processes in the Hudson River. Ph.D. Dissertation. Columbia University, New York, NY, 161 pp.

Helsinger, M.H. and G.M. Friedman. 1982. Distribution and incorporation of trace elements in the bottom sediments of the Hudson River and its tributaries. N.E. Environ. Sci. 1(1):33–47.

Hetling, L. 1976. Trends in wastewater loading, 1900 to 1976. Paper 14 *In* Hudson River Ecology, 4th Symp. Hudson River Environ. Soc., Bronx, NY.

Horne, W.S. (Study Manager). 1979. Hudson River basin. Level B: Water and related land resources study. Technical Paper 3. N.Y.S. Dept. of Environmental Conservation, Albany, NY. (in 3 vols.)

Houston, L. 1985. One ecosystem: land, wetlands, and the river. Proc. 7th Hudson River Ecology Symp., Hudson River Environ. Soc., New Paltz, NY. (In press).

Howells, G.P. and S. Weaver. 1969. Studies on phytoplankton at Indian Point, pp. 231–261 *In* 2nd Hudson River Ecology Symp. N.Y.S. Dept. of Environmental Conservation, Albany, NY.

Howells, G.P., E. Musnick, and H.I. Hirshfield. 1969. Invertebrates of the Hudson River, pp. 262–280 *In* 2nd Hudson River Ecology Symp. N.Y.S. Dept. of Environmental Conservation, Albany, NY.

Hudson River Policy Committee. 1969. Hudson River fisheries investigations, 1965–1968. Evaluation of a proposed pumped storage project at Cornwall, New York in relation to fish in the Hudson River. N.Y.S. Conservation Dept. Presented to Consolidated Edison Co. of NY, Inc.

Klauda, R.J., W.P. Dey, T.B. Hoff, J.B. McLaren, and Q.E. Ross. 1980. Biology of Hudson River juvenile striped bass, pp. 101-123 *In* Proceedings of the 5th Annual Marine Recreational Fisheries Symposium (H. Clepper, ed.). Sport Fishing Institute, Washington, D.C.

LMS. 1976. Environmental impact assessment. Hudson River water quality analysis. Prepared by Lawler, Matusky, and Skelly, Engineers, for Nat. Comm. on Water Quality. PB 251099, NTIS. Springfield, VA

LMS. 1977a. 1976 Hudson River aquatic ecology studies at Lovett Generating Station, Vol. 1. Chapters I–VII. Prepared by Lawler, Matusky, and Skelly, Engineers, for Orange and Rockland Utilities, Inc.

LMS. 1977b. Roseton Generating Station: Near-field effects of once-through cooling system operation on Hudson River biota. Prepared by Lawler, Matusky, and Skelly, Engineers, and by Ecological Analysts, Inc. for Central Hudson Gas and Electric Corp.

LMS. 1978a. Roseton and Danskammer Point Generating Stations. Hydrothermal analysis. Prepared by Lawler, Matusky, and Skelly, Engineers, for Central Hudson Gas and Electric Corp.

LMS. 1978b. Central Hudson Gas and Electric Corp. Annual Progress Report for 1974. Prepared by Lawler, Matusky, and Skelly Engineers. Pearl River, NY.

LMS. 1980. Evaluation of lower trophic level aquatic communities in the vicinity of the Bowline Point Generating Station—1971–1977. Prepared by Lawler, Matusky, and Skelly, Engineers, for Orange and Rockland Utilities, Inc.

LMS. 1981. 1980 Annual Progress Report (Roseton-Danskammer). Prepared by Lawler, Matusky, and Skelly, Engineers, for Central Hudson Gas and Electric Corp.

Malone, T.C. 1977. Environmental regulation of phytoplankton productivity in the lower Hudson estuary. Est. Coastal Mar. Sci. 5:157-171.

Malone, T.C. 1984. Anthropogenic nitrogen loading and assimilation capacity of the Hudson River estuarine system, USA, pp. 291–311 *In* The Estuary as a Filter (V.S. Kennedy, ed.). Academic Press, Orlando, FL.

Malone, T.C., P.J. Neale, and D. Boardman. 1980. Influences of estuarine circulation on the distribution and biomass of phytoplankton size fractions, pp. 249–262 *In* Estuarine Perspectives (V.S. Kennedy, ed.). Academic Press, New York, NY.

Marshall, N., C.A. Oviatt, and D.M. Skauen. 1971. Productivity of the benthic microflora of shoal estuarine environments in southern New England. Internationale Revue der Gesamten Hydrobiologie 56:947-956.

McFadden, J.T., Texas Instruments, Inc., and Lawler, Matusky, and Skelly, Engineers. 1977. Influence of Indian Point Unit 2 and other steam electric generating plants on the Hudson River estuary, with emphasis on striped bass and other fish populations. Prepared for Consolidated Edison Co. of NY, Inc.

McFadden, J.T., Texas Instruments, Inc., and Lawler, Matusky, and Skelly, Engineers. 1978. Influence of the proposed Cornwall pumped storage project and steam electric generating plants on the Hudson River Estuary, with emphasis on striped bass and other fish populations. Revised. Prepared for Consolidated Edison Co. of NY, Inc.

Mt. Pleasant, R.C. and H. Bagley. 1975. Towards purer waters. Progress Report. Environ. Qual. News. Conservationist 30(2):ii–iii.

Mueller, J.A., T.A. Gerrish, and M.C. Casey. 1982. Contaminant inputs to the Hudson-Raritan Estuary. NOAA Technical Memorandum OMPA-21. Boulder, CO.

Muenscher, W.C. 1937. Aquatic vegetation of the lower Hudson area, pp. 231–248 In A Biological Survey of the Lower Hudson Watershed. Suppl. to 26th Ann. Rep. N.Y.S. Conservation Dept.

Neale, P.J., T.C. Malone, and D.C. Boardman. 1981. Effects of freshwater flow on salinity and phytoplankton biomass in the lower Hudson estuary, pp. 168–184 In Proc. of the Nat. Symp. on Freshwater Inflow to Estuaries (R. Crossand and D. Williams, eds.). U.S. Fish and Wildl. Serv. Office of Biological Services. FWS/OBS-81/04. (2 vols.)

New York State Economic Development Board. 1975. Economic impact of regulating the use of PCB's in New York State. Prepared by Economic Development Board Staff, Albany, NY. 28 pp. (Mss.)

NOAA. 1982. Monthly normals of temperature, precipitation, and heating and cooling degree days 1951–1980. New York. (Climatography of the United States No. 81) U.S. Dept. of Commerce, National Oceanic and Atmospheric Administration, Environmental Data and Information Service, National Climatic Center, Asheville, NC. 18 pp.

NYSDEC. 1967. Periodic report of the Water Quality Surveillance Network. 1965 through 1967 water years. N.Y.S. Dept. of Environmental Conservation, Albany, NY. 390 pp.

NYU. 1978. Hudson River ecosystem studies. Effects of entrainment by the Indian Point power plant on biota in the Hudson River estuary. Prepared by Institute of Environmental Medicine, New York University Medical Center, for Consolidated Edison Co. of NY, Inc.

Odum, E.P. 1971. Fundamentals of Ecology, 3rd Ed. W.B. Saunders Co., Philadelphia, PA. 574 pp.

Odum, H.T. 1972. An energy circuit language for ecological and social systems: its physical basis, pp. 140–211 In Systems Analysis and Simulation in Ecology, Vol. 2 (B.C. Patten, ed.). Academic Press, New York, NY.

O'Reilly, J.E., J.P. Thomas, and C. Evans. 1976. Annual primary production (nannoplankton, netplankton, dissolved organic matter) in the Lower New York Bay. Paper 19 In Hudson River Ecology, 4th Symp. Hudson River Environ. Soc., Bronx, NY.

Pearce, J.B., D.J. Radosh, J.V. Carracciolo, and F.W. Steimle, Jr. 1981. Benthic fauna. NESA New York Bight Atlas Monograph 14. N.Y. Sea Grant Institute, Albany, NY. 79 pp.

Ristich, S.S., M. Crandall, and J. Fortier. 1977. Benthic and epibenthic macroinvertebrates of the Hudson River. I. Distribution, natural history and community structure. Est. Coastal Mar. Sci. 5:255–266.

Sanders, J.E. 1982. The PCB-pollution problem of the upper Hudson River from the perspective of the Hudson River PCB Settlement Advisory Committee. N.E. Environ. Sci. 1(1):7–18.

Schroeder, R.A. and C.R. Barnes. 1983. Trends in polychlorinated biphenyl concentrations in Hudson River water five years after elimination of point sources. U.S. Geological Survey. Water-Resources Investigations Report 83-4206. Albany, NY. 28 pp.

Schureman, P. 1934. Tides and currents in the Hudson River. U.S. Coastal and Geodetic Survey Spec. Publ. 180.

SERI. 1981. Solar Radiation Energy Resource Atlas of the United States. SERI/SP–642-1037. Solar Energy Research Institute, Golden, CO.

Sheppard, J.D. 1976. Valuation of the Hudson River fishery resources: past, present and future. Technical Report. N.Y.S. Dept. of Environmental Conservation, Bureau of Fisheries, Albany, NY. 50 pp.

Sheppard, J.D. 1983. Valuation of the Hudson River fisheries. Appendix B: Commercial fisheries; Appendix C: Recreational fisheries. N.Y.S. Dept. of Environmental Conservation, Bureau of Environmental Protection, Albany, NY. (Draft mss.)

Simpson, H.J., R. Bopp, and D. Thurber. 1974. Salt movement patterns in the Hudson. Paper 9 *In* Hudson River Ecology, 3rd Symp. Hudson River Environ. Soc., Bronx, NY.

Simpson, H.J., D.E. Hammond, B.L. Deck, and S.C. Williams. 1975. Nutrient budgets in the Hudson River estuary, pp. 616–635 *In* ACS Symposium Series No. 18. (T.M. Church, ed.). Marine Chemistry in the Coastal Environment.

Sirois, D.L. and S.W. Fredrick. 1978. Phytoplankton and primary productivity in the lower Hudson River estuary. Est. Coastal Mar. Sci. 7:413–423.

Stedfast, D.A. 1982. Flow model of the Hudson River estuary from Albany to New Hamburg, New York. U.S. Geological Survey, Water-Resources Investigations 81–55, Albany, NY. 69 pp.

Stewart, H.B. 1958. Upstream bottom currents in New York Harbor. Science 127:1113–1115.

Storm, P.C. and R.L. Heffner. 1976. A comparison of phytoplankton abundance, chlorophyll *a* and water quality factors in the Hudson River and its tributaries. Paper 17 *In* Hudson River Ecology, 4th Symp. Hudson River Environ. Soc., Bronx, NY.

Texas Instruments. 1974, 1975. Hudson River ecological study in the area of Indian Point. Annual Reports (for 1973 and 1974). Prepared by Texas Instruments for Consolidated Edison Co. of NY, Inc.

Texas Instruments. 1976a. A synthesis of available data pertaining to major physicochemical variables within the Hudson River estuary emphasizing the period from 1972 through 1975. Prepared by Texas Instruments for Consolidated Edison Co. of NY, Inc.

Texas Instruments. 1976b. Hudson River ecological study in the area of Indian Point. Thermal Effects Report. Prepared by Texas Instruments for Consolidated Edison Co. of NY, Inc.

Texas Instruments. 1977. 1974 year-class report for the multiplant impact study of the Hudson River estuary. Vol. II. Appendices. Prepared by Texas Instruments for Consolidated Edison Co. of NY, Inc.

Texas Instruments. 1979a. 1978 water quality data display. Prepared by Texas Instruments for Consolidated Edison Co. of NY, Inc.

Texas Instruments. 1979b. 1977 year-class report for the multiplant impact study of the Hudson River estuary. Prepared by Texas Instruments for Consolidated Edison Co. of NY, Inc. (Draft.)

Texas Instruments. 1980. 1979 water quality data display. Prepared by Texas Instruments for Consolidated Edison Co. of NY, Inc.

Texas Instruments. 1981. 1979 year-class report for the multiplant impact study of the Hudson River estuary. Prepared by Texas Instruments for Consolidated Edison Co. of NY, Inc.

Thomas, J.P., W. Phoel, J.E. O'Reilly and C. Evans. 1976. Seabed oxygen consumption in the lower Hudson estuary. NOAA, NMFS, Sandy Hook Lab, Highlands, N.J. 21 pp.

Tofflemire, T.J. and S.O. Quinn. 1979. PCB in the upper Hudson River: mapping and sediment relationships. Technical Paper No. 56. N.Y.S. Dept. of Environmental Conservation, Albany, NY. 140 pp.

Townes, H.K., Jr. 1937. Studies on the food organisms of fish, pp. 217–230 In A Biological Survey of the Lower Hudson Watershed. Suppl. to 26th Ann. Rep. N.Y.S. Conservation Dept., Albany, NY.

USGS. 1969, 1975. Water resources data for New York. Part 2. Water Quality Records. U.S. Geological Survey, Albany, NY.

USGS. 1977. Water resources data for New York. Water year 1976. U.S. Geological Survey Water Data Report NY–76–1. 615 pp.

USGS. 1978. Water resources data for New York. Water year 1977. U.S. Geological Survey Water Data Report NY–77–1. 566 pp.

USGS. 1979. Water resources data for New York. Water year 1978. U.S. Geological Survey Water Data Report NY–78–1.

USGS. 1980. Water resources data for New York. Water year 1979. U.S. Geological Survey Water Data Report NY–79–1. 538 pp.

USGS. 1981. Water resources data for New York. Water year 1980. U.S. Geological Survey Water Data Report NY–80–1. 310 pp.

Weinstein, L.H. (ed.) 1977. An Atlas of the Biologic Resources of the Hudson Estuary. Boyce Thompson Institute for Plant Research, Inc., NY. 104 pp.

3
Power Plant Operation on the Hudson River

WILLIAM H. MCDOWELL

The construction and operation of high-capacity power stations on waterways ranks among the most significant environmental alterations by technological society. In the U.S., the problems associated with power plant operation were recognized and addressed specifically by Section 316 of the Clean Water Act (CWA), which dealt with the issues of thermal discharges and intake structures for cooling water. To a large degree, the questions of environmental impact from power plants originated in the controversies surrounding the facilities built, or planned, on the Hudson to service the greater New York metropolitan area. The history of the 17-year controversy about six power stations provides us with several useful observations of various aspects of environmental management: 1) the evolution of awareness, on the part of both the general public and resource management agencies, of the large-scale effects that can be associated with power plants; 2) the development of research programs to accurately assess potential impacts; 3) the frustrating use of models to predict impacts long into the future; 4) the difficulties associated with bringing scientific assessment into the courtroom; and 5) the remarkable success of a mediation effort that led to a 10-year "ceasefire" between utility companies and regulatory agencies.

Description of the Power Plants and Their Operation

Power plants are found along the entire length of the lower Hudson River, from Manhattan to Albany (Figure 2.1). The largest generating stations are located in the middle reaches of the river (Table 3.1), from river mile (RM) 37 to RM 66 (Haverstraw Bay to Poughkeepsie). Indian Point Units Two and Three (nuclear) and the Roseton and Bowline plants (oil-fired) provide the bulk of power generated along the Hudson.

The oldest generating stations are those at 59th street on Manhattan, which began commercial operation in 1918. From 1950–1970, power plants of less than 500 MWe were constructed at Albany, Danskammer, and Lovett (Table 3.1). From 1970 – 1976, several larger power plants of greater than 500 MWe were constructed in the middle reaches of the river. These included the Roseton, Bowline, and Indian Point plants (Table 3.1).

Power plants operating along the Hudson remove water from the river for cooling purposes. Water is drawn from the river, passed through pumps and condensers, and discharged back into the river. This system of "once-through cooling" relies on the river to act as a heat sink to dissipate waste heat produced during the generation of electrical power. The major potential effects of once-through cooling systems include killing of organisms and changes in community structure or habitat.

Loss of organisms results from physical and chemical stresses encountered by smaller planktonic organisms that are drawn into the power plant ("entrained") in the cooling water stream. These stresses include changes in pressure, changes in temperature, contact with heat exchangers or other mechanical parts within the power plants, and exposure to chlorine and other chemicals added to the water. Death also results when larger organisms such as fish are trapped ("impinged") on the intake screens that prevent entry of debris into the cooling water stream. These losses can subsequently exert indirect effects on the community. Reduction in zooplankton abundance, for example, can cause decreases in the abundance of fish that rely on zooplankton for food. Other effects on the aquatic ecosystem include small changes in the temperature of the river, which may be important in determining the relative success of competing species (Coutant, 1972; Baslow and Logan, 1982).

The amount of water required for cooling purposes by the power plants along the Hudson is large relative to the net freshwater flow of the river (although not in terms of tidal flow). As shown in Table 2.1, the total maximum cooling water usage of all the plants along the Hudson is 9591 cfs; the average annual freshwater flow at Green Island, near Troy, N.Y., is 13,800 cfs (USGS, 1981). During summer, when average freshwater flows are much lower (11 to 17 m^3/s) (Figure 2.2), cooling water uptake rates may be greater than the net flow of fresh water in the estuary.

Cooling water use is greatest from RM 38 to RM 66. Although power plants are found along the length of the river, 88% of total cooling water usage is concentrated in this 28-mile stretch upstream from Haverstraw Bay (Figure 3.1; Table 3.1). Organisms with distributions centered in this area, or those with critical life stages found in this area, are particularly likely to be affected by power plant operation.

Table 3.1. Pertinent characteristics of power plants in the Hudson River estuary.[a]

Power plant	Location (mile point)	Total gross-rated capacity (Mwe)	Total cooling water flow (cfs)[b]	Plant temperature rise (°F)	Fuel	Operator	Yr of initial commercial operation
Albany, units 1–4	142.0	400	785	10.3	Fossil	Niagara Mohawk	1952–1954
Danskammer, units 1–4	66.5	480	705	17.0	Fossil	Central Hudson	1951–1967
Roseton, units 1, 2	66.0	1248	1429	17.8	Fossil	Central Hudson	1974
Indian Point							
Unit 1	43.0	285	709	12.0	Nuclear	Con. Ed.	1962[c]
Unit 2	43.0	906	1940	15.8	Nuclear	Con. Ed.	1973
Unit 3	43.0	1000	1940	17.1	Nuclear	PASNY	1976
Lovett, units 1–5	42.0	496	705	14.8	Fossil	Orange & Rockland	1949–1969
Bowline, units 1, 2	37.5	1244	1712	14.9	Fossil	Orange & Rockland	1972–1974
59th Street	5.0	132	375	6.7	Fossil	Con. Ed.	1918

[a] Adapted from EA (1978). Cooling water flow and plant temperature rise are design rather than operational values.
[b] Total cooling water flow is 9591 cfs; average daily freshwater inflow at Green Island has ranged from 882 cfs to 152,000 cfs (USGS, 1981); average monthly flow (1918–1980) at Green Island is 5583 cfs during August (McFadden, 1978; USGS, 1977–1980); average annual flow (1946–1980) at Green Island is approximately 13,800 cfs (USGS, 1981).
[c] Indian Point Unit 1 was withdrawn from commercial operation in 1974.

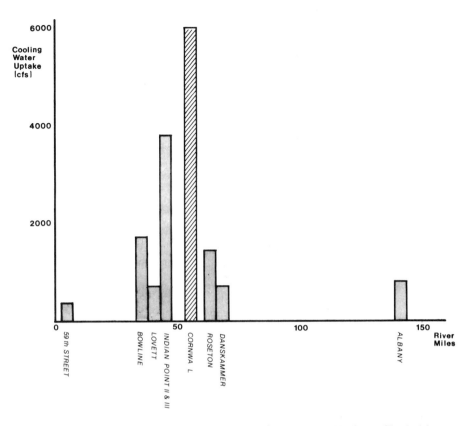

Figure 3.1. Uptake of river water by power plants on the Hudson. Shaded bars represent uptake by power plants in operation in 1982. The striped bar represents uptake of water by the proposed Cornwall Hydroelectric Plant.

Cooling water use and the temperature differential between cooling water at intake and discharge are inversely related in open-cycle cooling systems. Because the total heat exchange necessary for production of 1 kW of electricity is relatively constant, decreasing the flow of cooling water increases the temperature differential between intake and discharge. The temperature values given in Table 3.1 are the "design" temperature differentials. Actual temperature differentials are often higher as a result of reduced total flow of cooling water, as well as inefficiencies in the cooling system caused by build-up of material in the condenser tubes.

Closed-cycle cooling is an alternative method to the once-through cooling system used in the Hudson River power plants. Closed-cycle systems operate by transferring waste heat from power plant operation to an external cooling water stream that is open to the atmosphere. This is generally accomplished with "cooling towers." The towers can be of either "wet" or "dry" design (Hall et al., 1978). The dry towers operate in a fashion analogous to the operation of an automobile radiator. Heated water from the power plant passes through the interior of the "radiator," and is cooled by contact with the atmosphere. Wet cooling towers, which were proposed for construction on the Hudson River power plants, are not truly closed-cycle. Cooling water is drawn up into the cooling tower and trickles down a series of baffles. Heat is lost to the atmosphere due to evaporative cooling. Water is also lost from the system as part of the evaporative cooling process, and is replaced with river water. In addition, a small portion of the cooling water must continually be removed from the cooling water stream to prevent the accumulation of high concentrations of dissolved salts.

Closed-cycle cooling systems minimize the loss of fish due to impingement and entrainment. They are more expenseive to construct and operate than open-cycle cooling systems, however, and require more energy to operate. Cooling towers comprise 10 to 20% of total capital costs for power plant construction; about two or three times more than the capital costs of once-through cooling. Energy costs of operating wet cooling towers are normally less than 5% of plant output, while energy costs of dry cooling towers tend to be considerably greater—as much as 20% of a power plant's output (Hall *et al.*, 1978).

Chronology of Litigation and Hearings

In 1962, Consolidated Edison announced plans to construct a 2600-MW pumped storage reservoir at Cornwall-on-Hudson (Storm King Mountain; RM 56). During periods of off-peak electricity demand, water from the Hudson was to be pumped to a storage reservoir on Whitehorse Mountain; during peak hours, water was to be released from the reservoir to generate electricity. The pumped storage project was thus designed to act as a natural "storage battery" to provide increased generating capacity during periods of peak demand. Because of the need for peak-period electricity, Consolidated Edison considered the project worthwhile even though the electricity used to pump water up the mountain would be more than that generated when the water ran back down to the river. The projected average uptake of water by the Storm King (Cornwall) project was 18,000 cfs for 8 hr each day—approximately 50% more than the cooling water use of Indian Point Units Two and Three combined (Table 3.1). With the Storm King plant in operation, total uptake of water from the Hudson would have averaged 16,000 cfs, which is an amount greater than the average inflow of fresh water at Green Island.

The proposed Storm King pumped storage project started an environmental impact assessment (EIA) process that was to span several issues and many years, and involve federal agencies, state agencies, utility companies, and citizen groups. Two main issues formed the nucleus of the assessment process: the impacts of the Storm King storage project on the Hudson River ecosystem and the necessity of closed cycle cooling towers for Hudson power plants.

Following both the decision by Consolidated Edison to construct the Storm King pumped storage plant and initial licensing by the Federal Power Commission (FPC), the Scenic Hudson Preservation Conference petitioned to intervene in the licensing of the project. The FPC upheld the license for Storm King in 1964, but the license was set aside by the Court of Appeals in 1965 (Sandler and Schoenbrod, 1981). In an historic decision, the Court of Appeals stated that:

> The Commission's renewed proceedings must include as a basic concern the preservation of natural beauty and national historic sites . . . On remand, the Commission should take the whole fisheries question into consideration before deciding whether the Storm King project is to be licensed. (*Scenic Hudson Preservation Conference* v. *Federal Power Commission*, 354 F.2d 608 [2d Cir. 1965], Cert. denied. *Sub. nom. Consolidated Edison v. Hudson River Preservation Conference*, 384 U.S. 841 [1966].)

This decision resulted in a sharp increase in research on fisheries of the Hudson River. Major emphasis was placed on the striped bass, and the effects that passage of juvenile fish and eggs through the turbine system of the pumped storage plant would have on bass populations.

Two years of hearings followed the *Scenic Hudson v. FPC* decision. In August 1968, the FPC Hearing Officer recommended that the license for Storm King be granted (Sandler and Schoenbrod, 1981). In October 1968, objections to the license were raised by New York City, which alleged possible damage to the Catskill Aqueduct. In August 1970, the FPC relicensed the Storm King project after rejecting New York City's claims. The New York State Department of Environmental Conservation subsequently granted water quality certification for the project in 1971. This approval was challenged, revoked, reinstated, and finally upheld in March 1973. In March 1974, construction of the plant began on a limited basis. Construction was terminated in July 1974 with the opening of new FPC hearings. The hearings resulted from a successful appeal by the Hudson River Fishermen's Association (HRFA) to the Court of Appeals. In 1975, the Court of Appeals authorized further fisheries studies and halted the construction of Storm King pending an FPC ruling. In 1977, the FPC hearings were temporarily suspended (Sandler and Schoenbrod, 1981).

During this same time period, three other environmental controversies were brewing on the Hudson—all three connected with once-through cooling systems. Passage of the National Environmental Policy Act (NEPA) occurred in 1969. Consolidated Edison was in the process of constructing Indian Point Unit Two under a permit granted from the Atomic Energy Commission (AEC) in 1966. With the passage of NEPA, any federally funded or permitted project was required to undergo an EIA. In 1970, the AEC adopted regulations stipulating that atomic power plants need only consider radiological issues under NEPA. This regulation was overturned by the U.S. Court of Appeals, *Calvert Cliffs' Coord. Com. v. United States Atomic Energy Commission,* 449 F.2d 1109 (D.C. Cir. 1971). In compliance with this decision, the AEC staff evaluated the environmental effects of Indian Point Unit Two. After extensive hearings, AEC issued an operating license contingent upon installation of a closed-cycle cooling system no later than May 1, 1978 (Sandler and Schoenbrod, 1981).

Consolidated Edison appealed the terms of the license, and in April 1974, the date for installation of closed-cycle cooling was advanced to May 1, 1979. The Licensing Board also authorized extension of the installation date if new research showed that closed-system cooling was not necessary to protect the Hudson River fisheries. Indian Point Unit Three received a similar license, with the provision that a closed-cycle cooling system be installed by September 15, 1980. Licensing hearings regarding the design and necessity of closed-cycle cooling continued, and the date for installation of cooling towers was once again extended—this time to May 1, 1982 (Sandler and Schoenbrod, 1981).

Another cooling tower controversy had begun for Central Hudson Gas and Electric Company (CHG&E) and Orange and Rockland Utilities (O&R). In 1970, the U.S. Army Corps of Engineers issued a permit to CHG&E authorizing dredging of the Hudson River and construction of a barge-unloading facility and cooling water intake for the Roseton Plant. In 1971, O&R was issued a permit to construct a pier, to dredge, and to install a discharge pipe at the Bowline Point Plant. Both permits were issued without preparation of a final EIA, and were challenged by the HRFA in 1972. The HRFA claimed that the impact of water withdrawal on fish populations should be addressed in an EIA, because scientific testimony presented in the Indian Point hearings suggested that closed-cycle cooling systems were necessary to mitigate potential harm to the fish population (*HRFA v. Central Hudson,* 72 Civ. 5459 [CMM] [S.D.N.Y.] and *HRFA v. Orange and Rockland,* 72 Civ. 5460 [CMM] [S.D.N.Y.]). In January 1974, the Court ordered the Corps to prepare a final environmental impact statement by July 1974. The Corps did not meet this court-imposed deadline; it was not until August 1977 that a draft environmental impact statement was produced, and not until January 1981 that a final impact statement was issued (Sandler and Schoenbrod, 1981).

The third controversy was linked to the CWA, Public Law 92-500, signed into law October 18, 1972. With this law, the Environmental Protection Agency (EPA) was given the authority to regulate pollutant discharge in the nation's waterways. The National Pollutant Discharge Elimination System (NPDES) was established by EPA to grant permits allowing the discharge of pollutants under specified conditions. Thermal pollution is included among NPDES concerns, and thus utilities operating power plants were required to obtain permits regulating the design and operation of cooling systems. Two sections, 316(a) and 316(b), were particularly germane to regulation of power plant operation. Section 316(a) stipulates that thermal effluent limitations may be relaxed if the limitations are "more stringent than necessary to assume the protection and propagation of a balanced, indigenous population of shellfish, fish, and wildlife . . ." Section 316(b) states that "the location, design, construction, and capacity of cooling water intake structures (shall) reflect the best technology available for minimizing adverse environmental impact."

In 1975, EPA Region II (which includes New York State) issued proposed NPDES permits requiring closed-cycle cooling systems at Indian Point One and Two, Bowline, and Roseton. The utilities involved (Consolidated Edison, Power Authority of the State of New York, CHG&E, and O&R) requested and were granted an "adjudicatory hearing" to examine the utilities' claim that closed-cycle cooling systems were not necessary pursuant to section 316(a) of the CWA. With the start of hearings, the requirement to install closed-cycle cooling systems was automatically stayed. Intervenors in the adjudicatory hearings included HRFA, Commonwealth of Massachusetts, New York State Department of Environmental Conservation, and the Orange County Chamber of Commerce, among others. In July 1977, the utilities submitted their prepared written testimony, based primarily on research they had sponsored since 1968. Witnesses for the utilities were cross-examined on their testimony from December 1977 to December 1978 by EPA attorneys and scientists. In spring 1979, EPA witnesses presented their written testimony, and the utilities commenced cross-examination of EPA witnesses in July 1979. Cross-examination had not been completed by December 1980, when an out-of-court settlement was reached (Sandler and Schoenbrod, 1981).

On December 19, 1980, the utilities, EPA, New York State Department of Environmental Conservation, the Attorney General of the State of New York, and several environmental groups agreed to settle their disputes over the Storm King project and installation of cooling towers. The settlement was hailed as a "Peace Treaty for the Hudson" by the *New York Times,* and it may hold considerable promise as a model agreement for similar disputes. The major provisions of the settlement agreement are as follows: 1) no cooling towers needed to be built on the Indian Point, Bowline, and Roseton plants for at least 10 years; 2) the Cornwall pumped storage project at Storm King Mountain would not be built; 3) the utilities would contribute a $12 million endowment to the Hudson River Foundation, a newly created research foundation; 4) the utilities would spend $2 million annually on a biological monitoring program on the Hudson; and 5) mitigative measures would be taken to decrease the impact of power plant operation on fish populations. These mitigative measures include plant shutdown during peak spawning periods (which will eliminate some entrainment), and installation on the cooling water intakes of angled screens and dual-speed pumps that are designed to minimize impingement. In addition, the utilities have funded the establishment of a hatchery that will attempt to stock the Hudson River with 600,000 striped bass fingerlings (3 inches in length) each year.

Scientific Research Sponsored by the Utilities

Goals of the presettlement research programs sponsored by the utilities are described in detail by McFadden *et al.* (1977). The overall goal of the research was to provide information on the striped bass and other fish of the Hudson sufficient to demonstrate the impact (or lack thereof) of once-through cooling on fish populations. The specific goals included: 1) basic ecosystem studies; 2) understanding striped bass ecology and natural history; 3) estimation of entrainment and impingement mortality; and 4) estimates of power plant impact on the "equilibrium level" of the striped bass production using simulation models. The following sections describe the research conducted to attain each of these goals.

Basic Ecosystem Studies

Basic ecosystem studies conducted as part of the impact assessment process included quantitative and qualitative surveys of various communities, including those of phytoplankton, zooplankton, fish, benthic invertebrates, and rooted vegetation. No concise goals for "basic ecosystem studies" appear to have been formulated. As a consequence, most of the available information consists of species lists, with relatively little information on the abundance or spatial distribution of particular populations. Fish populations represent an exception to this rule—extensive data on spatial and temporal changes in fish populations were collected for striped bass and white perch, in particular. The specific information found in various reports is given in Chapter 7.

Understanding the structure and function of biological communities in relation to the abiotic environment is central to a "basic ecosystem study." Proportionately little research was directed toward understanding the functional aspects of the ecosystem—the cycles of nutrients and energy that support "important" species such as fish.

As described in the previous section, the lack of emphasis on understanding the river as an ecosystem was due in part to historical accident. Litigation brought by environmental groups placed immediate emphasis on the effects of power plants on fish populations, especially striped bass. Striped bass have considerable economic value for sport and commercial fisheries, and they are one of the major predators in the Hudson River. Thus, they are a reasonable focus for investigation and meet several of the criteria proposed by Coutant (1975) for choosing "representative important species."

Given the focus on striped bass and a few other species, which resulted from litigation, a very narrow view of the ecosystem was taken in the impact assessment process. Population dynamics of striped bass in the Hudson were studied in a virtual ecological vacuum, making it extremely difficult to establish the causes of fluctuations in population levels of the striped bass, much less evaluate the effects of power plant operation on striped bass populations.

Research on striped bass in other areas has shown that several mechanisms may be responsible for fluctuations in striped bass populations. Mihursky et al. (1981) investigated the effects of hydrologic regime, water quality, and the abundances of zooplankton and phytoplankton on the various life stages of striped bass in Chesapeake Bay. They found that strong year-classes of juvenile striped bass were correlated with cold winters followed by above-average spring runoff, and were not related to the abundance of spawning stock, eggs, or early larval stages. They concluded that control of striped bass breeding success was regulated by the abundance of zooplankton at the time of first larval feeding, which was in turn affected by hydrologic regime.

Hydrologic regime is also important in determining the success of juvenile striped bass in the Sacramento-San Joaquin estuary. In this instance, however, flow conditions in the estuary affect striped bass success by direct cropping of larvae. During periods of high riverine flow, striped bass are carried downstream to favorable nursery grounds in Suisun Bay (Chadwick et al., 1977). At low flows, young striped bass are less widely distributed in the estuary, and more likely to be drawn into the diversion intakes that pump water to agricultural lands south of the estuary.

In both the San Joaquin and Chesapeake Bay estuaries, strong year-classes of young-of-the-year striped bass show strong correlation with flow regime. Despite this apparent similarity, the underlying relationships that produce the observed correlation are radically different. In the San Joaquin estuary, flow is related to direct cropping of organisms, while in Chesapeake Bay flow affects the abundance of phytoplankton and consequently zooplankton, which are the major food of striped bass larvae. The results of these two studies show the potential dangers inherent in using results of multiple correlation analyses in impact assessment. Without understanding the mechanisms responsible for a correlation between two variables, predictions regarding the response of a system to changes in external variables will be unreliable.

Multiple linear regression analysis provided the backbone of much of the data analysis conducted by utility-sponsored researchers. The basic question asked was, "What environmental parameters determine the abundance of young-of-the-year striped bass in August?" Abundance indices were based on catch per unit effort from beach seines during July and August. McFadden *et al.* (1977) examined seven variables using this approach: 1) weighted average freshwater inflow; 2) weighted average spring freshwater inflow minus weighted average winter flow; 3) rate of warming of surface water from 12 to 16°C; 4) rate of warming of surface water from 16 to 20°C; 5) power plant water withdrawal rate; 6) predator abundance (juvenile bluefish and older striped bass); and 7) egg production index. For 1965–1975, it was found that predator index, egg production index, and rate of temperature change (16 to 20°C) explained 79% of the variability in young-of-the-year abundance. Power plant water withdrawal rates did not explain a significant fraction of the abundance of young-of-the-year, from which it was concluded that power plant operation had no significant effect on young-of-the-year abundance.

Ricker (1975) cautions that depending on the number of variables, a minimum of 15 to 25 years of data is necessary for reliable multiple linear regressions. Research on the Hudson subsequent to McFadden *et al.*'s (1977) study supports Ricker's contention. Klauda *et al.* (1980) reports that for the period 1965 through 1977, striped bass year-class success was most consistently related to average daily freshwater inflow. The discrepancy between his analysis and that of McFadden demonstrates the tenuous nature of conclusions based on multiple linear regression analysis. As of 1980, Texas Instruments, a major consultant to the utilities, had abandoned its regression analysis in favor of simulation modeling (Klauda *et al.*, 1980).

The observation that year-class success is related to freshwater inflow to the estuary does little in and of itself to clarify the mechanisms regulating striped bass populations, and hence the effects of power plant operation on striped bass. The studies previously cited on the San Joaquin estuary (Chadwick *et al.*, 1977) and Chesapeake Bay (Mihursky *et al.*, 1981) demonstrate that radically different processes may result in a correlation between flow and striped bass success. Unfortunately, data collected by the utilities do not provide any insight into the causal relationships between freshwater flow and success of young-of-the-year striped bass. Sampling of zooplankton abundance and phytoplankton production was sporadic and confined largely to studies of the effects of entrainment in power plant cooling water streams (see Chapter 7). No attempt was made to examine primary production (e.g., along the length of the river) or to maintain an annual sampling regime that could be related to striped bass success or abiotic variables such as temperature regime and freshwater inflow. These two problems—lack of coherent sampling design and lack of continuity from year to year—make the data base on basic ecosystem processes quite weak.

Much of the data on basic ecosystem processes is buried in reports on entrainment, impingement, or fisheries, and has not received adequate review. A case in point is that of studies on primary production under-

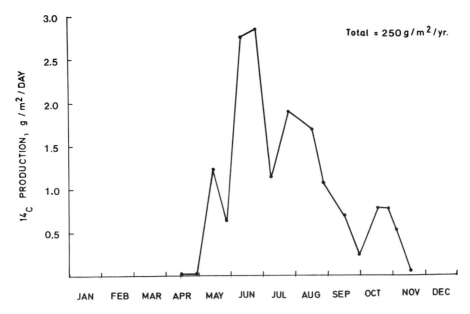

Figure 3.2. ^{14}C primary production at Bowline Point power plant (adapted from Ecological Analysts, Inc., 1981, Table B-3).

taken during 1976 as part of an analysis of the effects of entrainment on riverine biota. Primary productivity (^{14}C method) was measured on planktonic samples taken before and after passage of cooling water through the Bowline Generating Station (Ecological Analysts, Inc., 1981). Results of the study show no appreciable effects on primary productivity due to entrainment. Data presented in the body of the report are extraordinarily high compared to average values for temperate estuaries (up to 20 mg C/m²/day, Ecological Analysts, Inc., 1981 (Figure 4-1)), and are almost certainly incorrect. An appendix of the report (Table B-3) contains raw data that appear to be more reliable. These data can be used to estimate annual primary production for this reach of the river (Figure 3.2). The numbers agree reasonably well with those collected by other workers in other parts of the river (Sirois and Fredrick, 1978). This simple exercise indicates that further analysis of the existing data sets could provide much useful information for future decision-making. Unfortunately, due to the lack of a comprehensive and consistent sampling regime, it does not appear that data on primary production of the river as a whole are available.

Striped Bass Ecology and Natural History

Studies of the ecology and natural history of striped bass in the Hudson were more successful than the basic ecosystem studies described in the previous section. There are several reasons for this success—the most important being a clear statement of research goals and a large sampling effort. Much of the research described in this section was conducted as part of the extensive study of fish populations in the Hudson River, which formed the core of research conducted by consultants hired by the utility industry.

The major goal of the Hudson River fisheries studies was to determine standing crops in specific river regions (Q. Ross, pers. comm.). This section examines three projects from the overall study in detail: 1) a study of the mortality and growth of young-of-the-year striped bass (Dey, 1981); 2) a study of the movements of older (age II and above) striped bass (McLaren *et al.*, 1981); and 3) a study of the relative importance of Hudson River stocks to the Atlantic coast fishery (Berggren and Lieberman, 1978). Each of these research projects has been presented as part of the testimony in various hearings, and has also been published in peer-reviewed journals.

Mortality rates of striped bass larvae and young-of-the-year juveniles have a strong influence on adult population levels in subsequent years. An understanding of the mechanisms that govern larval and juvenile mortality is thus very important in determining the effects of power plant operation, or other anthropogenic stresses, on striped bass populations. In the Hudson River, Texas Instruments conducted an extensive sampling program during 1975 and 1976 (Dey, 1981). Striped bass eggs, larvae, and early juveniles were collected weekly from 12 sampling regions with Tucker trawls and epibenthic sleds. Juvenile striped bass were collected using beach seines and epibenthic sleds. Weekly estimates of standing crops and average size of each life stage in the entire river were obtained by summing data collected from each of the 12 sampling regions. Weekly growth and mortality rates of striped bass larvae were calculated from length-frequency histograms. The effects of water temperature and flow on growth and mortality of larval stages and juveniles were analyzed using correlation analysis.

The results of this study show that for 1975 and 1976, mortality of larvae and juvenile striped bass were high during June and lower during July and August (Figure 3.3). Similar trends (i.e., rapidly decreasing mortality rate in early life history stages) are observed in other animals that lay large numbers of eggs (Hutchinson, 1978), as well as in striped bass of the Potomac estuary (Polgar, 1977). Dey (1981) observed distinct differences in the abundance and growth of larvae and juveniles during 1975 and 1976, which he attributed to an unusually sharp decline in temperature during May. He has hypothesized that this decrease in temperature caused the apparent cessation of spawning that was reflected in the bimodal abundance of eggs and yolk-sac larvae of striped bass (Figure 3.4). Dey (1981) observed that the much smaller spawn in early June 1976 appeared to be the source of all the juvenile fish surviving until August.

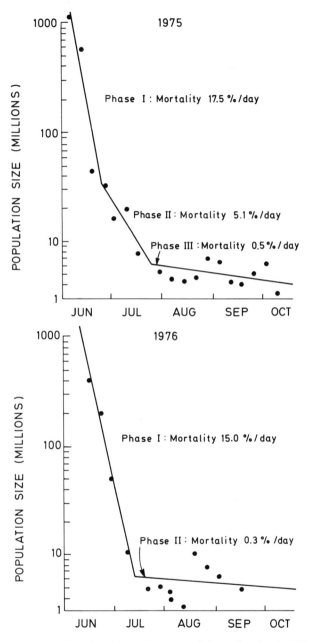

Figure 3.3. Mortality of striped bass larvae and juveniles in the Hudson River estuary, 1975–1976 (from Dey, 1981).

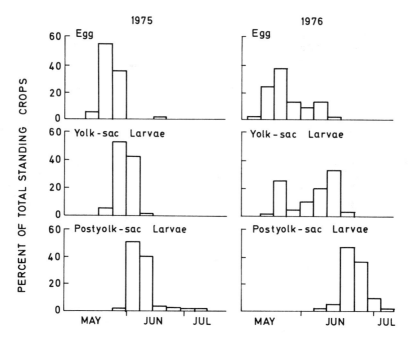

Figure 3.4. Temporal pattern in abundance of striped bass eggs, yolk-sac larvae, and postyolk-sac larvae in the Hudson River estuary, 1975–1976 (from Dey, 1981).

From these results, the author concluded that temperature regime was an important factor regulating the growth and abundance of striped bass in the Hudson. This conclusion was further supported by a positive correlation between instantaneous growth rates of striped bass and water temperature during summer for the period 1965–1976 (Dey, 1981). Boreman (1983) reached the same conclusion in a simulation study of striped bass eggs and larval development.

The mechanisms by which temperature affects growth and mortality of striped bass in the Hudson are not known. Direct effects on metabolic processes may be responsible; indirect effects such as changes in population levels of prey organisms may also be responsible (Dey, 1981). Taken at face value, a positive correlation between growth of striped bass and river temperature appears to indicate that increases in the temperature of the Hudson from the operation of power plants would be beneficial to the striped bass population. However, if indirect effects are responsible for the observed correlation—for example, zooplankton densities also increase with temperature, providing more food for the striped bass—then the effects of power plant operation are much harder to predict. Another point to consider is the importance of average conditions versus extreme events. If the normal range in river temperatures is exceeded due to heat generated by power plants, then one or two days of extremely high temperatures might decrease the growth of larval fish and greatly increase their mortality rate. The research carried out by the utilities (Dey, 1981) clearly establishes the importance of temperature in the life history of striped bass, but it is inadequate to predict the effects of increased temperatures associated with power plant operation.

McLaren et al. (1981) studied the seasonal movements of older (age II and above) striped bass in the Hudson during 1976 and 1977. A major goal of their study was to describe the migration of bass in and out of the Hudson in order to determine the possible importance of striped bass from the Hudson to the Atlantic coastal fishery. Their study was also designed to provide information regarding the movement of spawning stocks within the Hudson during the breeding season.

Striped bass were collected and tagged by McLaren et al. (1981) on the spawning grounds from the Tappan Zee Bridge (RM 26; Figure 2.1) upstream to RM 69 (above Newburgh; Figure 2.1). Approximately 5000 fish were tagged during 1976 and 1977. Rewards were offered for the return of tags and information on the date, location, and method of capture. Several hundred tags were returned during 1976 and 1977. Tagged fish were found throughout the Hudson in both years, and substantial numbers were recovered outside the Hudson. Most of the fish leaving the river traveled northeast into Long Island Sound. Fish were caught as far north as Newburyport, Massachusetts and as far south as Slaughter Beach, Delaware.

Sampling with gill nets and haul seines to obtain the fish used in the tagging study provided information about the movement of older fish in the Hudson prior to summer migrations. Males arrived on the spawning grounds earlier than females. In addition, a higher proportion of fish near the principal spawning grounds (around RM 60) were found to be sexually mature than those captured further downstream.

The results of McLaren et al. (1981) show that after spawning, many striped bass leave the Hudson and begin a northern migration similar to that observed in fish leaving Chesapeake Bay (Mansueti, 1961). The major difference in the migration patterns observed for Chesapeake and Hudson stocks is the sex ratio of migrating fish. In the Hudson, no bias toward either females or large size was observed in migrating fish (McLaren et al., 1981). In contrast, fish leaving the Chesapeake are primarily larger females (Kohlenstein, 1980). The preponderance of females found along the Atlantic coast (Schaefer, 1968) suggests that the Chesapeake is the major source of migratory striped bass entering the coastal fishery. Although Hudson stocks do contribute to the coastal fishery, they are most important in New York Bight and Long Island Sound. Thus, any effects of power plants on striped bass of the Hudson River probably will have most impact on the coastal fisheries in close proximity to the mouth of the Hudson.

The final utilities-sponsored research project discussed here is that of Berggren and Lieberman (1978). The major purpose of their study also was to determine the importance of Hudson River stocks in the coastal fishery. In contrast to the mark-recapture technique used by McLaren *et al.* (1981) and described above, Berggren and Lieberman (1978) used innate "tags" to determine the birthplace of fish captured along the Atlantic coast.

Striped bass (4 to 12 years old) were collected on the Atlantic coast (Maine to North Carolina) and at spawning sites in three estuaries—Hudson River, Chesapeake Bay, and Roanoke River. Assuming that fish spawn in the river in which they were born, the birthplace of the oceanic fish can be determined by comparing their identifying characteristics with those of the three estuarine populations. Obviously, the accuracy of such a process depends on the reliability of the characteristics used to identify the origin of oceanic fishes. Discriminant analysis was used by Berggren and Lieberman (1978) to determine what morphological and meristic characters would best serve as identifying characteristics. They found that snout length, internostril width, distance between focus and first annulus, distance between first annulus and second annulus, and number of rays in various fins were the characteristics that most readily discriminated among fish of the three estuaries.

The probable origin of fish collected off the Atlantic coast during 1975 was primarily the Chesapeake Bay. Berggren and Lieberman (1978) estimated that approximately 6% of the total coastal fishery originated in the Hudson. Although higher proportions of Hudson stock were found in areas adjacent to the mouth of the Hudson, even in these areas fish from Chespeake Bay were more abundant. Thus, on the basis of their research, it appeared that fish from the Hudson do not play an important role in the coastal fishery.

The methods of data analysis employed by Berggren and Lieberman (1978) received close scrutiny by staff members at Oak Ridge National Laboratory as part of the Nuclear Regulatory Commission hearings. Van Winkle and Kumar (1982) obtained the original data set used by Berggren and Lieberman (1978) for their reanalysis. Using the maximum likelihood method as well as the discriminant function method, they obtained results similar to those of the original authors when the entire data set was used in the analysis. They also examined the data by year-classes and sex, and obtained somewhat different results. Because 1970 was a very strong year-class in Chesapeake Bay, the catch of striped bass in the coastal fishery was dominated by these fish (44% of the total oceanic catch in 1975 was comprised of five-year-old females; Berggren and Lieberman, 1978). Dominance by this one year-class from the Chesapeake tended to overwhelm the possible contribution of Hudson River fish from other year-classes. When the origin of each sex and age classification was considered in a separate analysis, Van Winkle and Kumar (1982) found that as many as 49% of the fish in a particular age class were probably from the Hudson. Highest proportions of total catch with origins in the Hudson were obtained for the 1965 year-class; most years, however, showed a much smaller contribution from the Hudson. Van Winkle and Kumar (1982) suggested that a similar study should be undertaken as soon as possible, because by 1982, the fish of the strong 1970 year-class in the Chesapeake would no longer be an important part of the coastal fishery.

In the case of Berggren and Lieberman (1978), the adjudicatory proceedings provided substantiation of the results obtained by the utilities' consultants, as well as further insights into the origins of the striped bass fishery of the Atlantic coast (Van Winkle and Kumar, 1982). In this instance, research conducted as part of the EIA process entered the scientific mainstream. Reanalysis of the data as part of the adjudicatory process (Van Winkle and Kumar, 1982) probably resulted in more careful scrutiny than most data ever receive in the course of normal peer review. The interplay between science and the law thus appears to have had a positive impact on both in this particular case. Unfortunately, the same cannot be said for other research projects conducted by the utilities. Much of the data has not been thoroughly analyzed or published in peer-reviewed journals, and may never be.

Entrainment and Impingement

Concern over the effects of drawing organisms through a power plant's cooling system became a major issue in EIA during the early 1970s. In the 1960s, EIA tended to focus primarily on the problems associated with waste heat loading or "thermal pollution." When it became apparent that original fears of damage to aquatic ecosystems from waste heat loading would not be as catastrophic as originally feared, attention shifted towards the impact of entrainment and impingement of aquatic biota in cooling water flows (Hanson et al., 1977). Recognition of the potential magnitude of problems of entrainment and impingement was fostered by a realization of the numbers of fish involved—up to many millions per year at a single plant (Hanson et al., 1977)—and the graphic display of the problem that dumps filled with striped bass carcasses presented (Boyle, 1969).

Estimation of the effects of entrainment and impingement has received considerable study on the Hudson River, as well as on many other power plant sites. Numerous conferences and several edited volumes have dealt with this aspect of power plant-induced mortality in fish populations (e.g., Jensen, 1974, 1976, 1981; Schubel and Marcy, 1978; Van Winkle, 1977). The results of some of these studies will be mentioned briefly in this section, but the interested reader is urged to consult these references for consideration of entrainment and impingement mortality in areas other than the Hudson River.

During summer, average freshwater flows in the Hudson reach or drop below the total rate at which cooling water is withdrawn from the river by power plants (Table 3.1), and this has some effect on the amount of water in this tidal stretch of the estuary. The potential impact of mortality due to entrainment and/or impingement of fish and other organisms is substantial, and much of the impact assessment research was focused on this subject in studies by the utilities and their consultants. Most of the data (field and modelled) are to be found in unreduced form in the annual reports produced by various utility companies; these reports are catalogued in Chapter 7. The published reports included here are primarily those authored by scientists at Oak Ridge National Laboratory, and are based on their preparation for the Nuclear Regulatory Commission hearings. The raw data, however, were obtained by consultants to the utilities; primarily Lawler, Matusky, and Skelly (LMS), Ecological Analysts, Inc. (EA), and Texas Instruments.

The effects of entrainment on a fish population generally have been studied in three phases: 1) measurement of total numbers of entrained fish; 2) determination of the entrainment mortality fraction (the ratio of the number of organisms killed to the number entrained; Beck *et al.* 1978); and 3) estimation of the effects that this mortality will have on the abundance of fish in a particular year-class. The first two phases—determination of total entrainment and entrainment mortality fraction—are discussed in this section, while the final phase is discussed in the following section.

The total number of fish entrained in the cooling water inflow to a power plant is a function of many factors, including density of fish, their age and size, design of the cooling water intake, velocity of the cooling water inflow, and flushing rate in the vicinity of the intake. Mortality of entrained fish is a function of residence time, temperature, and mechanical stress within the cooling water system. Biocides such as sodium hypochlorite, often used to control fouling organisms within condensing tubes, may also cause fish mortality (Morgan and Carpenter, 1978). In general, younger fish and planktonic forms (eggs, ichthyoplankton) are most susceptible to entrainment (McFadden, 1977). Because larvae of many fish, including striped bass, show vertical diurnal migrations, time of day may affect the susceptibility of larvae to entrainment (McFadden, 1977).

A major obstacle to accurate determination of the mortality rate suffered by entrained fish is the sampling mortality associated with using nets to filter entrained organisms from a high-velocity current. Intake and discharge velocities can be as high as several feet per second in a power plant (McFadden, 1977), causing considerable mortality of entrained fish.

Sampling mortality is thus a major problem in the estimation of entrainment mortality fractions.

Significant technical advances allowing improved estimation of entrainment mortality fractions occurred with the introduction of "larval tables" to sample entrained organisms from cooling water intakes and discharges (Christensen *et al.*, 1981). Originally designed by LMS and refined by EA, the larval table allows the passage of a cooling water sample through collecting nets at relatively low velocities (McGroddy and Wyman, 1977). With this improvement in sampling gear, much lower and presumably more accurate estimates of entrainment mortality fractions were made. For example, in 1974, entrainment sampling using nets resulted in an estimated entrainment mortality fraction of 0.73 at the Indian Point power plant for postyolk-sac larvae of striped bass. At the Bowline and Roseton plants, use of larval tables resulted in values of 0.09 and 0.25, respectively, for the estimated entrainment mortality fraction of postyolk-sac larvae (McFadden, 1977). In contrast to earlier expectations, the best estimates of mortality rates showed that up to 75% or more of entrained fish larvae survived passage through a power plant cooling system. Although there continued to be minor disagreement between the consulting firms and scientists retained by EPA and the Nuclear Regulatory Commission over the best estimates of entrainment mortality fractions, this is an instance in which improved methodology resulted in a consensus of opinion regarding one aspect of the effects of power plant operation on fish of the Hudson River (Christensen *et al.*, 1981).

The effects of sampling regime on estimation of entrainment mortality fraction have been studied quantitatively by Vaughan and Kumar (1982). To aid in the design of sampling programs at power plants, they examined the interrelationships among several variables, including: 1) number of fish collected at intake and discharge stations; 2) fraction alive in the intake sample; 3) mean entrainment mortality fraction; 4) minimum detectable entrainment mortality; and 5) width of the confidence interval about estimated entrainment mortality. They observed that minimum detectable entrainment mortality and the confidence interval of estimated entrainment mortality are related to both sample size at the intake and the fraction alive in the intake sample (Vaughan and Kumar, 1982). The relationship between these parameters is then presented in a series of graphs, such as that shown in Figure 3.5. The results shown in Figure 3.5 demonstrate that there is a rapid initial decline in minimum detectable entrainment mortality as sample size increases. An asymptote is reached (dependent on intake survival) beyond which increased sample size is ineffective in reducing the minimum detectable entrainment mortality (Figure 3.5). The results obtained by Vaughan and Kumar (1982) have direct practical application in the design of sampling regimes for studies of entrainment mortality. They also demonstrate in a quantitative manner the importance

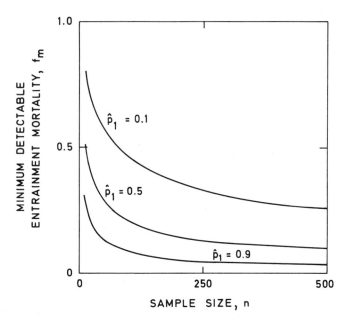

Figure 3.5. Minimum detectable entrainment mortality (f_m) plotted against sample size (n) for three levels of observed intake survival ($\hat{p}_1$), $\delta = 1$, and $\alpha = 0.05$ (from Vaughan and Kumar, 1982).

of improved sampling techniques, such as the larval table, for accurate estimation of entrainment mortality in ichthyoplankton. In this instance, then, scientists working in the regulatory arena have helped to improve the effectiveness of the sampling regimes used in studies of the environmental impact of power plant operation.

Fish that are too large to pass through protective screens as they are drawn into a power plant cooling water intake are said to be "impinged." Mortality rates of impinged fish are a function of species, size, life history stage, and water velocity. Direct mortality can occur due to abrasion and suffocation; delayed mortality may result from exhaustion, internal hemorrhaging, or scale loss (Hanson et al., 1977). Loss of scales or other trauma may result in increased susceptibility to disease and parasitism. Accumulation of debris on intake screens may trap and entangle fish, as well as alter approach velocity at the intake (Hanson et al., 1977). Researchers originally assumed that mortality of both impinged and entrained fish was complete. With more extensive sampling, it is now generally recognized that not all fish subjected to impingement and entrainment die.

Impingement of fish by power plants along the Hudson was measured by periodically collecting fish from the "traveling screens" which continuously remove accumulated debris from screens at cooling water intakes. Fish not sampled were returned to the river. Impingement rates for striped bass are shown in Table 3.2 for 1975.

Many more white perch than striped bass are killed annually by Hudson River power plants. During 1974, for example, more than three million perch were impinged (Barnthouse and Van Winkle, 1981). The implications of impingement mortality for white perch populations are discussed by Barnthouse and Van Winkle (1981). They used as a measure of impact the impingement conditional mortality rate (Ricker, 1975) defined as "the fraction of the vulnerable population that would be killed by impingement in the absence of mortality from all other sources . . ." (Barnthouse and Van Winkle, 1981). The impingement conditional mortality rate is numerically equivalent to the average fractional reduction in year-class abundance due to impingement; and it also reflects the differential impact of impinging fish of different ages, because losses of fish in each year-class are treated separately. No attempt was made to extrapolate from estimates of loss in a particular year-class to declines in the long-term abundance of white perch. The use of conditional mortality rates in modeling fish population dynamics on the Hudson is discussed in more detail in a later section.

Based on raw data collected by the consulting firms, Barnthouse and Van Winkle (1981) calculated the conditional mortality rates of white perch due to impingement on a monthly basis at the six major power plants on the Hudson. Data used in their analysis included estimates of the abundance of each year-class, monthly counts of impinged fish, and

Table 3.2. Estimates of absolute numbers of striped bass killed by impingement at each plant, all plants combined, and post-1970 units during 3-month intervals, 1975[a]

Plant	Jan.-Mar.	Apr.-June	July-Sept.	Oct.-Dec.	Annual
Bowline	42,204	32,042	188	6922	81,356
Lovett	3088	998	403	835	5324
Indian Point Unit 1	938	108	17	—[b]	1063
Indian Point Unit 2	5134	3524	18,638	5135	32,431
Indian Point Unit 3	—[b]	—[b]	—[b]	—[b]	—[b]
Indian Point—all units combined	6072	3632	18,655	5135	33,494
Roseton	89	96	643	560	1388
Danskammer	151	467	733	1235	2586
Post-1970 units	47,427	35,662	19,469	12,617	115,175
All plants combined	51,604	37,235	20,622	14,687	124,128

(From McFadden (1978, Table 9.4-2).
[a]Assumes 100% mortality.
[b]Not operating.

monthly estimates of total mortality of white perch vulnerable to impingement. Due to the lack of reliable figures for monthly mortality rates, a high and low value were used in two separate analyses. The values used were annual conditional mortalities of 0.8 (assumed by McFadden and Lawler, 1977) and 0.5. Data on the number of white perch in various age classes killed by impingement are given by Van Winkle et al. (1980).

Results of the study for all power plants combined are shown in Table 3.3. Conditional mortality rates of white perch (i.e., fractional reduction in year-class strength) range from 10 to 59%, depending upon the year in question and various assumptions that are made. For the 1974 year-class, it appears that impingement probably accounted for a 20% reduction in year-class strength (Barnthouse and Van Winkle, 1981).

Analysis of conditional mortality rates on a plant-by-plant basis (Figure 3.6) shows that the Indian Point plants accounted for an inordinately large fraction of total impingement mortality. Indian Point (Units One, Two, and Three combined) was responsible for the death of more white perch than all the other five plants combined (Barnthouse and Van Winkle, 1981) (Figure 3.6). Impingement mortality is particularly high during win-

Table 3.3. Estimates of total conditional impingement mortality rates (m_1) and impingement exploitation rates (in parentheses) for 1974 and 1975 year-classes of Hudson River white perch. Estimates were computed using all combinations of assumptions about initial population size, natural mortality, and number of years of vulnerability.[a]

No. of yr of vulner-ability	Yr-class	Initial population Size					
		Low natural mortality rate		Best estimate natural mortality rate		High natural mortality rate	
		Low	High	Low	High	Low	High
2	1974	0.309 (0.165)	0.446 (0.200)	0.177 (0.094)	0.255 (0.114)	0.095 (0.051)	0.137 (0.061)
	1975	0.166 (0.082)	0.245 (0.099)	0.116 (0.057)	0.172 (0.069)	0.077 (0.038)	0.115 (0.046)
3	1974	0.387 (0.172)	0.588 (0.209)	0.221 (0.099)	0.336 (0.119)	0.119 (0.053)	0.181 (0.064)
	1974	—	—	—	—	—	—

(From Barnthouse and Van Winkle (1981, p. 203).
[a] Total conditional impingement mortality rate calculated using equation 3 in Barnthouse and Van Winkle (1981). Total conditional impingement mortality rates are equal to fractional (or percentage) reductions in year-class strength due to impingement assuming no compensation. Exploitation rate calculated by dividing the total number of white perch impinged in a year-class during the entire period of vulnerability on the initial size of the young-of-the-year population at start of the period of vulnerability.

Entrainment and Impingement

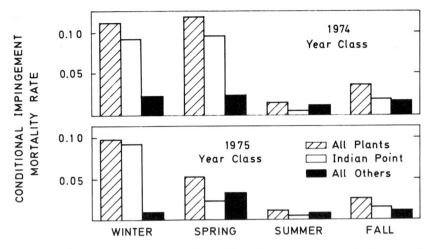

Figure 3.6. Seasonal comparison of conditional impingement mortality rates for all plants combined, for Indian Point (all units combined), and for all other plants combined (from Barnthouse et al., 1982).

ter (Figure 3.6), when cold temperatures reduce the mobility of white perch.

High rates of impingement at Indian Point appear to be related to the natural history of white perch. Perch overwinter in the lower and middle estuary (McFadden, 1977) and tend to congregate in the deepest areas of the river channel (Texas Instruments, 1974). Indian Point is located in the middle of the preferred overwintering area, and the cooling water intake is located in the river channel. The plant thus appears well situated for the impingement of white perch.

Barnthouse and Van Winkle (1981) argue that the large impingement mortality of white perch at Indian Point should be mitigated by reducing the number of impinged fish or increasing the survival rate of those that are impinged. While this argument is valid, a point of more fundamental importance is the extent to which their results point out the problem inherent in after-the-fact impact assessment. After a thorough study of the spatial and temporal distribution of white perch in the Hudson River, it becomes evident that large numbers of white perch are being killed simply because the Indian Point power plant was built on the "wrong" (from the perspective of white perch) side of the river. If it had been built on the other side of the river, away from the main channel, the cooling water intake would not be drawing perch from their preferred spot in the channel. Likewise, if it had been built in another area of the estuary, away from the overwintering sites of the white perch, losses due to impingement would be dramatically reduced (Figure 3.6). Basic, irreversible decisions about the location and design of the plant were made before thorough environmental studies were completed (in fact, most of the plant was completed prior to enactment of NEPA). Millions of dollars were spent on environmental research in an effort to justify the existing placement of power plants, rather than on research to choose the most environmentally sound site before construction began.

Ability to Detect Reduction in Year-Class Strength

An important consideration in impact assessment is the ability to detect impact in the presence of a high level of background "noise." It is intuitively obvious, of course, that high variability in the reference values for a parameter of interest (e.g., year-class strength in white perch) makes detection of changes difficult. Van Winkle et al. (1981) and Vaughan and Van Winkle (1982) examined this problem quantitatively using data on the Hudson to estimate year-class strength in white perch. They found no significant trends in year-class strength for the period 1972–1977. Using data gathered during 1972–1977 as the reference or "baseline" condition, they then examined the extent to which declines in year-class strength in subsequent years could be detected, given the variability associated with the "baseline" condition. Two versions of the analysis were published — Van Winkle et al. (1981) and a corrected version (Vaughan and Van Winkle, 1982). The results of the corrected version are discussed here.

The results of Vaughan and Van Winkle (1982) are not encouraging. With the actual data on white perch impingement, a minimum of 20 years of additional data collection would be necessary to detect a reduction in year-class strength of 50%. The relationship between minimum detectable reduction and the number of years of impingement data is shown explicitly in Figure 3.7. For populations with high variability (coefficients of variation 100%; Figure 3.7), reductions of less than 50% can never be detected with a statistical power (probability of correctly rejecting the null hypothesis that there is no reduction in year-class strength) greater than 0.25 (Figure 3.7).

This sort of quantitative analysis has very important implications for the practice of impact assessment. Clearly, the demonstration of impact in a population with a high natural coefficient of variation is impossible within the time frame imposed by most adjudicatory hearings. "I'll let you know in 20 years" is not an adequate answer for issues such as the environmental impact of power plant operation. Other, indirect methods must be used to assess the likely effects of power plant operation. One such method is the creation of a simulation model based on field studies of the distribution of fish and the direct mortality associated with power plant operation. The use of models in the impact assessment process on the Hudson is described in the next section.

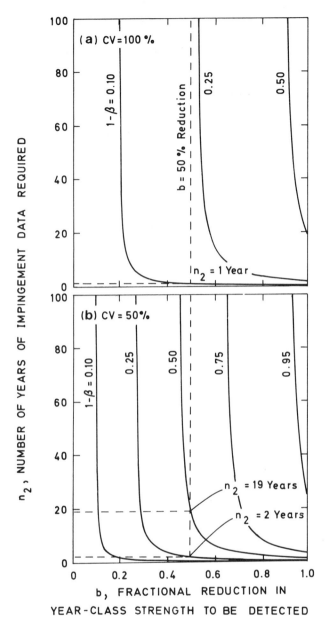

Figure 3.7. Number of years of additional impingement data (starting in 1978) required to detect a specified fractional reduction in mean year-class strength of young-of-the-year white perch in the Hudson River. Curves are drawn for $a = 0.05$ over a range of powers $(1 - B)$, for $n_1 =$ five years of pre-1978 data and for two values of the coefficient of variation (CV): (a)CV = 100% and (b)CV = 50%. Values of n_2 on the figure are the number of additional years of data required for the specified statistical powers to detect a 50% reduction in mean year-class strength. Detectable fractional reduction in year-class strength is 0.44 for CV = 50% and is 0.88 for CV = 100%. (from Vaughan and Van Winkle, 1982).

Models Used in Power Plant Impact Assessment*

Most of the controversy generated during the EPA and Nuclear Regulatory Commission hearings revolved around the appropriate use of models in impact assessment. As described in the previous section, high variability in baseline data on fish populations makes it virtually impossible to detect anything less than catastrophic changes in year-class strength without 10 or 20 years of additional data. A model then becomes a logical necessity, whether it be a simple conceptualization or a highly elaborate simulation model. In this section, we present the models of effects on striped bass populations developed primarily by Oak Ridge National Laboratory (ORNL) scientists as technical advisors to the regulatory agencies, and by the utilities' consultants (Quirk, Lawler, and Matusky [QLM], later becoming Lawler, Matusky, and Skelly [LMS], and Texas Instruments [TI]).

Viewed in retrospect, the development and use of impact assessment models for the Hudson may be divided into three phases (Barnthouse *et al.*, 1984): 1) an initial, fairly simple phase; 2) a period of quite complex models; and, in the end, 3) a move back toward simpler descriptions. Modelling efforts were also tightly coupled to data collection; and for a time, field research was directed toward verification of, or providing input for, a number of increasingly complex models. Figure 3.8 and Table 3.4 present a general chronology of these efforts. From 1968 to approximately 1977, most models developed tended to invoke increasing numbers of mechanisms and refined data inputs to explain the movements of bass populations and project their potential vulnerability to entrainment and impingement. However, for various reasons (below), none of these could predict power plant effects with much certainty. Beginning in 1975, a trend toward simpler, more empirical models was evident. These proved to be the easiest models to work with, and were in fact used in developing the Hudson River Settlement Agreement (Barnthouse *et al.*, 1984).

Phase 1

The earliest models were based on relatively simple assumptions regarding the relationship of river flow, fish abundance, and power plant water consumption. The first documented model assessed the impact of the proposed Cornwall plant at Storm King Mountain on striped bass populations (Hudson River Policy Committee, 1968), and involved comparison of the average daily water withdrawal rate with the average tidal flows:

$$\left\{ \begin{array}{l} \% \text{ Reduction in} \\ i^{\text{th}} \text{ life stage} \end{array} \right\} = \frac{(N_{1,i})(V_w)(f)}{(N_{2,i})(V_t)} (100),$$

---

*This section of the chapter is authored by Karin Limburg.

Models Used in Power Plant Impact Assessment

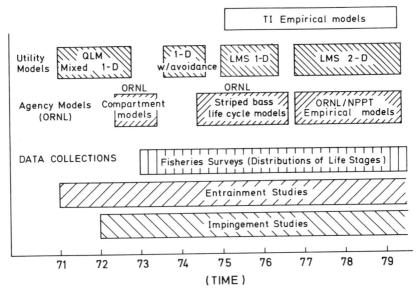

Figure 3.8. Time chart of Hudson River fisheries impact models and concurrent, relevant data collections. (Note: data reports were published with about a two-year time lag following sampling.)

where $N_{1,i}$ = weekly abundance of the ith life stage (eggs, larvae, or y-o-y) in zone of withdrawal,
$N_{2,i}$ = weekly abundance of ith life stage in total cross section,
V_w = average volume withdrawn during pumping (18,000 cfs),
V_t = average tidal flow at Cornwall (100,000 cfs),
f = fraction of day when pumping occurs (8 hr/day or 1/3).

The results gave percent reductions in the immediate vicinity of the Cornwall plant. These were multiplied by factors expressing the ratios of near-field to river-wide standing crops to obtain estimates of river-wide reduction (Christensen and Barnthouse, pers. comm.).

This model projected that less than 0.7% of the eggs, 2.9% of the larvae, and up to 6.2% of young-of-the-year (y-o-y) fish would be vulnerable, leading the Committee to conclude:

> This study indicates that the operation of the Cornwall plant alone would have negligible effects on the fisheries of striped bass and other species occurring in the estuary. (Hudson River Policy Committee, 1968, p. 6)

However, as was apparent to early workers at ORNL in 1972 (Christensen *et al.*, 1981), the use of average tidal flow (V_t) did not account for the fact that passively borne organisms would wash back and forth in front of the plant with repeated tidal cycles. If, instead, the net downstream flow

Table 3.4. Models used to assess impacts of entrainment and impingement on the Hudson River striped bass population. Dashed lines separate models into different periods of development.

	Models	
Proceedings	Utilities	Agencies
(No hearing) 1968		Cornwall Entrainment Model[a]
Indian Point Unit 2 licensing hearing (1972–1973)	QLM completely mixed model[b] QLM 1-D model[d]	AEC/ORNL compartment model[c]
FPC Cornwall hearing (1974)	QLM 1-D model with transport-avoidance factors[e]	—
Indian Point Unit 3 final environmental statement (1975)	—	ORNL 1-D transport model[f] ORNL striped bass life-cycle model[f]
Indian Point Unit 2 license (1976) extension hearing	LMS 1-D model[g]	ORNL striped bass life-cycle model[h]
Corps of Engineers Bowline study (1977)	—	LMS 1-D model[i] ORNL striped bass life-cycle model[i]
EPA NPDES permit hearings (1977-1980)	LMS 2-D Real time life-cycle model[j] TI empirical entrainment and models[j]	NPPT/ORNL empirical transport model[k] ORNL empirical im-impingement models[k]

The author thanks L.W. Barnthouse for providing the information in this table.
[a]Hudson River Policy Committee (1968).
[b]Lawler (1972a).
[c]U.S. Atomic Energy Commission (1972).
[d]Lawler (1972b).
[e]Lawler (1974).
[f]Van Winkle et al. (1974); U.S. Nuclear Regulatory Commission (1975).
[g]Consolidated Edison Co. of NY, Inc. (1975).
[h]U.S. Nuclear Regulatory Commission (1976).
[i]Barnthouse et al. (1977).
[j]McFadden (1977); McFadden and Lawler (1977).
[k]Boreman et al. (1982); Barnthouse et al. (1982).

(11,000 cfs) had been used, calculated impacts would have increased roughly 9-fold (C. Hall, pers. comm.).

No further models were presented until 1972, when both utility consultants and advisors to the U.S. Atomic Energy Commission (AEC) produced models to predict impacts of one of the nuclear plants, Indian Point

Unit 2 (Lawler, 1972a, 1972b; USAEC, 1972). The first of these (Lawler, 1972a), presented at the Atomic Safety and Licensing Board (ASLB) hearings in April of that year, assumed a uniform distribution of vulnerable striped bass life stages. Impacts were calculated by: 1) comparing water volumes withdrawn by the power plant with the total estuarine volume; and 2) employing a life-cycle model to project y-o-y mortality over a single year of operation (Barnthouse et al., 1984). The assumption of uniformly distributed fish was later relaxed in a model presented in October 1972 at the same hearings (Lawler, 1972b). The latter model, referred to as QLM 1-D, developed longitudinal-advective and -dispersive transport equations to move eggs and larvae past the power plant. Long-term (>10 yr) projections of impacts to striped bass populations were also possible.

Meanwhile, scientists at ORNL were developing their own models. One early model, described by Hall (1977), divided the estuary from Croton to Albany into nine sections. In each of these, changes in populations of striped bass were ascribed to the processes of birth, natural death, movements into and out of the section, and entrainment-induced mortality where appropriate. Early life stages were assumed to be homogeneously distributed in each section and to move "planktonically," i.e., drift with the currents. The number of eggs, larvae, and juveniles dying from entrainment was a function of their densities in the section, the quantity of water withdrawn from any nearby power plant (assumed to operate at full capacity), and the expected mortality of individuals passing through the plant (assumed to be 100%). The model predicted a 20% entrainment mortality over the first four months of life with the Lovett, Danskammer, and Indian Point Units One and Two operating, as was the situation in 1973 (Hall, 1977). Sensitivity analyses, which included the operation of all on-line, under construction, or planned power stations, drove the estimated entrainment mortality up to 70 to 80% (Hall, 1977).

Another, somewhat more complicated, model was appended to the Final Environmental Statement for Indian Point Unit Two (USAEC, 1972). This model was based on the principle of two-layered estuarine transport: freshwater would carry eggs and larvae downstream to the zone of two-layered bidirectional flow, where some of the organisms would sink into the lower layer and be transported back upriver by the incoming tide. Those individuals that were brought upriver at night would migrate up toward daylight, following their zooplanktonic prey, and would be subject to downstream transport once again. Because the Indian Point intakes are located around the salt front in the estuary, the life stages vulnerable to entrainment could be exposed repeatedly. The ORNL model captured the essence of this process, without using complicated physical transport equations (Barnthouse et al., 1984).

Both the more complex ORNL and QLM 1-D models employed sets of weighting coefficients, referred to as "W-factors" or "migration factors," to account for differential susceptibility of life stages to power plant withdrawal (e.g., avoidance of the intake by juveniles). These coefficients were calculated with different assumptions about the effectiveness of the intake's withdrawal ability; QLM assumed that the best estimate of intake concentration of eggs and larvae was derivable from surface concentrations near the intake, whereas ORNL used a depth-averaged concentration estimate. The former resulted in a lower estimate of entrainment susceptibility than the latter (Christensen et al., 1981).

Another set of factors, now called "f-factors," were used to account for differential survival of entrained organisms (entrainment mortality fraction), as well as for their non-uniform distribution in the vicinity of the power plant (Barnthouse et al., 1984). Again, these factors gave different emphasis to entrainment mortality, with ORNL's factors assuming greater mortality (Christensen et al., 1981).

However, the greatest conceptual difference between the ORNL and QLM models, in fact between all successive agency and utility models, lay in their respective formulations of fish population growth. Specifically, the utilities' consultants invoked the theory of compensatory growth, which holds that as populations are reduced in size, individuals' survival increases because food is relatively more abundant, predation is relatively decreased, waste products do not accumulate, etc. The QLM-LMS models expressed early life stage mortality with the following form (QLM, 1974; Swartzman et al., 1978):

$$K_{E'} = K_E + (K_E - K_O) * \left[\frac{C_k - C_s}{C_s}\right]^3$$

where K_E = equilibrium mortality (per day),
K_O = minimum rate of decay (per day),
C_k = total concentration of life stage in segment k (mass or numbers/volume),
C_s = carrying capacity of segment k (mass or numbers/volume).

The cubic term served to accentuate the response of population mortality (and hence survival as well) to changes in population density. This had a large effect on predictions of long-term impact. Whereas the ORNL model projected that 30 to 50% of each year-class would suffer entrainment mortality, the QLM 1-D model predicted a stabilization, after one decade of equilibration, at a population level only 3.5 to 6% lower than that prior to operation of the power plant.

The ASLB decided (Atomic Safety and Licensing Board, 1973) that the issue of entrainment impacts had not been satisfactorily resolved, largely because of uncertainty in parameter estimates and the questionable validity of the assumption of compensatory growth. Both the ORNL and QLM models were then run with similar conservative input assumptions and produced similar results (Barnthouse et al., 1984). Based on these conservative impact estimates, the ASLB ordered Con Edison to construct at once a closed-cycle cooling system. The utility appealed the decision on the grounds that the costs of such mitigation were not justified by the expected benefits. The Atomic Safety and Licensing Appeals Board (ASLAB) upheld Con Edison's appeal in 1974; furthermore, it considered the QLM analysis to be more accurate than ORNL's. ORNL was ordered to reassess its analysis, and the required closed-cycle cooling system was delayed (Barnthouse et al., 1984).

Phase 2

The years immediately following the 1974 ASLAB decision witnessed the generation of several more models and revisions of previous analyses. QLM resubmitted its one-dimensional model, with modifications, for the FPC's Cornwall hearing in 1974 (Lawler, 1974). At about the same time, ORNL prepared its own one-dimensional transport model and striped bass life-cycle model (Van Winkle et al., 1974) and applied it in the Indian Point Unit Three Final Environmental Statement (FES) (U.S. Nuclear Regulatory Commission, 1975; Eraslan et al., 1976). The ORNL model projected the combined impact estimates for all power plants, with different runs either including or deleting the operation of the proposed Cornwall station (Christensen et al., 1981). New, more refined data on spatiotemporal distributions of striped bass life stages were used. During the calibration runs, actual transport rates of eggs and larvae were found to be slower than predicted by ORNL's model, so terms were included in ORNL 1-D to partially counter hydrodynamic transport. Analogous terms were also added to the QLM 1-D model in 1974 (Christensen et al., 1981) (Table 3-4).

During 1975–1977, three implementations of the ORNL striped bass life-cycle model and a new version of the one-dimensional model (LMS 1-D) from the now reorganized consultants were presented at hearings for Indian Point Units Two and Three and Bowline Point (Table 3.4). In 1977, the utilities presented testimony at the EPA hearings for NPDES permits; the testimony (McFadden, 1977; McFadden and Lawler, 1977) included a new, even more complex analysis in two dimensions (LMS 2-D Real Time Life-Cycle model, or LMS 2-D).

For LMS 2-D, the river was longitudinally divided as before to estimate hydrodynamic transport; additionally, each segment was split into two vertical layers in order to estimate vertical movements of eggs and larvae. LMS 2-D simulated river movement and transport of organisms over each tidal cycle, and could be run for extended periods to estimate growth and mortality of striped bass from eggs to juveniles over the course of a growing season. Results of LMS 2-D were then used to predict long-term (40 year) impacts on the entire Hudson striped bass population. A parallel approach was developed by Texas Instruments and referred to as the "equilibrium reduction equation" (ERE), because it estimated the reduction in size of the fish population that is assumed to have been in equilibrium (steady state) prior to operation of the power plants in question (but not *all* of the Hudson power plants). The model took the form of a so-called Ricker stock-recruitment equation (after Ricker, 1954), which predicts the amount of young fish recruited to the spawning stock as a function of the size of their parents' stock (McFadden and Lawler, 1977):

$R = \alpha \cdot P \cdot e^{-\beta P}$,
$\alpha = E \cdot e^{(-k_1 T)}$,
$\beta = k_0 \cdot T$,
R = number of recruited (1-year-old) adults,
E = egg production per spawning adult,
P = number of parental fish (females),
T = time of recruitment,
k_1 = sum of all unit mortality rates for density-independent mortality of all prerecruitment age fish,
k_o = density-dependent unit mortality rate for prerecruits.

Alpha (α) is termed the "reproductive reserve" of the stock and represents mortality factors that act independently of the density of recruits. Beta (β) is referred to as a compensatory term that keeps the stock size "fine-tuned" to an equilibrium level. McFadden and Lawler (1977) went on to state that alpha ". . . becomes a measure of compensatory reserve, since what it measures, by implication, is the amount of compensatory or density-dependent mortality (beta) required to bring the *actual* (sic) survival ratio of recruits to spawners to unity on a long-term basis (p. 1-IV-12)." The basic equation was then modified to include a number of factors, such as a multiple-age spawning stock, density-dependent influences, and increased temperatures from cooling cycle effluents.

This model, and the LMS 2-D model, received a great number of criticisms by numerous witnesses for EPA (e.g., Levin, 1979; Slobodkin, 1979; Levins, 1979; Fletcher and DeRiso, 1979; Ricker, 1979; Policansky, 1979; and others). These ranged from criticism of the nature and appropriateness of the models ("borrowing" a fishery stock-recruitment model for the present purposes) to the lack of evidence for compensatory mechanisms, and to the total exclusion of community- and ecosystem-level interactions. Among the most serious criticisms were the charges that the curve-fitting techniques used to estimate the alpha and beta parameters were scientifically unsound.

Christensen *et al.* (1982) demonstrated one very useful technique for validating the utilities' estimates—one that bears mentioning. Alpha's value had been chosen as 4.0; a lower value would have resulted in a much higher estimate of power plant impact. Because it is almost physically impossible to verify alpha by field measurements (due to the enormous sampling requirements over a long time period), a numerical technique was used. A Leslie-type matrix was used to calculate fish population changes over time. Alpha was arbitrarily specified; and from it, beta was calculated at the start of the simulation. Both were incorporated into a term describing the probability of survival of 0-year fish into the one-year class (i.e., recruitment probability). An index of catch per unit effort (CPUE) was generated (such an index had been the basis of the utilities' parameter estimation technique), and random error was introduced into it (as the utilities claimed was true for their estimated CPUE). The model was run for 50 simulated years to equilibrate, followed by 3120 more simulated years. The latter yearly results were divided into 120 groups of replicates of 26 years each (because the length of the utilities' CPUE record was 26 years). Each replicate was analyzed in the same manner that the utilities had used on the actual Hudson CPUE set, ultimately yielding 12 data sets, each containing 22 simulated stock- recruit data points per replicate. Finally, the same methods used by the utilities to estimate alpha and beta were employed on the 1440 data sets.

The model was run with a series of specified alpha values under different assumptions and randomizations. Not surprisingly, the curve-fitting techniques failed to estimate the correct value for alpha. When alpha was set at 2.5, alpha was usually overestimated; when set > 5, alpha was usually underestimated, and above 10, estimates of alpha were almost always underestimated. Thus, the analysis of Christensen *et al.* (1982) appeared to dismiss effectively the utilities' contention that their choice of alpha was conservative.

Phase 3

It was obvious to all participants that substantial grounds for conflict continued to exist. Further analyses of analyses (e.g., Barnthouse *et al.*, 1977; Swartzman *et al.*, 1978) revealed less obvious differences in the utility and agency models. Swartzman *et al.* (1978), in reviewing all of the major models used by ORNL and LMS, concluded that:

> The high variability in data . . . and the simplicity of assumption and lack of information in such areas as density-dependent mortality make *no existing model useful as a predictor of power plant impact*. . . . The most striking result [of comparing the models by means of a simulator] is the ability to demonstrate insignificant impact (low PR[1]) through the use of d-d mortality functions with parameters used in the LMS models. (Swartzman *et al.*, 1978, pp. 142–144)

Other factors determining major differences in model projections included the number of segments into which the river was divided, choice of

---

[1]PR, percent reduction of fish stock due to entrainment and impingement; d-d, density-dependent.

entrainment factors, and choice of transport avoidance factors; all of these differences led to lower LMS impact predictions relative to ORNL (Swartzman *et al.*, 1978). The use of hydrodynamic data collected for a given year, as input to other years for predicting fish spatial distribution, was deemed a failure as well.

Dissatisfaction with the state of modelling prompted both sides to try a different approach—one that first appeared in a Texas Instruments annual report (Texas Instruments, 1975). The models made maximal use of the large data sets from the utilities' ongoing biological monitoring program. Known as direct impact assessment techniques, the models calculated an index of impact called the conditional mortality rate, m, from classical fishing theory (Ricker, 1975):

$$m = 1 - e^{-Ft},$$
$F =$ instantaneous rate of power plant mortality (analogous to fishing mortality),
$t =$ period when young fish are vulnerable to entrainment and impingement.

The method was far simpler (and cheaper!) to apply than other models and eliminated ambitious predictions of future stock reductions (Barnthouse *et al.*, 1984), and was applied by utility and agency consultants alike. When both groups evaluated the same 1974 and 1975 data sets, their estimates of impact were closer than ever before (McFadden and Lawler, 1977; Barnthouse *et al.*, 1982; Boreman *et al.*, 1982). By the time the Settlement Agreement was being negotiated, direct impact assessment models were most trusted, and were used extensively to evaluate various schemes for mitigating the power plants' combined impacts.

The results of several of the different kinds of models are given in Table 3.5. Compared to the first stage of modelling (Part A, Table 3.5), the estimates by both sides of power plant induced-mortality were much closer (Parts B and C) in the later phases.

Concluding Remarks

Environmental impact assessment (EIA) is the illegitimate offspring of an unhappy union between law and science. The very nature of the two disciplines argues against their compatibility. Science and scientists deal hierarchically with observations, hypotheses, generally accepted principles, dogma, and ultimately "laws." Very few research efforts produce laws—those seemingly indisputable truths such as the Second Law of Thermodynamics. Self-criticism, skepticism, and an underlying belief that one can never be absolutely certain of anything characterize most scientists. Lawyers, however, deal with the inverse of the scientist's universe. The laws are already known; and it is the job of the lawyer to make the facts fit the law, to force the universe to work under previously defined constraints. Lawyers do not deal in probability levels—a case is either won or lost, rather than probably won with a confidence level of 95%.

Christensen et al. (1981) argue forcefully that the quality of science was improved by focusing attention on the specific issues addressed in the adjudicatory hearings regarding operation of Hudson River power plants. Unquestionably, a clearly defined focus greatly enhances the chances that meaningful results will be produced in any scientific endeavor. Examination of the research undertaken by utility consultants shows that the most successful projects, such as that examining the relative contribution of Hudson River fish to the Atlantic coast fishery, were the ones with clear-

Table 3.5. Comparison of impact predictions from different models.

A. Life-cycle models (striped bass only)

Model	% Reduction of population	Young-of-year compensation	Long-term % reduction in 1-yr-old fish No. of yr of plant operation				Source
			5	10			
QLM 1-D[a]	2.07	High	2.71	4.01			Lawler (1973),
	3.42	Low	5.68	7.48			cited in
	3.13	None	5.55	12.00			Swartzman et al., (1977)
			7	10	40		
LMS 2-D[b]	1.21	High	1.33	1.68	2.18		Lawler (1975),
	1.26	High	1.38	1.75	2.26		cited in
	2.44	Low	2.81	3.91	6.99		Swartzman et
	3.14	Low	3.61	5.03	8.99		al. (1977)
	4.47	Low	5.13	7.11	12.46		
			5	10	20	40	
ORNL 1-D[c]	10	None	10	14	17	18	USNRC
	25	None	25	33	38	42	(1975), cited
	50	None	50	62	70	75	in Swartzman et al. (1977)
ORNL (Hall)[d]	20	None	(1973 conditions, May-Aug.)				Hall (1977)
	70–80	None	(all power plants on-line)				
	50–60	None	(all power plants on-line)				

Table 3.5. *Continued*

B. Empirical transport model for 1974 and 1975 year-classes of four fish species[e]

Species	Yr-class	% Entrained	% Impinged	% Combined
Striped bass	1974	11.1–14.5	1.1–9.2	12.1–22.4
	1975	18.2–18.4	0.4–3.5	18.5–21.3
White perch	1974	10.9–11.7	11.9–44.6	21.5–51.1
	1975	13.0–13.6	11.5–24.5	23.0–34.8
Atlantic tomcod	1975	5.2–8.4	0.6–3.0	5.8–11.1
American shad	1974	13.6	0.1–0.5	13.7–14.1

C. Conditional mortalities for 1974 and 1975 year-classes, estimated with LMS 2-D with no compensation[f]

Species	Yr-class	% Entrained	% Impinged	% Combined
Striped bass	1974	4.46	3.88	8.34
	1975	6.22	2.15	8.37

[a]QLM 1-D, one-dimensional model developed by Quirk, Lawler, and Matusky; hydrodynamic transport equations modelled with longitudinal segmentation, and coupled to population model of striped bass.
[b]LMS 2-D, two-dimensional model by Lawler, Matusky, and Skelly; included a vertical dimension in each river segment.
[c]ORNL 1-D, one-dimensional (longitudinal) striped bass life-cycle model developed by Oak Ridge National Labortory for NRC.
[d]Percent reduction in this model refers to fish in first 4 months of life.
[e]From Boreman *et al.* (1982), cited in Barnthouse *et al.* (1984).
[f]From McFadden and Lawler (1977).

est focus. It may be argued, however, that adjudicatory hearings are not the most effective means of establishing scientific focus. Nor is the impact assessment circus an appropriate arena for good science.

Too often, the wrong questions are asked by the law (Barnthouse *et al.*, 1984). The legal system should not expect scientists to provide definitive numerical answers to the question "What is the ultimate impact of once-through cooling on striped bass populations in the Hudson?" The capabilities and limitations of science should be established at the outset, and these limitations should be reflected in the types of questions that are asked of scientists and scientific research in the courtroom.

There are numerous ways by which the impact assessment process might be transformed to better serve the public interest. Most important among them would be a restructured system of funding and a commitment to support smaller numbers of detailed long-term studies, rather than more numerous short-term studies. The program developed in Maryland to examine environmental effects of power plant siting is a distinct improvement over more conventional arrangements, in which the utilities pay to have the data collected, edit final reports to their specifications, and generally control the scientific process. In Maryland, an independent research team is maintained by the state government. They are responsible to the state, rather than the utility company, and thus enjoy a much higher level of autonomy and much less rewriting or outright censorship than that occurring in most other areas. This sort of system has many advantages over that employed in most states, and that followed by the Federal government under NEPA.

An important aspect of impact assessment that warrants additional discussion is the use of "representative and important" species as the specific focal point of an investigation. Studies on the Hudson do not provide a good test case for the usefulness of the "representative and important species" concept in impact assessment. The focus on striped bass populations of the Hudson was due, in large part, to public awareness and interest rather than a specific decision on the part of the utilities or EPA.

As noted in the opening section of this chapter, research efforts by the utilities were directed toward fish populations in 1965, when *Scenic Hudson Preservation Conference v. Federal Power Com.* held that "the whole fisheries question" should be taken into consideration before licensing of the Storm King hydroelectric plant. Subsequent litigation undertaken by the HRFA brought continued focus on fish populations, in particular those of the striped bass. Consideration of one or several species, rather than the entire ecosystem, was reaffirmed by EPA's endorsement of the "Representative Important Species" (RIS) concept for the NPDES hearings.

Coutant (1975) argues that the RIS concept is a pragmatic means by which the intractable problem of studying all aspects of an ecosystem can be reduced to manageable proportions. He describes several criteria that should be considered in the designation of representative and important species, including economic importance, vulnerability to the particular action under consideration, and ecological importance, in the sense that some species may be particularly important to community structure and function. Striped bass, white perch, and Atlantic tomcod were considered to be among the representative and important species for the Hudson. Although they may have been adequate choices, there appears to have been only nominal consideration given to taxa other than fish. This is discussed further in Chapter 6.

References

Atomic Safety and Licensing Board. 1973. Initial decision in the matter of Consolidated Edison Company of New York, Inc. (Indian Point Station Unit No. 2). Atomic Safety and Licensing Board of the U.S. Atomic Energy Commission. Sept. 25, 1973.

Barnthouse, L.W., J.B. Cannon, S.W. Christensen, A.H. Eraslan, J.L. Harris, K.H. Kim, M.E. LaVerne, H.A. McLain, B.D. Murphy, R.J. Raridon, T.H. Row, R.D. Sharp, and W. Van Winkle. 1977. A selective analysis of power plant operation on the Hudson River with emphasis in the Bowline Point generating station. ORNL/TM-5877 (Vols. 1 and 2), Oak Ridge National Laboratory, Oak Ridge, TN.

Barnthouse, L.W. and W. Van Winkle. 1981. The impact of impingement on the Hudson River white perch population, pp. 199–205 In Issues Associated with Impact Assessment. Proceedings of Fifth Annual Workshop on Entrainment and Impingement (L.D. Jensen, ed.). EA Communications, Sparks, MD.

Barnthouse, L.W., W. Van Winkle, J. Golumbek, G.F. Cada, C.P. Goodyear, S.W. Christensen, J.B. Cannon, and D.W. Lee. 1982. Impingement impact analyses, evaluations of alternative screening devices, and critiques of utility analyses relating to density-dependent growth, the age structure of the Hudson River striped bass population, and the LMS real-time life-cycle model. ORNL/NUREG/TM-385/V2. Oak Ridge National Laboratory, Oak Ridge, TN.

Barnthouse, L.W., W. Van Winkle, and D.S. Vaughan. 1983. Impingement losses of white perch at Hudson River power plants: magnitude and biological significance. Env. Mgmt. 7(4):355-364.

Barnthouse, L.W., J. Boreman, S.W. Christensen, C.P. Goodyear, W. Van Winkle, and D.S. Vaughan. 1984. Population biology in the courtroom: the Hudson River controversy. Bioscience 34(1):14–19.

Baslow, M.H. and D.T. Logan. 1982. The Hudson River ecosystem. A case study. Report to the Ecosystems Research Center. Cornell University, Ithaca, NY. 460 pp.

Berggren, T.J. and J.T. Lieberman. 1978. Relative contribution of Hudson, Chesapeake, and Roanoke striped bass, *Morone saxatilis*, stocks to the Atlantic coast fishery. Fish. Bull. 76(2):335–345.

Boreman, J., L.W. Barnthouse, D.S. Vaughan, C.P. Goodyear, S.W. Christensen, K.D. Kumar, B.L. Kirk, and W. Van Winkle. 1982. Entrainment impact estimates for six fish species inhabiting the Hudson River estuary. ORNL/NUREG/TM-385/V1. Oak Ridge National Laboratory, Oak Ridge, TN.

Boreman, J. 1983. Simulation of striped bass egg and larva development based on temperature. Trans. Am. Fish. Soc. 112:286–292.

Boyle, R.H. 1969. The Hudson River, a Natural and Un-natural History. W.W. Norton and Co., Inc., New York, NY. 304 pp.

Chadwick, H.K., D.E. Stevens, and L.W. Miller. 1977. Some factors regulating the striped bass population in the Sacramento-San Joaquin Estuary, California, pp. 18-351 In Proceedings of the Conference on Assessing the Effects of Power-Plant-Induced Mortality on Fish Populations (W. Van Winkle, ed.). Pergamon Press, New York, NY.

Christensen, S.W., W. Van Winkle, L.W. Barnthouse, and D.S. Vaughan. 1981. Science and the law: confluence and conflict on the Hudson River. Environmental Impact Assessment Review 2:63–88.

Christensen, S.W., C.P. Goodyear, and B.L. Kirk. 1982. An analysis of the validity of the utilities' stock-recruitment curve-fitting exercise and "prior estimation of beta" technique, Vol. III. The Impact of Entrainment and Impingement on Fish Populations in the Hudson River Estuary. ORNL/NUREG/TM-385/V3. Oak Ridge National Laboratory, Oak Ridge, TN.

Coutant, C.C. 1972. Biological aspects of thermal pollution II. Scientific basis for water temperature standards at power plants. CRC Critical Reviews in Environmental Control 3(1):1–24.

Coutant, C.C. 1975. Temperature selection by fish—a factor in power plant impact assessments, pp. 575–597 In Environmental Effects of Cooling Systems at Nuclear Power Stations. Proceedings of a Symposium in Sweden, August 26–30, 1974. IAEM-SM-187/11.

Dey, W.P. 1981. Mortality and growth of young-of-the-year striped bass in the Hudson River estuary. Trans. Am. Fish. Soc. 110:151–157.

EA. 1978. Roseton and Danskammer Point Generating Stations 316(a) demonstration. Prepared by Ecological Analysts, Inc. for Central Hudson Gas and Electric Corp.

Ecological Analysts, Inc., 1981. Bowline Point Generating Station entrainment abundance and survival studies. 1979 Annual Report with Overview of 1975–1979 Studies. Prepared for Orange and Rockland Utilities, Inc.

Eraslan, A.H., W. Van Winkle, R.D. Sharp, S.W. Christensen, C.P. Goodyear, R.M. Rush, and W. Fulkerson. 1976. A computer simulation model for the striped bass young-of-the-year population in the Hudson River. ORNL/NUREG-8. Oak Ridge National Laboratory, Oak Ridge, TN.

Fletcher, R.I. and R.B. Deriso. 1979. Appraisal of certain arguments, analyses, forecasts, and precedents contained in the utilities' evidentiary studies on power plant insult to fish stocks of the Hudson River estuary. University of Washington, Seattle, WA. Prepared for the U.S. EPA. Submitted in the EPA-Utilities Adjudicatory Hearings No. C/II-WP-77-01, Exb. 218.

Hall, C.A.S. 1977. Models and the decision making process: the Hudson River power plant case, pp. 345–364 In Ecosystem modelling in theory and practice (C.A.S. Hall and J.W. Day, Jr., eds.). John Wiley & Sons, New York, NY.

Hall, C.A.S., R. Howarth, B. Moore, III, and C.J. Vorosmarty. 1978. Environmental impacts of industrial energy systems in the coastal zone. Ann. Rev. Energy 3:395–475.

Hanson, C.H., J.R. White, and H.W. Li. 1977. Entrapment and impingement of fishes by power plant cooling-water intakes: an overview. Mar. Fish. Rev. 39:7–17.

Hudson River Policy Committee. 1968. Hudson River fisheries investigations (1965–1968). Report to Consolidated Edison Co. of NY, Inc. (2 vols.)

Hutchinson, G.E. 1978. Introduction to Population Ecology. Yale University Press, New Haven, CT. 260 pp.

Jensen, L.D. (ed.) 1974. Proceedings of the Second Workshop on Entrainment and Intake Screening. EPRI. Proj. RP-49 Report 15. Palo Alto, CA.

Jensen, L.D. (ed.) 1976. Third National Workshop on Entrainment and Impingement. Ecological Analysts, Inc., Melville, NY.

Jensen, L.D. (ed.) 1981. Issues associated with impact assessment. Proceedings of Fifth National Workshop on Entrainment and Impingement. Ecological Analysts, Inc., Middletown, NY.

Klauda, R.J., W.P. Dey, T.B. Hoff, J.B. McLaren, and Q.E. Ross. 1980. Biology of Hudson River juvenile striped bass, pp. 101-123 In Proceedings of Fifth Annual Marine Recreational Fisheries Symposium (H. Clepper, ed.). Sport Fishing Institute, Washington, D.C.

Kohlenstein, L.C. 1980. On the proportion of the Chesapeake Bay stock of striped bass that migrates into the coastal fishery. Trans. Am. Fish. Soc. 110:168–179.

Lawler, J.P. 1972a. The effect of entrainment at Indian Point on the population of the Hudson River striped bass. Written testimony presented on April 5, 1972 before the U.S. Atomic Energy Commission in the matter of Consolidated Edison Co. of NY, Inc. (Indian Point Station, Unit 2).

Lawler, J.P. 1972b. The effect of entrainment at Indian Point on the population of the Hudson River striped bass. Written testimony presented on October 30, 1972, before the U.S. Atomic Safety and Licensing Board in the matter of Consolidated Edison Co. of NY, Inc. (Indian Point Station, Unit No. 2), USAEC Docket 50-247.

Lawler, J.P. 1974. Effect of entrainment and impingement at Cornwall on the Hudson River striped bass population. Testimony presented to the Federal Power Commission in the matter of Cornwall, USFPC Project No. 2338, October 1974.

Levin, S.A. 1979. The concept of compensatory mortality in relation to impacts of power plants on fish populations. Written testimony prepared for the U.S. Environmental Protection Agency, Region II.

Levins, R. 1979. Community structure and population change. Written testimony prepared for the U.S. Environmental Protection Agency, Region II.

Mansueti, R.J. 1961. Age, growth, and movements of the striped bass, *Roccus saxatilis*, taken in size-selective fishing gear in Maryland. Chesapeake Sci. 2:9–36.

McFadden, J.T. (ed.) 1977. Influence of Indian Point Unit 2 and other steam electric generating plants on the Hudson River estuary, with emphasis on striped bass and other fish populations. Prepared for Consolidated Edison Co. of NY, Inc.

McFadden, J.T. and J.P. Lawler (eds.). 1977. Supplement I to: Influence of Indian Point Unit 2 and other steam electric generating plants on the Hudson River estuary with emphasis on striped bass and other fish populations. Prepared for Consolidated Edison Co. of NY, Inc.

McFadden, J.T. (ed.) 1978. Influence of the proposed Cornwall pumped storage project and steam electric generating plants on the Hudson River estuary, with emphasis on bass and other fish populations. (Revised) Prepared for Consolidated Edison Co. of NY, Inc.

McGroddy, P.M. and R.L. Wyman. 1977. Efficiency of nets and a new device for sampling living fish larvae. J. Fish. Res. Bd. Can. 34:571–574.

McLaren, J.B., J.C. Cooper, T.B. Hoff, and V. Lander. 1981. Movements of Hudson River striped bass. Trans. Am. Fish. Soc. 110:158–167.

Mihursky, J.A., W.R. Boynton, E.M. Sletzer-Hamilton, and K.U. Wood. 1981. Freshwater influences on striped bass population dynamics, pp. 149–167 *In* Proc. Natl. Symp. on Freshwater Inflow to Estuaries. U.S. Fish and Wildlife Service. FWS/OBS-81/04.

Morgan, R.P., II and E.J. Carpenter. 1978. Biocides, pp. 95–134 *In* Power Plant Entrainment: A Biological Assessment (J.R. Schubel and B.C. Marcy, eds.) Academic Press, New York, NY.

Polgar, T.T. 1977. Striped bass ichthyoplankton abundance, mortality, and production estimation for the Potomac River population, pp. 110–126 *In* Proceedings of the Conference on Assessing the Effects of Power-Plant-Induced Mortality on Fish Populations (W. Van Winkle, ed.). Pergamon Press, New York, NY.

Policansky, D. 1979. Prepared testimony. Submitted on behalf of the Hudson River Fishermen's Assn. U.S. Environmental Protection Agency, Region II. Adjudicatory Hearing Docket No. C/II-WP-77-01.

QLM. 1974. Documentation for mathematical models of the Hudson River striped bass population. Volume I: Transport Model. Quirk, Lawler, and Matusky Engineers, Tappan, NY. 263 pp.

Ricker, W.E. 1954. Stock and recruitment. J. Fish. Res. Bd. Can. 11:559–623.

Ricker, W.E. 1975. Computation and interpretation of biological statistics of fish population. Fish. Res. Bd. Can. Bull. 191. 382 pp.

Ricker, W.E. 1979. Notes on certain of the testimonial documents that pertain to the effects of power plants on striped bass in the lower Hudson River and estuary, *In* Volume 2 of Fletcher and DeRiso (1979).

Sandler, R. and D. Schoenbrod (eds.). 1981. The Hudson River power plant settlement. Materials prepared for a conference. New York University, New York, NY.

Schaefer, R.H. 1968. Sex composition of striped bass from the Long Island surf. N.Y. Fish and Game J. 15:117–118.

Schubel, J.R. and B.C. Marcy, Jr. (eds.). 1978. Power plant entrainment: A biological assessment. Academic Press, New York, NY.

Sirois, D.L. and S.W. Fredrick. 1978. Phytoplankton and primary production in the lower Hudson River estuary. Est. Coastal Mar. Sci. 7:413–423.

Slobodkin, L.J. 1979. Critique of the utilities' striped bass density-dependence arguments and research policy and programs. Written testimony prepared for the New York State and Massachusetts Attorneys General.

Swartzman, G.L., R.B. DeRiso, and C. Cowan. 1978. Comparison of simulation models used in assessing the effects of power-plant-induced mortality on fish populations. UW-NRC-10. Center for Quantitative Science, College of Fisheries, University of Washington, Seattle, WA.

Texas Instruments. 1974. Hudson River ecological survey in the area of Indian Point. 1973 Annual Report. Consolidated Edison Co. of NY, Inc.

Texas Instruments. 1975. Multiplant impact study of the Hudson River estuary. First Annual Report. Prepared for Consolidated Edison Co. of NY, Inc.

U.S. Atomic Energy Commission (USAEC). 1972. Final environmental statement related to operation of Indian Point nuclear generating plant, Unit No. 2, (Vols. 1 and 2.) Docket No. 50-247.

USGS. 1977. Water resources data for New York. Water year 1976. U.S. Geological Survey Water Data Report NY-76-1. 615 pp.

USGS. 1978. Water resources data for New York. Water year 1977. U.S. Geological Survey Water Data Report NY-77-1. 566 pp.

USGS. 1979. Water resources data for New York. Water year 1978. U.S. Geological Survey Water Data Report NY-78-1.

USGS. 1980. Water resources data for New York. Water year 1979. U.S. Geological Survey Water Data Report NY-79-1. 538 pp.

USGS. 1981. Water resources data for New York. Water year 1980. U.S. Geological Survey Water Data Report NY-80-1. 310 pp.

U.S. Nuclear Regulatory Commission. 1975. Final environmental statement related to operation of Indian Point nuclear generating plant, Unit No. 3. (Vols. I and II.) NUREG-75/002 and NUREG-75/003.

U.S. Nuclear Regulatory Commission. 1976. Final environmental statement for facility license amendment for extension of operation with once-through cooling, Indian Point Unit 2. NUREG-0130.

Van Winkle, W., B.W. Rust, C.P. Goodyear, S.R. Blum, and P. Thall. 1974. A striped bass population model and computer programs. ORNL/TM-4578, ESD-643, Oak Ridge National Laboratory, Oak Ridge, TN. 200 pp.

Van Winkle, W. (ed.) 1977. Assessing the Effects of Power-Plant Induced Mortality on Fish Populations. Pergamon Press, New York, NY.

Van Winkle, W., L.W. Barnthouse, B.L. Kirk, and D.S. Vaughan. 1980. Evaluation of impingement losses of white perch at the Indian Point nuclear station and other Hudson River power plants. ORNL/NUREG/TM-361. Oak Ridge National Laboratory, Oak Ridge, TN.

Van Winkle, W., D.S. Vaughan, L.W. Barnthouse, and B.L. Kirk. 1981. An analysis of the ability to detect reductions in year-class strength of the Hudson River white perch (*Morone americana*) population. Can. J. Fish. Aquat. Sci. 38:627–632.

Van Winkle, W. and K.D. Kumar. 1982. Relative stock composition of the Atlantic coast striped bass population: further analysis. ORNL/TM-8217. 31 pp.

Vaughan, D.S. and K.D. Kumar. 1982. Entrainment mortality of ichthyoplankton: detectability and precision of estimates. Environ. Management 6:155–162.

Vaughan, D.S. and W. Van Winkle. 1982. Corrected analysis of the ability to detect reductions in year-class strength of the Hudson River white perch (*Morone americana*) population. Can. J. Fish. Aquat. Sci. 39:782–785.

4
PCBs in the Hudson

KARIN E. LIMBURG

When former New York State Commissioner of Environmental Conservation Ogden Reid warned consumers on August 8, 1975 against eating fish from the Hudson River and Lake Ontario because of contamination with polychlorinated biphenyls (PCBs), relatively few people were even aware of the existence of these compounds. Less than one decade later, PCB has become a household term with notoriety at par with DDT (1,1,1-trichloro-2,2-bis(p-chlorophenyl)ethane); moreover, a burst of scientific research has revealed PCB dispersal around the globe and throughout the biosphere. The discovery and scientific concern about such wide distribution of PCBs lent needed impetus for Congress to pass the Toxic Substances Control Act in 1976 (Letz, 1983). In the Hudson, the problem of PCBs went unrecognized for decades, since PCBs were considered nonreactive. That problem became compounded year after year until, finally, regulatory agencies were forced to recognize the severity of this toxic hazard.

The presence of PCBs in the Hudson in greater quantity than anywhere else in the U.S. has given rise to one of the best-documented series of studies on the upper and lower Hudson ecosystems. It has also led to the lengthy process of deciding on the best measures for dealing with the problem. The magnitude of PCB contamination of the Hudson ecosystem, together with the socioeconomic implications, make this issue one of major importance in the documentation of Hudson River environmental impacts.

Background on PCBs: History in Hudson Through 1976

PCBs: Description and History of Use

PCBs represent a class of organic chemical compounds characterized by two joined phenyl rings that are subsequently chlorinated to greater or lesser degree. With 10 possible sites for chlorination on this 12-carbon ring system, 209 isomers are possible, although only 102 are likely to occur in industrial manufacture (DiNardi and Desmarais, 1976) (Figure 4.1).

PCBs have been manufactured in West Germany, France, Italy, Japan, the U.S., and the Soviet Union (Ahmed, 1976). PCBs were manufactured in the U.S. from 1929 until 1977 solely by Monsanto Corp., and were sold under the trade name of Aroclor. Products were graded on the basis of percent chlorination. With the name Aroclor 12xx, the "12" represents the number of carbon atoms and the "xx" denotes some percent level of chlorination of the product. One notable (and important for the Hudson) exception is their Aroclor 1016 (A-1016), which in fact has 12 carbons that are more than 41% chlorinated. (A major difference between A-1016 and a similar compound, A-1242, has to do with the relative amounts of contamination by polychlorinated dibenzofurans [PCDFs]; A-1016 virtually lacks contamination, while A-1242 contained approximately 1.5–2.0 μg PCDFs per gram PCBs [Buckley, pers. comm.].)

First described in the late nineteenth century (Schmidt and Schultz, 1881), the propitious dielectrical properties of PCBs were known by 1930 (Kimbrough, 1974). PCBs are highly stable compounds that are resistant to heat and fire, and are excellent electrical insulators. These properties have suited them to a wide variety of uses. The electrical industry has not only used these compounds in capacitors and transformers, replacing the volatile mineral oils previously employed; they have also been used as sealants, as additives to many plastic materials (including those com-

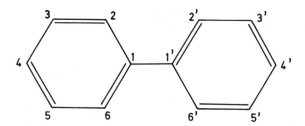

Figure 4.1. Basic structure of biphenyl molecule. Each intersection of lines represents a carbon atom; double lines indicate double bonds. Chlorine atoms may substitute for hydrogen atoms at any of the numbered positions; this numerical system is used in nomenclature.

monly found in the home), as flame retardants, in "carbonless" carbon paper, as adhesives, and in many more products (Kramer, 1975).

Distribution and Toxicity of PCBs

PCBs were not recognized as environmental contaminants until 1966, when Sören Jensen, a Swedish chemist, discovered the compounds in fish that were being screened for pesticides (Jensen, 1966). Following that, many more discoveries of PCBs in the environment were made worldwide. Because of their low solubility in water and their high affinity for particulates, PCBs enter rivers and estuaries via the hydrologic cycle, and tend to accumulate in bottom sediments.

At least two mechanisms exist for PCB accumulation in the biosphere: these are termed *bioaccumulation* (or bioconcentration) and *biomagnification*. The first is the tendency for organisms to extract a compound present in ambiently low concentrations and to store it cumulatively in their bodies. This has been repeatedly demonstrated for PCBs, particularly in aquatic ecosystems. PCBs are lipophilic and accumulate in body fat. Bioaccumulation rates are calculated as (concentration of compound in organism)/(concentration in environment). The second term, biomagnification, refers to the tendency of a compound such as PCB to be passed along a trophic chain in a cumulative manner. Hence, each successive consumer may receive a larger dose than the previous one. White-tailed eagles and seals have been shown to biomagnify PCBs (Jensen *et al.*, 1969).

Summaries of toxicity work may be found in Stalling and Mayer (1972); Kimbrough (1974); volumes 1 and 24 of *Environmental Health Perspectives;* and Letz (1983), among others. After hundreds of studies, some patterns of PCB toxicity have emerged. Key factors determining effects now appear to be structure-activity relationships and disposition and persistence of isomers in the environment (Bickel, 1982; DHEW, 1978). It is now known that PCB homologs having greater than five chlorine atoms are those that tend to bioaccumulate, while the less-chlorinated ones are metabolized and excreted. The congener[1] 2,4,5,2',4',5'-hexachlorobiphenyl (6CB) is the single most abundantly found form in PCB residues (Bickel, 1982). Furthermore, different congeners can vary widely in eliciting toxic responses in different species. Kimbrough *et al.* (1978) and Letz

---

[1]Several terms are used (often erroneously) in describing families of compounds such as PCBs. A *congener* is a single member of a family of related chemical substances. An *isomer* is one of two or more molecules having the same number and kind of atoms, but differing in respect to the arrangement or configuration of the atoms. Lastly, congeners of PCBs may be classified in a *homologous series;* that is, a series of organic compounds differing by some multiple (in this case, chlorine atoms). Thus, one may refer to mono-, di-, tri-, etc. chlorobiphenyls as *homologs*.

(1983) refer to work that compared the effects of different isomers. For example, the congener 3,4,5,3',4',5'-6CB was found to be far more toxic to both birds and mammals than most other isomers; however, it seems to be more quickly eliminated from the body than other 6CBs, at least in humans (Wolff et al., 1981). Even more toxic are isomers of tetrachlorodibenzofuran (TCDF) and tetrachlorodibenzodioxin (TCDD), which are contaminants and byproducts of commercial PCB mixtures, respectively. The degree to which these are associated with various Aroclors ranges from 2.0 ppm for A-1248 to less than 1.0 ppb for A-1016 (National Research Council, 1979). The processes of PCB metabolite formation are not understood well enough to predict which metabolites are likely to be toxic or mutagenic. Carcinogenic risks to humans are still largely unknown, although induction of certain enzyme systems, which in turn might lead to disease due to increased metabolism of endogenous or exogenous materials, is a well-established effect of PCB exposure (Letz, 1983).

PCB Discharges into the Hudson

The General Electric (G.E.) Co., with two facilities for capacitor production located at Ft. Edward and Hudson Falls along the upper Hudson, was by far the largest discharger of PCBs. The Ft. Edward plant began to use these compounds in capacitors in 1946, and in 1952 the Hudson Falls plant followed suit (R. Arisman, General Electric Co., Hudson Falls, pers. comm.). Figure 4.2 shows the location of these facilities. There appears to be no way of obtaining actual data on discharges of PCBs prior to 1966, although it can be assumed that the average annual loadings were at least on the order of those from the late 1960s, or roughly 5 metric tons (mt)/yr (Sofaer, 1976). However, G.E.'s records of PCBs purchased for the two plants were available for the years 1966–1975 (Sofaer, 1976). Estimates of purchases back to 1957 are possible by extrapolation from a linear regression ($r^2 = 0.88$) of the existing records of G.E.'s purchases (dependent variable) against Monsanto Corp.'s nationwide sales of PCBs for capacitors—data for which can be found in Hutzinger et al. (1974). G.E.'s purchases, including these estimates, add up to approximately 60×10^3 mt for the period 1957–1975, as shown in Figure 4.3. The graph indicates a stepped-up volume of capacitor production and, presumably, PCB discharges during the period after 1964 until 1975. In 1971, G.E. began substituting A-1016 for A-1242 (Sofaer, 1976).

Discovery and Hearings

Despite the large amounts entering the river from these two facilities, the chemical went largely unnoticed among all of the other industrial and municipal discharges of the region. It was not until the late 1960s, when PCB contamination had gained widespread recognition within the scien-

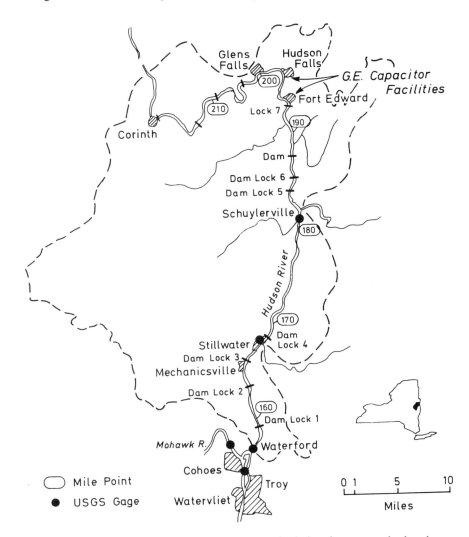

Figure 4.2. Map of upper Hudson River watershed showing towns, locks, dams, and location of General Electric Co.'s capacitor plants (from Tofflemire and Quinn, 1979).

tific community, that the chemicals began to be looked for in the American environment. In a 1970 investigation by local fishermen, with laboratory analysis by the Wisconsin Alumni Research Foundation (WARF) Institute, striped bass from the Hudson were reported to contain an average of 4 ppm PCB in their flesh; their eggs averaged 11.4 ppm (Boyle and Highland, 1979).

Meanwhile, G.E. had applied for a federal permit to discharge wastes into the Hudson in November 1971 (Sofaer, 1976). This application was

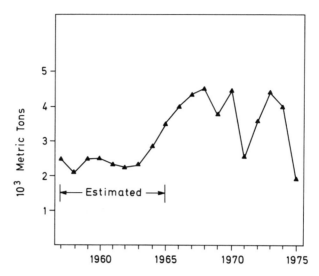

Figure 4.3. Purchases of PCBs by General Electric Co. for use at its Hudson and Fort Edward, NY capacitor manufacturing plants. Data from 1966 are from Sofaer (1976); earlier values are estimated by regression of GE data (1966–1975) against data from Hutzinger et al. (1974) on PCBs sales by Monsanto.

later revised in 1973 to comply with the new National Pollutant Discharge Elimination System (NPDES) established under the 1972 amendments to the Federal Water Pollution Control Act (FWPCA). In the same year, New York's Department of Environmental Conservation (DEC) overtook jurisdiction of the discharge permitting system and received G.E.'s application that August. The application stated that on the average, 30 pounds of "chlorinated hydrocarbons," explained in a footnote as PCBs, were discharged daily. Apparently, this was not regarded by the state as a violation of either the FWPCA or New York's own Environmental Conservation Law (ECL), although the latter specifically prohibits discharges that are not in compliance with water quality standards and criteria. As stated in the law, toxic wastes are prohibited "in amounts that will be injurious to fishlife or which in any manner shall adversely affect the flavor, color, or odor thereof, or impair the waters for any best usage," (6 NYCRR 701.5) for any class of waters as defined by section 301 of the ECL (Sofaer, 1976).

The state government, becoming aware of the dispersal of PCBs in the river and of the potential ramifications, began fish investigations of its own in 1972. Largemouth bass were recorded with PCB concentrations of 0.38 to 14.62 ppm; white perch had 0.38 to 15.81 ppm and striped bass contained 3.70 to 49.63 ppm (Sofaer, 1976). Beginning in December 1974, the DEC's Division of Pure Waters began a monitoring program for sediments and water column PCB concentrations in the upper Hudson River basin. A-1016 was found in sediment core samples at concentrations up to 1850 ppm downstream of the G.E. plants, together with significant concentrations of A-1221 and A-1254. Later, in 1975, the DEC began a statewide fish sampling program, again including stations below the G.E. plants. Concentrations of A-1242 and A-1016 were much higher than previously found.

Another important event took place in October 1973 that drastically altered the distribution of PCBs in the Hudson. This was the removal by Niagara Mohawk Power Corp. (R. Mt. Pleasant and M. Brown, DEC, pers. comm.) of an already flood-damaged dam at Ft. Edward, directly below the discharges. Following the removal, nearly 1 million m^3 of sediments that had accumulated for scores of years in a pool behind the dam were washed downstream. It is estimated that more than 70% of the PCB burden in the sediments behind the former Ft. Edward Dam was gone by 1979 (NYSDEC, 1979). Included in this loss were sediments that the state Department of Transportation (DOT) had dredged out of the river from 1974–1978 near Rogers Island in order to maintain a shipping channel (Tofflemire et al., 1979). Geochemists at Lamont-Doherty Geological Observatory have been able to trace the effects of the dam removal all the way to New York Harbor (Bopp et al., 1982).

It should be noted that the significance of the dam removal to the PCB problem was not recognized until several years after the fact. This lack of perception was due, in part, to the fact that information collected by one state agency division was not always shared with others, so that discussion and interpretation of the data collected were severely hampered. This was the situation that Ogden Reid found in 1975 when he assumed the job of Commissioner of Environmental Conservation (Severo, 1975a).

Also in 1975, federal researchers from the Environmental Protection Agency (EPA) contacted Reid and informed him of their own studies, summarized in Nadeau and Davis (1974). In 1974, these researchers had sampled water and sediment at Hudson Falls, together with a site at the Ft. Edward discharge site and at three sites—each located successively at quarter-mile intervals downstream. Fish were seined where possible at these sites, and snails were simultaneously sampled. Measurable concentrations of PCBs (A-1016 and A-1242) were found at all sites in the water and sediments; especially high concentrations were found at the outfall of the Ft. Edward G.E. plant. More startling were the PCB concentrations found in biota, identified as A-1254 and A-1248. A northern rock bass contained 350 ppm; shiners averaged 78 ppm, perch averaged 17 ppm, and gastropods had up to 27 ppm averaged at one station. From these results, Nadeau and Davis calculated bioaccumulation rates. All exceeded 1000-fold; for the rock bass at station 3 (0.5 miles downstream from Ft. Edward), the concentration factor was 117,000:1. The authors presented a conceptual model of biological transport of PCBs throughout that stretch of the river in which the perch and shiners were identified as important links (as grazers of periphyton and as food for game fish).

The news came as a surprise to Commissioner Reid, who had not been previously advised of the problem by his staff. On August 8, 1975, Reid put out a public alert warning people against eating striped bass from the Hudson and salmon from the Great Lakes (Severo, 1975b). At the same time, a public relations representative for G.E. was quoted as saying, "We don't put PCBs into the (Hudson) river intentionally. But we live in a real world, and it is going to take time to solve the problem" (Severo, 1975b).

Public response was immediate, and concern was voiced by political representatives of neighboring states as to the possible consequences of PCB contamination to east coast fisheries (Severo, 1975c). Reid further discovered that the G.E. plants had been given water quality certifications in 1971 and 1973 by the New York State DOT and the DEC's Division of Pure Waters, respectively. However, there were still grounds for the state to initiate an administrative proceeding against G.E. under the ECL. The hearings were conducted from the end of 1975 to September 1976, and were presided over by Abraham Sofaer. Sofaer wrote a lengthy interim opinion (Sofaer, 1976), which summarily dealt with the information at hand.

As with virtually every other instance of major environmental litigation concerning the Hudson (other chapters in this book), the condition of the major commercial and recreational fish stocks was of utmost concern. Among other things, because PCB levels in fish were found to exceed the U.S. Food and Drug Administration (FDA) limits, they could not be shipped for interstate sale. Unfortunately for G.E.'s case, a hired expert witness from the consulting firm of Ecological Analysts, Inc. testified that Hudson River fish that had been sampled by his firm and analyzed by another laboratory in Tennessee showed PCB concentrations in all fish that were below the 5 ppm limit set by the FDA. Upon cross-examination, this witness admitted to errors and omissions in the data. He then mentioned PCB levels in four fish that were above 5 ppm, ranging from 66 to 143 ppm. The G.E. lawyer offered to consent to a withdrawal of the data, and G.E. had a complete reworking of its Hudson River samples. Several of the reanalyzed fish showed levels of PCBs in excess of 100 ppm (Boyle and Highland, 1979).

Based on the thousands of pages of testimony, profiled reports, and data collected, it was determined that G.E. had undeniably been responsible for the presence of tons of PCBs in the river (NYSDEC, 1979). Sofaer found G.E. guilty of violating two state environmental quality standards by discharging wastes injurious to fish and by impairing fishing—a protected use of the waters. On record was Sofaer's judgment that the violations were the result not only of corporate abuse, but also failure of the regulatory agencies to recognize and remedy it (Sofaer, 1976).

The Settlement

In order to determine what steps should be taken to deal with the estimated 0.5 million pounds of PCBs in the river and estuary, a second round of hearings was convened following Sofaer's interim report. In September 1976, exactly one year after the initial hearings began, the DEC and G.E. reached a settlement agreement. G.E. agreed to pay a lump sum of $3 million to the DEC for monitoring the presence of PCBs and for determining the need for and implementing remedial action if necessary. In addition, the company agreed to fund $1 million of its own in-house research: 1) to look for suitable alternatives for capacitor materials; 2) to conduct research and pilot studies on the removal and treatment of PCBs from the river; and 3) to fund another $200,000 worth of research on the environmental hazards of up to three potentially hazardous substances, as specified by the Commissioner of Environmental Conservation—the new one being Peter Berle. In return, the settlement provisions did not construe that G.E. had "violated any law or regulation or otherwise committed a breach of duty at any time. . . . No amount of the settlement contribution by General Electric constitutes a fine or penalty" (NYSDEC, 1976). The DEC agreed to contribute another $3 million, making a grand total for the remedial action fund of $7 million.

In addition to monetary terms, G.E. agreed to cease using PCBs in its capacitors by July 1, 1977 and no more than 1 lb/day after the date of signing the agreement. Lastly, a Settlement Advisory Committee was established to watch over and administer the funds. The formation of this "watchdog" committee was a stipulation of the environmental interests before signing the settlement (Sanders, 1984). The Committee, still in existence as of this writing, over the years has been comprised of independent scientists, environmentalists, and staff representing the DEC, EPA, DOT, N.Y. Department of Health (DOH), and the U.S. Army Corps of Engineers.

Scientific Assessment of the PCB Problem in the Hudson

As mandated by the PCB settlement agreement, the state established a broad program to evaluate the distribution of PCBs—a program that involved investigators from consulting firms, universities, and federal and state agencies. Because of the all-pervasive nature of PCB distribution, and because the material had been deposited in the upper river, a total ecosystem approach to the problem was needed as never before. There-

Table 4.1. Summary of major studies conducted to assess the extent of PCB distribution in the Hudson ecosystem, 1970–1983.

Type of assessment	Yr conducted	Geographic location	Investigators
I. Scientific Studies: Impact Assessment			
A. Physical studies			
1. Monitoring river flow, sediment, and PCB transport	1977–present (1983)	Upper Hudson (Green Island and other stations	US Geological Survey
2. PCB areal mapping	1977	Upper Hudson	Normandeau Assoc., DEC, Gahagan & Bryant
3. Sediment PCB analyses			
PCBs in estuarine sed.	1976	Troy-Tappan Zee Bridge Estuary (main stem)	EPA
PCBs in estuarine sed.	1977	Upper Hudson	Lamont Doherty
PCBs in upper Hudson sed.	1973	Upper Hudson	EPA
PCBs in upper Hudson sed.	1974–1977	Upper Hudson	DEC Div. Pure Waters
PCBs in upper Hudson sed.	1977	Upper Hudson	NUS (in conjunction with DEC Bur. Water Research)
PCBs in upper Hudson sed.	1983	Upper Hudson	Rensselaer Polytechnic Inst.
4. Bedload sediment transport	1977	Upper Hudson	Weston Environ. Consultants
5. Groundwater concentrations	1977	Upper Hudson	Pure Waters, O'Brien & Gere
6. Wastewater	1977	Upper Hudson	DEC Bur. Water Research
7. Air-water particulate interchange	1978	Upper Hudson	DEC Bur. Air Research, Boyce Thompson Inst.
8. Air monitoring	1976–1982	Ft. Edward and dump and river sites	Studies by R. Boyle, EPA, and DEC prior to GE hearings
B. Ecological studies			
1. Fish monitoring	1970–1975	Upper Hudson and estuary	
	1976–1983	Entire river and estuary	DEC, with analyses by O'Brien & Gere and Raltech

2. Macroinvertebrate monitoring	1977–1983	Entire river and estuary	NYS Dept. of Health
3. Aquatic food chain dynamics	1978–1979	Estuary	NYU Medical Center
4. Lower trophic levels	1978–1979	Estuary	SUNY Stony Brook, Marine Science Res. Ctr.
5. Plant- and farm-product uptake	1978–1983	Ft. Edward vicinity and nearby riverside locations	Boyce Thompson Inst. (BTI), DEC Bur. Water Res.; PCB analyses by Raltech and BTI
6. Microbial degradation of PCBs	1982–1984	Upper HR sediments	GE Co.
C. Modelling studies			
1. Physical			
PCB sediment transport model	1978–1979	Upper Hudson (Ft. Edward to Troy Dam)	Lawler, Matusky and Skelly Engineers, Inc.
2. Ecological			
Estuarine food web fate and transport	1978–1979	Estuary	Hydroscience, Inc.
Lower food chain kinetics of PCBs	1978–1979	Estuary	Fordham University
II. Engineering Related to Remedial Action			
A. Dredging feasibility study	1977–1980	Upper Hudson	Malcolm Pirnie Inc., DEC Bur. Water Res.
B. Sediment treatment	1976–1978	Upper Hudson	GE Co.
C. Landfills and dumps	1977–1979	Upper Hudson	Weston, DEC Div. of Solid Waste
D. Public water supply	1978–1979	Waterford (Upper HR), Poughkeepsie (estuary)	NYS Dept. of Health, O'Brien & Gere Horstman, 1977
E. Assessment of non-dredging remedial action alternatives	1977	Upper Hudson	
III. Sociological Assessment			
A. Public opinion survey on reactions to PCB cleanup plan	1981	Washington County	RA Hayslip Assoc.

From DEC (1979) and this study.

fore, the assessment process brought forth teams of researchers from diverse disciplines.

The main components of the studies are discussed in NYSDEC (1979) and Horn *et al.* (1979), and are included in Table 4.1. Many will be dealt with separately below. The studies may be divided into scientific assessment and engineering feasibility projects. The scientific assessments were initiated in 1976 following the settlement, and many are ongoing at this writing. These investigations have focused on the extent of PCB burdens in physical and biological components of both the aquatic and adjacent terrestrial systems, although some assessment of transport mechanisms and rates has been initiated as well. Of primary concern have been the questions of mobility of the PCBs into the estuary, the possible health threats imposed by substantial estuarine contamination, and how best to remove such threats indefinitely (NYSDEC, 1979; Hetling *et al.*, 1979). The PCB Settlement Advisory Committee, together with the DEC, has overseen the execution of these projects.

PCBs in Sediments

Upper River. Samples of the sediments had been collected initially by the DOT in 1974. In 1975, the DEC and DOT undertook a more comprehensive survey of sediments from above Glens Falls down to Waterford; especially high concentrations (up to 3707 ppm) were measured near the Thompson Island Dam between Ft. Edward and Ft. Miller (Hullar *et al.*, 1976). In September 1976, the DEC collected more core and grab samples between Ft. Edward and Troy, and the consulting firm of Lawler, Matusky, and Skelly took a few sediment samples in early 1976. These were analyzed by the DOT and another consulting firm, O'Brien and Gere (Tofflemire and Quinn, 1979).

Following the PCB settlement, contracts were awarded to several groups to identify and quantify the physical presence of PCB (Table 4.1) (Tofflemire and Quinn, 1979). Normandeau Associates, Inc. undertook a major sediment sampling program in 1977 under the guidance of DEC and the Settlement Advisory Committee. Their assessment included mapping from aerial photographs, grab sampling, and coring down to 8 ft over 171 transects from Ft. Edward to Troy. Additionally, the DOT used a clamshell dredge at one channel site to take grab samples. Malcolm Pirnie, Inc. took samples from the remnant deposits (remains of built-up sediments that appeared along the banks when the river elevation dropped as a result of removal of the Ft. Edward Dam) in November 1977 and April 1978. Sediments were probed in the summer of 1978 by the DEC to identify precisely the location of 40 "hot spots," defined as areas with sediments containing 50 ppm PCB or greater (Tofflemire and Quinn, 1979). Also, in October 1981, eight cores from the upper Hudson were collected and

Scientific Assessment of the PCB Problem in the Hudson

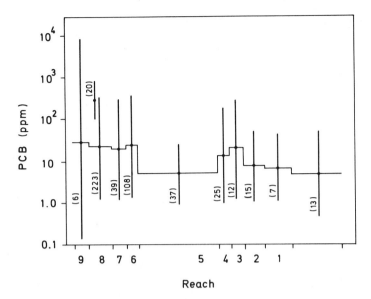

Figure 4.4. Average PCB concentration in sediments of the upper Hudson River. Log of concentrations are plotted against river reach, also showing locations of dams and locks (from Tofflemire and Quinn, 1979).

analyzed by EPA and reported by EPA's Edison, N.J. lab (Johnson, 1981).

No further surveys were undertaken until the summer of 1983, when a team of investigators contracted by EPA returned to the area with the DEC Bureau of Water Research and attempted a partial replication of the 1978 sampling. This was done in order to determine whether the contaminated sediments had been significantly moved or scoured (NUS, 1983).

Figure 4.4 shows the average concentrations of PCBs in the upper river basin in 1978 (Tofflemire and Quinn, 1979). The figure and the surveys revealed a high degree of local variation in PCB concentrations, even over very short distances. Concentrations were also found to be approximately lognormally distributed with respect to sampling frequency (Tofflemire and Quinn, 1979). Samples were analyzed according to grain size and textural characteristics. North of the Thompson Island dam, the upper river sediments contain large quantities of sawdust and wood chips that came from past timber transport and pulp manufacture. PCBs were found in greatest concentrations, often exceeding 50 ppm, in sediments consisting of muck and/or wood chips, which have high adsorptive capacities for the substances (Hetling *et al.*, 1978).

Heavy metals analyses and ^{137}Cs isotope dating were carried out on some of the sediment cores. PCBs were positively correlated with ^{137}Cs (a fallout product from atomic bomb testing) and lead. Dating with cesium was used as a rough meter of past PCB discharge rates, and as such indicated that peak discharges (mainly from the G.E. capacitor plants) occurred during the 1960s (Tofflemire and Quinn, 1979; Bopp *et al.*, 1982).

Lower River. Sediment sampling of the tidal portion of the Hudson (south of Troy) was carried out by geochemists from Lamont Doherty Laboratory between 1977 and 1979, during which time over 80 cores were taken (Bopp, 1979; Bopp *et al.*, 1981). More than 150 analyses for A-1242, A-1016, and A-1254 were performed. ^{137}Cs, ^{134}Cs, and ^{60}Co were concurrently measured and were used as chronological tracers in the sediments. As in the upper Hudson, ^{137}Cs was absent prior to 1954 and records the time of atomic testing. ^{134}Cs- and ^{60}Co-radioactive products released in trace amounts from the Indian Point nuclear power plants (RM 41) can be used to label sediments that passed these plants. The main findings included identifying a PCB concentration gradient of A-1242 decreasing downstream (Figure 4.5); at the same time, the relative proportion of A-1254 (a more highly chlorinated product) to A-1242 increased in the same direction. From this, and from analyses of peaks in chromatograms, it was concluded that the lower chlorinated PCBs were eluting from the particulate into the dissolved phase, with volatilization a secondary avenue of loss from the sediments (Bopp *et al.*, 1981).

A budget of PCBs in the estuary was also made (Bopp *et al.*, 1981). An estimated 76 mt were distributed in three types of deposition sites: 1) 65% were estimated to have gone into areas of rapid deposition (often up to 20–40 cm yr^{-1}), mainly in New York Harbor; 2) another 25% was incor-

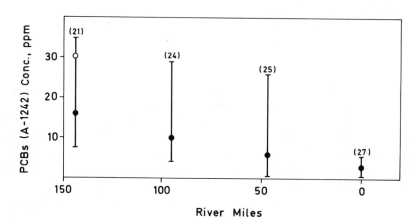

Figure 4.5. Plot of Aroclor 1242 concentrations vs. distance in the Hudson estuary (numbers decrease toward the estuary mouth). Ranges are indicated by vertical bars; numbers of samples indicated in parentheses above. Open circle at RM 145 is an average concentration that includes one sample with unusually high PCB concentrations in the top 60 cm (data from Bopp *et al.*, 1981).

porated into sediments in quiet coves and broad shallows; and 3) only 10% went into the channel and subtidal banks. These deposition sites account for 10%, 25%, and 65%, respectively, of the total area of the estuary.

Finally, an attempt was made to estimate the relative contribution of lower Hudson sources to the total PCB input to New York Harbor (Bopp et al., 1981). These sources are largely sewage and sewage treatment plant discharges, 70% of which occur in the harbor itself. Based on analyses of chromatogram peak height ratios in samples taken from Yonkers sewage, New York City sewage sludge, and upstream (RM 53 and 43) and downstream (RM −1.5 and 0.1) sediment samples, the investigators estimated that around 25 to 30% of the PCB inputs derive from south of the Narrows (RM −5). This indicates that up to 75% of the PCBs found in the southern end of the river originate above Troy. This is an important point of which more will be said later.

PCBs in the Water Column

The U.S. Geological Survey (USGS) has monitored the entire Hudson for PCBs since 1977, although data have been taken since 1982 only at stations above the Troy Dam (M. Brown et al., 1984). Roughly 150 upriver and 25 downriver (estuarine) depth-integrated samples were collected annually and analyzed for PCBs and suspended sediments. River discharge is monitored at four upriver stations: Glens Falls, Schuylerville, Stillwater, and Waterford.

Turk (1980), Turk and Troutman (1981), and Schroeder and Barnes (1983) discussed PCB sediment-water relationships and downstream transport. Taken together, the reports illustrate how understanding PCB transport has changed with additional monitoring, making possible time-series analysis. In the earlier two reports, the investigators identified two mechanisms by which PCBs entered the water column. At low-to-moderate flows (50 to 500 or 600 m^3/s), PCBs appeared to be released at a constant rate from the sediments. This is reflected by the inverse relationship of PCB concentrations (measured in 1977; Turk, 1980) to flow at rates less than about 600 m^3/s, as shown in Figure 4.6a. Higher flows have enough force to scour and resuspend the sediments. PCBs are then mobilized in an exponential fashion, as seen in the right-hand side of the graph in Figure 4.6a.

By 1982, it was clear that PCB transport rates were declining, particularly at low flows (Figure 4.6b). Schroeder and Barnes (1983) demonstrated that recent data (post-1980) do not support the original, empirically derived low-flow relationship. Schroeder and Barnes suggested that the decline in PCB concentrations during non-scour conditions was due, in part, to the below-average flow regimes, which allowed uncontaminated sediments to settle out over the PCB-contaminated bottoms. An-

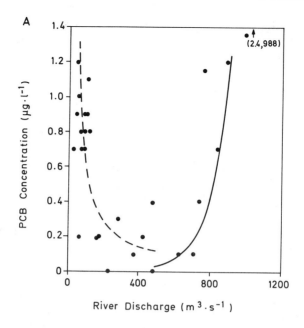
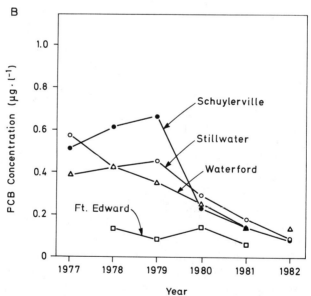

Figure 4.6. Relation of PCB concentration (μg/l) to river discharge, Hudson River at Stillwater (from Turk, 1980). (b) Declines in low-flow (< 400 m³/s) PCB concentrations, 1977–1982, for four upper Hudson River stations. Ft. Edward is located upstream from the Thompson Island Dam, whereas all other stations are located downstream (data from U.S. Geological Survey).

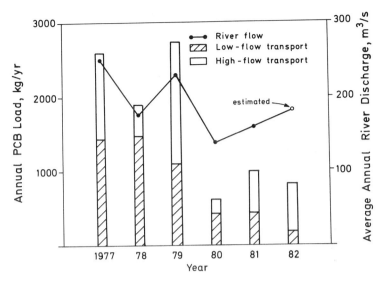

Figure 4.7. River discharges (line) and PCB loadings (bars) measured and calculated, respectively, at Stillwater. Stippled shadings indicate volume of PCBs transported at low velocity (< 400 m³/s) flows (adapted from M. Brown et al., 1984).

other cause for the decrease may be the loss of lighter chlorinated congeners from the system through volatilization.

M. Brown et al. (1984) calculated annual rates of PCB transport with river flows measured at Stillwater for the years 1977–1982; these are presented in Figure 4.7. The bars show the proportions of chemical transported due to the two mechanisms: low-flow release and high-flow scouring. Although average annual flow rates mask the daily variation that in fact, governs the extent of scour, it is interesting to note that higher flow years have proportionately more PCBs transported via resuspension (compare years 1980 and 1982). A 100-year flood occurred in the upper Hudson in 1976, reflected in the average annual flow of 626 m³/s at the Green Island gauging station (USGS, 1977). Presumably, much greater amounts of PCBs were scavenged from the sediments at that time. The spring of 1983 also brought unusually high waters to the upper river, and high sediment yields were evident in May (M. Brown, DEC Bureau of Water Research, pers. comm.).

PCBs in the Air and Landfills

The DEC carried out an extensive air monitoring program from late 1976 through 1977. Five sites were monitored between Glens Falls and Ft. Miller, with PCB analyses carried out by the DOH. Prior to 1977, the two G.E. factories were the main sources of aerial contamination. Other sources include landfills and dump sites, river water, and remnant deposits. Following the cessation of PCB usage by G.E., ambient air concentrations declined more than 3-fold from 1.0 to 0.3 $\mu g/m^3$ of air in the Ft. Edward vicinity (Hetling *et al.*, 1978).

More recent monitoring has been carried out by researchers from the Boyce Thompson Institute in conjunction with plant foliar uptake studies (E.H. Buckley, Boyce Thompson Institute, pers. comm.). Air samples were collected at a dump site in 1980 and 1981, and later at sites near the river and remnant deposits. Experimental and field work (Tofflemire *et al.*, 1981; Buckley and Tofflemire, 1983) suggest that volatilization rates at the dam outfalls are up to 10 times higher than average due to turbulent mixing.

Weston Environmental Consultants assessed the amounts of PCBs in upriver dredge spoil sites, landfills, and dumps, as well as transfers from these sites to ground-waters and surface waters (Hetling *et al.*, 1978). In 1977, dredge spoil areas were estimated to contain some 70 mt of PCBs, while landfills and dumps held roughly 320 mt. Yearly losses from both of these were estimated by Tofflemire and Quinn (1979) as approximately 1450 kg to the air and 630 kg to water systems.

PCBs in the Biota

Biological assessment of PCB fate and transport in the Hudson ecosystem has been carried out for nearly 15 years. Of all the biological monitoring projects, the most extensive has been the DEC fish survey, which has spanned eight years and the entire length of the Hudson. Other biological assessment and/or monitoring programs have included: 1) a long-term (1977–1983) survey of macroinvertebrates by the DOH (Simpson, 1982); 2) wildlife and bird surveys by the DEC (NYSDEC, 1981a, 1981b; 1982a); 3) monitoring of snapping turtles (Stone *et al.*, 1980); 4) an incidental study of Atlantic tomcod (Klauda *et al.*, 1981; Smith *et al.*, 1979); 5) studies of PCB dynamics in phyto- and zooplankton by researchers at the Louis Calder Conservation and Ecology Study Center of Fordham University (e.g., Roscigno and Punto, 1982; Brown *et al.*, 1982); 6) related studies on phytoplankton, zooplankton, and fish by scientists at New York University Medical Center (NYU Medical Center, 1982; Califano, 1981; Peters and O'Connor, 1982; Pizza and O'Connor, 1983) and the State University at Stony Brook, N.Y. (e.g., Powers *et al.*, 1982); and 7) studies on uptake by wild and cultivated plants (Buckley, 1982; Buckley and Tofflemire, 1983). Researchers at the Columbia, Mo. National Fisheries Research Laboratory also compared PCB and other contaminant levels in striped bass in the Hudson with fish collected from two other eastern seaboard estuaries (Merhle *et al.*, 1982). Freshwater clams also have been used as *in situ* monitoring organisms for PCBs (Werner, 1981).

Measurements, Mechanisms, and Trends: Fish. Several thousand fish, representing dozens of resident and migratory species and several trophic levels, have been collected by gill-netting, electroshocking, and purchases from commercial fishermen (who are otherwise not permitted to fish the Hudson). For fish greater than 150 mm in length, an entire, scaled side ("fillet") is analyzed; smaller fish are analyzed whole. Most of the analyses have been performed by Hazleton Raltech, Inc. (Brown et al., 1985).

Armstrong and Sloan (1980, 1981), Sloan and Armstrong (1981), and Sloan et al. (1983) reported on trends in the fish data that are also summarized by Brown et al. (1985). With the cessation of the G.E. discharges, PCB burdens in monitored fish declined rapidly until about 1981. More recent observations indicate that concentrations in some resident species (including pumpkinseeds and brown bullheads) have stabilized or increased, with most of the increase caused by more heavily chlorinated congeners (R. Sloan, pers. comm.). Armstrong and Sloan (1980, 1981) estimated rates of decline with a simple model. By this method they calculated a net depuration rate[2] of 1.3 years for A-1016 and 2.9 years for A-1254. This is consistent with the latter compound's greater chlorination and, therefore, lesser tendency to volatilize. Armstrong and Sloan (1980, 1981) also found a strong correlation between lipid content and PCB body burdens for resident fish species. Some migratory species (e.g., Atlantic tomcod and blue crab) also showed this pattern, whereas others (e.g., shad and striped bass) did not (Sloan and Armstrong, 1981). The authors concluded that most migratory species do not spend enough time in contaminated waters to absorb much material. Brown et al. (1985) also suggest that for anadromous species, variability in data may be caused by intraspecific variation in migratory behavior; it is postulated that striped bass occur as subpopulations that exhibit a gradient from year-round residency to in-migration solely during spawning.

Positive correlations between fish PCB concentrations and levels of dissolved PCBs in ambient waters were found (Armstrong and Sloan, 1981; Sloan and Armstrong, 1981; Brown et al., 1985). A major suggestion, then, is that body burdens in fish would be expected to decline until a

---

[2]The depuration rate refers to the rate at which the concentration of PCBs in the fish population decreases, and may be calculated as:

$$\frac{dC}{dt} = -kC,$$

$$\frac{[C_t]}{[C_{t_o}]} = e^{-kt}$$

where $[C_{t_o}]$ = initial concentration,
$[C_t]$ = concentration at some time $t > t_o$,
k = a constant.

major scouring event releases larger quantities of PCBs from the sediments. High spring floods in 1983 occasioned more scouring than in the preceding years (Limburg, 1984), and may have been responsible for the increased PCB burdens observed in some of the resident fish species. Some trends are summarized in Table 4.2. Recently reported PCB levels in striped bass confirm the declining trend (Horn et al., 1983). The average 1983 PCB level in legal-size striped bass was reported for the first time to be below the then pertaining FDA limit of 5 ppm (note: 1 ppm = 1 μg/g); larger fish had even lower values, which was a surprising result.

Merhle et al. (1982) compared striped bass collected in the Hudson, Potomac, and Nanticoke estuaries. PCB, lead, and cadmium concentrations were significantly higher in Hudson River fish; PCB levels were about 10 times greater. Tests performed on vertebrae gave Hudson River fish lowest scores on strength, stiffness, toughness, and rupture point. Although the relative importance of PCBs is not clear, the authors suggested a mechanism by which organic contaminants may adversely affect vertebral development.

Klauda et al. (1981) reported PCB concentrations in Atlantic tomcod liver tissues, collected in 1978, that averaged 37.5 μg/g (range, 10.9 to 98.2 μg/g). Of 13 fish sampled, three had livers appearing normal, three had "hemorragic" livers, and the remainder showed small pustules and tumors.

Lower Trophic Relationships. PCBs in lower trophic organisms were studied with a range of field and experimental methods. The DOH has monitored PCBs in aquatic macroinvertebrates since 1977. Caddisfly larvae have been monitored at five upriver stations; multiplate samplers, set up on buoys at eight upper and lower river stations, have sampled diverse

Table 4.2. Changes in PCB concentrations in selected Hudson River fish.

Species	Mean PCB concentration (ppm)		Location
	1977 (78)	1981 (82)	
Pumpkinseed[a]	1019	362	Upper Hudson
Brown bullhead[a]	2510	428	Upper Hudson
Goldfish[a]	6761	310	Upper Hudson
Largemouth bass[a]	6010	1000	Upper Hudson
Largemouth bass[b]	145.3	10.2	Stillwater
Largemouth bass[b]	29.5	1.0	Catskill (estuary)
Striped bass[b,c]	(9.9)	(2.6)	Estuary

From Brown et al. (1985).
[a]PCB concentrations expressed per unit lipid in fish.
[b]PCB concentrations expressed per unit wet-weight tissue.
[c]Median values used for striped bass.

invertebrates (Simpson, 1982). Collections have been made at five-week intervals from June to September, and samples were analyzed for A-1016 and A-1254; in 1983, samples were analyzed for individual PCB congeners (Bush et al., 1984). In another field study, micro- and macrozooplankton were collected at 13 lower Hudson sites by investigators at New York University Medical Center (NYU Medical Center, 1982). Experimental work with ^{14}C-labelled PCB was carried out to study partitioning and uptake kinetics of PCBs in phytoplankton in the presence and absence of contaminated sediments (Roscigno and Punto, 1982), and for PCB uptake in zooplankton and fish in PCB-contaminated water and/or food (Peters and O'Connor, 1982). Powers et al. (1982) found growth rates and photosynthetic rates to be reduced in phytoplankton exposed to A-1254 desorbed from fine particles.

Until 1981, the results of the macroinvertebrate study indicated a similar declining trend in PCB concentrations. Between 1979 and 1980, caddisfly larvae concentrations decreased 34%, multiplate residues decreased 27%, and water concentrations decreased 43% (Brown et al., 1985). In 1981, unexpected significant increases in A-1016 concentrations in both caddisfly larvae and multiplate residues were measured (Sloan et al., 1983); the increase is inconsistent with other biological trends and is not well understood. Zooplankton showed wide variation in PCB concentrations due to species specificity, location, and season. Microzooplankton in particular showed a geographic trend, having highest concentrations (11.12 μg/g) at Albany and lowest ones (0.85 μg/g) in the lower estuary (NYU Medical Center, 1982, cited in Brown et al., 1985).

The macrobenthic monitoring and planktonic experiments provided estimates of bioconcentration rates in lower trophic levels. From 1977–1980, average bioconcentration factors of 16,800 and 106,500 times ambient water concentrations were reported for multiplate residues (diverse macroinvertebrates) and caddisfly larvae, respectively (NYSDEC, 1982b). The importance of uptake through direct adsorption of particles vs. consumption has been a matter of debate. Peters and O'Connor (1982) reported bioconcentration factors of 1×10^4 for amphipods and mysid shrimp via food pathways, but 2 to 4×10^4 via direct sorption of dissolved and particulate PCBs; all equilibrated within 40 hr. Results of their experiments led the authors to conclude that knowledge of both water concentrations and octanol-water partition coefficients is essential for prediction of potential for food chain transport. Feeding studies of *Gammarus* with *Myriophyllum* and *Chlorococcus* did not provide any evidence for food chain magnification (Peters and O'Connor, 1982).

In another study that combined batch experiments with a simulation model, Brown et al. (1982) investigated PCB partitioning equilibria and bioaccumulation in plankton. The experimental results indicated that partitioning equilibria is partly size-dependent, while bioaccumulation varies with species. The kinetics of sorption-desorption were modelled using the experimental data for calibration, and were included in a nine-compartment pelagic model of soluble PCB interacting with various phyto- and zooplanktonic size classes, as well as suspended solids and detritus. The simulations, running over 96-hr periods, indicated that PCBs could accumulate at levels above those predicted by equilibrium partitioning when feeding exceeded desorption and excretion. "Unless referenced to a specific set of biological and environmental conditions, the importance of direct partitioning from water vs. food uptake appears to be a moot topic" (Brown et al., 1982, p. 29).

Evidence has also recently been presented (J. Brown et al., 1984) showing that upper Hudson sediment PCBs have undergone considerable microbial transformation. In 1984, G.E. researchers investigated the characteristics of a number of sediment cores that had been preserved from the original upper Hudson sediment survey in 1977. These cores, collected mainly in the vicinity of the Thompson Island pool, were analyzed by means of capillary gas chromatography. In addition, portions of each core sample were investigated for bacterial composition. The sediments were found to have highly altered PCB compositions relative to what was originally discharged. Evidence was found of dechlorination in deeper sediments, which was suggestive of microbial transformation. According to J. Brown et al. (1984), most of the originally discharged PCBs consisted of A-1242 mixtures; much of what was observed in sampled biota bore greater resemblance to A-1254, which was discharged in relatively small amounts. At least two classes of aerobic and three classes of anaerobic microbial dechlorinators have been identified, the former existing in upper sediments and the latter in deeper anoxic sediments (J. Brown et al., 1984). Based on the diversity of vertical and horizontal distribution patterns of measured PCB congeners, the microbial populations appear to exhibit considerable spatial heterogeneity. In contrast, Bopp et al. (1984) reported no significant alteration of PCB congeners in the lower Hudson.

Terrestrial Organisms. When the PCB situation came to light, it was feared that the chemical would be transported through the food chain to organisms that eventually would be consumed by people. Therefore, in addition to monitoring aquatic organisms, the DEC also conducted and funded research on terrestrial plants and animals.

In conjunction with the air monitoring program, Boyce Thompson Institute has monitored a variety of wetland and terrestrial (including both endemic and cultivated) plants since 1978. PCBs were found to be taken up mainly from the air; they volatilize from the soil and adsorb onto plant surfaces, mainly leaves (NYSDEC, 1979). Roots also absorb PCBs, but probably less than 1% is translocated to other parts of the plant (Buckley, 1982). Different species also exhibit different accumulation rates, but all served as good integrators over time of the PCBs that were volatilized (Buckley, 1982). Aspen, goldenrod, and sumac very near an old dump site accumulated PCBs in excess of 600 times the mean background levels. It has been suggested that plant leaves, analogously to benthic organisms in contaminated sediment, can be used successfully to monitor atmospheric transport of such volatile compounds as the less-chlorinated PCB congeners (Buckley, 1982; Buckley and Tofflemire, 1983).

Wildlife has been monitored by the DEC as part of a statewide toxics program (NYSDEC, 1981a, 1981b; 1982a). PCBs have been found to accumulate in apex or near-apex predators such as mink, herring, gulls, harbor seals, kestrels, etc. Liver, brain, and fatty tissues were the sites of greatest accumulation. One great horned owl found near Catskill (RM 120) had a brain PCB concentration of 357 ppm; however, most of the other organisms assayed had concentrations one to two orders of magnitude lower. A notable exception was the snapping turtle; this species was shown by Stone *et al.* (1980) to concentrate very high levels of PCBs in fat, liver, and eggs. The average PCB concentration in fat of 11 snapping turtles was 2990.6 ppm (Stone *et al.,* 1980). Turtle soup prepared by these investigators from Hudson River snapping turtles yielded about 62 mg of PCBs in an average portion.

Integration: The Role of Models in PCB Assessment

All of the field studies showed that PCBs, largely originating from the GE facilities, had undeniably spread to contaminate every aspect of the Hudson and surrounding ecosystems. Figure 4.8 is a conceptual model of PCB dispersal throughout the Hudson ecosystem, drawn with energy circuit language symbols (Odum, 1971). Estimates are approximates only, and are quite uncertain for biota since total biomass of various species and ecosystem components has never been assessed. Thomann and St. John (1979) estimated the total PCB burden in estuarine biomass to be around 100 kg, or less than 1% of that amount entering the estuary. It is obvious from such a diagram that the portion of PCBs sequestered in sediments is enormously greater than that in biota. This better represents a storage of chemical rather than a sink, in terms of potential future contamination that might reach humans.

Out of these and other studies, two questions emerged: 1) How long would these compounds remain in the system? and 2) What would the effects be? To help obtain an estimate for the first question, two numerical models were developed that synthesized much of the information available at the time. One modelled the physical transport of PCBs from the concentrated deposits upriver down to the federal dam at Troy, and could thus provide estimates of transport to the estuary. The other model simulated aspects of biological transport through the estuarine food web, placing emphasis on predicting levels in striped bass. The physical model was developed by Lawler, Matusky, and Skelly (LMS), Engineers, and the ecosystem model was developed by Hydroscience, Inc. A brief description and discussion of results follows.

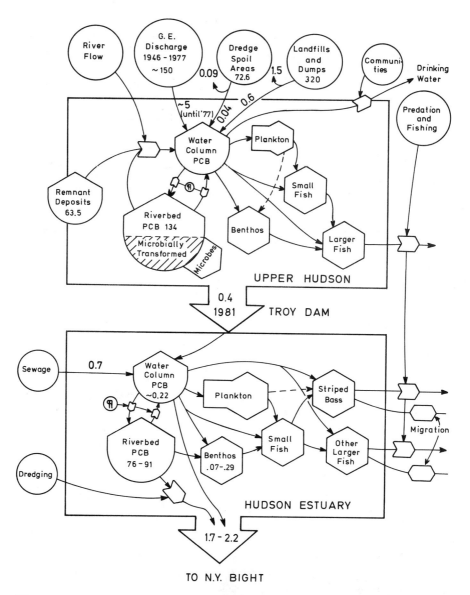

Figure 4.8. Diagram of PCBs in the Hudson ecosystem, showing relationships of biotic and abiotic components (symbols from Odum, 1971). Storages are metric tons; flows are metric tons/yr. Most of the estimates are from Tofflemire and Quinn (1979); sewage estimate is from Bopp *et al.* (1981); and contribution to New York Bight is from Mueller *et al.* (1982); concentration in benthos was calculated from data in Sloan *et al.* (1983).

Physical Sediment and PCB Transport Model (LMS, 1978, 1979; Apicella and Zimmie, 1978)

LMS used a standard transport model, HEC-6 (Hydrologic Engineering Center, Davis, Calif.), together with a model that kept track of PCBs. A schematic of the program is presented in Figure 4.9. As input, the model used data on river flow, basin morphology, grain size distribution, and PCB concentration and distribution. The system modelled was defined as the 40-mile (64 km) stretch from Lock 7 above the Thompson Island Dam to the Troy Dam. This was divided into eight segments by dams and locks (Figure 4.9b). River flow was the driving force expressed in an energy equation. This, in turn, was used in a submodel that calculated sediment scour and deposition, coupled with another submodel that calculated movement of the upper sediment bed. Bed change and sediment load for each segment was computed every 0.25 miles (0.4 km) and provided input to the PCB Inventory Model. Predictions could be made over both time and space.

Time-series analysis of USGS river flow data revealed a strong 20-year cyclical component (Apicella and Zimmie, 1978); more will be said about

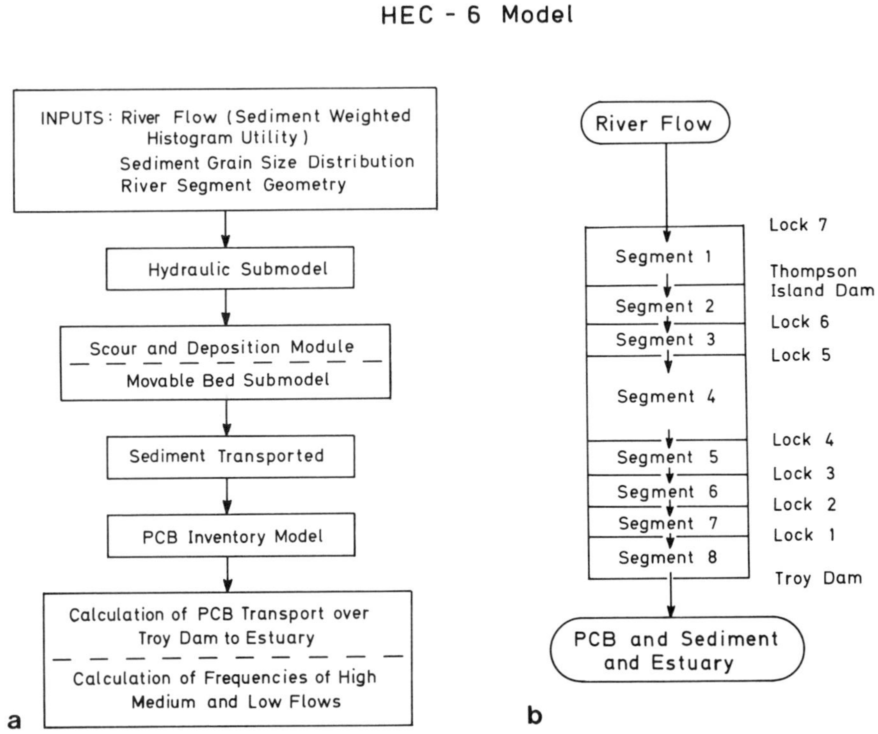

Figure 4.9. Schematic of LMS sediment and PCBs transport model. (a) Generalized computer flow diagram. (b) Idealized river segments modelled.

this later. Therefore, it was considered reasonable to use river flow data from the preceding 20 years (1958–1977) as the input. Initial conditions were based on field measurements in each of the reaches (LMS, 1978).

The model was used to simulate several possible management alternatives, including no action, no action except for channel-dredging maintenance, and dredging the hot spots in the river with concentrations ≥ 50 ppm. For the no-action simulation, the model was able to reproduce measured medium- and high-flow sediment and PCB loadings fairly well, but it significantly underestimated transport of PCBs during low flow. Therefore, at low flows, the model was forced with a transport rate of 6 to 8 pounds (2.7 to 3.6 kg) per day of PCBs. This forcing resulted in increasing the estimated 20-year transport of PCBs to the estuary by 23%, from 130,000 lb (59,000 kg) to 160,000 lb (73,000 kg) (Apicella and Zimmie, 1978).

For the other management alternatives, HEC-6 was recalibrated as new, better estimates were made available (LMS, 1979). Also, in 1976, DOT had dredged some 26,800 m^3 for channel maintenance; in 1977–1978, DEC dredged 37,600 m^3 from the river's east channel near Rogers Island; in 1978, it also excavated 10,750 m^3 of higher contaminated spoil from a remnant deposit (3A) just south of Hudson Falls and stabilized another one with rip-rap (EPA, 1981a). This action was reflected in the model with prediction of lower transport than previously reported. Figure 4.10 shows 20-year projections of PCB loading at the Green Island (Troy) Dam. Upper and lower bounds to PCB loadings are given by a simulation based on no further mitigative action, and another based on hot spot dredging combined with removal of two major remnant shoreline deposits, respectively. The new no-action simulation predicted an average annual loading of 7200 pounds (3272 kg). Hot spot dredging was predicted to cause a 20% drop in this rate, while hot spot dredging in combination with remnant deposit removal or capping would reduce the load to 1864 kg/yr or a 43% reduction over the simulated 20-year period.

Ecosystem Fate Model (Hydroscience, 1979)

The purpose of developing the ecosystem model was to have a means of synthesizing the large amount of data collected on PCB burdens in estuarine biota so as to understand its ultimate fate. It should be noted that "fate" in a toxicological sense connotes four subprocesses: 1) transport (hydrodynamic processes); 2) transfers (volatilization, sedimentation); 3) source and sink identification; and 4) transformations (chemical and biological processes). Furthermore, the model was to aid in predicting PCB body burdens following an eventual reduction in water column concentration. Although never explicitly stated, a primary practical motive appears to have been the forecasting of the time horizon over which striped bass PCB levels could be expected to exceed FDA standards. This would, in

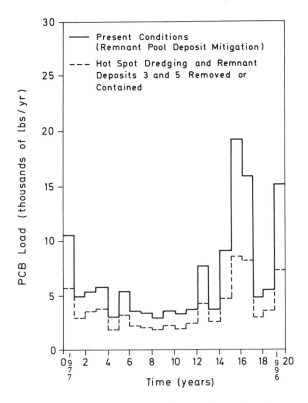

Figure 4.10. 20-year projection of PCB loadings at the Federal Dam at Troy, NY as predicted by LMS model (LMS, 1979).

effect, predict the length of the ban on commercial and recreational fishing for this important species. Two models were developed for this purpose, as shown in Figure 4.11.

The "food web" model described six biological compartments: phytoplankton, micro- (< 600 μg) and macrozooplankton, benthic fauna, large benthic detritivores and predators, and small (< 30 cm) fish. In addition, detritus was modelled as a state variable that interacted with most of the other compartments. The idealized estuary was fairly coarsely divided into six segments starting at the federal dam at Troy (RM 154) and extending into the apex waters of the New York Bight (RM -40). The major physical processes included in each segment were: advection in and out, tidal dispersion, and settling (these parameters were for the planktonic and detritus compartments). Growth (primary production for phytoplankton; consumption and assimilation for others), death and respiration, and mortality due to predation or grazing were the major biological processes. PCBs were followed throughout the system; the model was forced with assigned concentrations of PCBs in the water column and sediments.

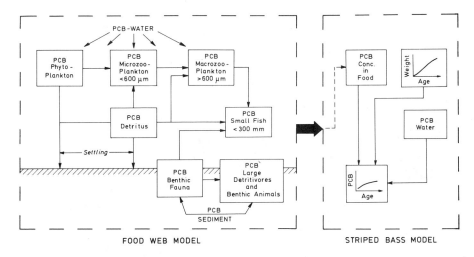

Figure 4.11. Schematic of ecosystem and striped bass models developed by Hydroscience, Inc. to study ecosystem fate and transport of PCBs (from Hydroscience, Inc., 1979).

The second model was coupled to the food web model and evaluated PCB transfers through 12-year cohorts of striped bass. Some of the calculated values of biomass and PCB burdens from the food web model were used as inputs to this. The physical environment was modelled simply as two segments: the estuary and coastal waters. Striped bass moved into the estuary to spawn, and young up to 2.5 years of age remained there. After that, the fish migrated in and out of the estuary each year. Prey of the young fish ($\leq$ 3 years) included micro- and macrozooplankton; older striped bass were assumed to eat the "small fish" of the food web model.

Parameters that were highly critical in the models were the PCB uptake and excretion rates, because these determined the body burdens over time. The modellers tried different scenarios of uptake and excretion; in general, excretion rates were much lower (order of magnitude) than uptake rates, and in many scenarios they were set to zero. This had the effect of emphasizing the importance of food chain transformation of PCBs. It should be noted that this was a critical part of the calibration process, since data on rates (and even good estimates of biomass) were scarce. The models were run and PCB burdens were calculated and compared with actual values, as these were the only measurements made.

Under a scenario of zero PCB excretion from the food web (except by phytoplankton), and with increased uptake rates by all compartments, the striped bass cohort model predicted essentially a logistic-type curve for cumulative PCB contamination through a fish's life span, reaching a maximum at about nine years and tapering somewhat thereafter. The simulation calibrated for median levels of PCBs in striped bass used bass PCB excretion rates half the magnitude of those originally assigned.

Under these assumptions, food chain PCB levels did decline with a drop in water concentration in the upper estuary; however, the decline became less marked down-stream, and finally coincided with the "present condition" simulations in the offshore segment. Small estuarine fish never dropped below the newer FDA "action levels" of 2 ppm, and so always provided some source to the striped bass. A major conclusion was that average bass PCB concentrations could only be sufficiently lowered with a water concentration reduction to 25 times less than present (1978) levels (i.e., to 2 to 4 ppb dissolved).

Relative Successes and Failures of the Models

Because both the physical and ecosystem forecasting models were completed in 1978, the passage of time and careful monitoring programs have provided enough data to allow some trend comparisons.

It is interesting to note that both models were critically reviewed as part of the draft environmental impact statement prepared for a remedial action plan by the EPA (EPA, 1981a). The reviewers were highly critical of the LMS work, but were quite favorable toward Hydroscience's model, blaming the former for using a less than state-of-the-art approach and praising the other for incorporating a state-of-the-art analytical framework. It was pointed out that HEC-6, the model chosen by LMS for the physical transport calculations, is designed for alluvial stream beds, while most of the PCBs in the Hudson are associated with fine organic materials; and such river beds have different sedimentation dynamics compared to alluvial beds. The model was also criticized for not including volatilization and navigational dredging effects. By contrast, the reviewers found the ecosystem model to have ". . . a solid scientific basis as well as sound engineering practicality" (EPA, 1981a, p. C-3). Both modelling reports recommended further monitoring studies to reduce uncertainties; however, only the recommendation by Hydroscience was mentioned in the review.

The ecosystem model (Hydroscience, 1979) also had several shortcomings. Although assumptions were clearly stated, they were not always justified, as for example with the arbitrarily set PCB excretion and uptake rates. Temporal trend data for the biomass and detritus were not even mentioned, although temperate estuarine functions are generally extremely seasonal (the Hudson being no exception). While mentioning some of the data scatter, other sources of uncertainty appeared to be ignored. Extremely poor estimates of biomass and, thus, PCB levels in some compartments—the benthic infauna and predators—lend a good deal of uncertainty to model predictions, especially since these may be important food chain links. Sensitivity testing to vary their levels was not discussed in the Hydroscience report. Finally, the work of LMS was not even mentioned, let alone linked to the estuarine ecosystem model. This was extremely unfortunate, but may have been the result of not coordinating task efforts. The PCB concentrations in the water going over the Troy Dam, as predicted by the physical transport model, could have provided a natural coupling to the downstream ecosystem model, which was hampered by a lack of year-to-year variation in PCB levels.

How well have the models done to predict the five years that have elapsed since their publication? It would appear that for all its limitations, the physical transport model has had greater success than the ecosystem one. The main reason for this was probably not the choice of correct mechanism (since low-flow mechanisms were unknown and entirely underestimated, and also because scour and deposition were given undue importance), but rather the LMS' time-series analysis of the USGS stream flow records. This revealed a highly significant 20-year cycle with five discrete, yearly subharmonics (Apicella and Zimmie, 1978). Therefore, by starting simulations at the appropriate phase of the cycle, LMS was able to make predictions that have thus far been fairly correct. Figure 4.12 shows a plot of the cumulative PCB transport from both the original and updated versions of LMS "no action" simulation, together with the cumulative estimated transport to date at Waterford (for all intents and purposes comparable to Troy). Also included is an estimate from a far simpler model that assumes an exponential decay of stored sediment as

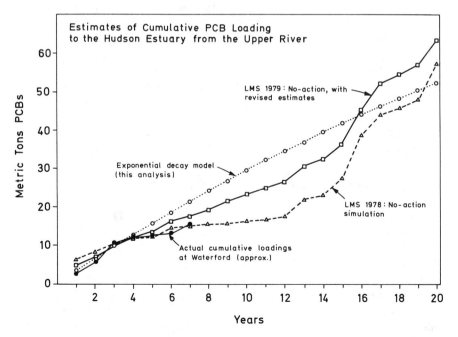

Figure 4.12. Plot of simulated, cumulative PCBs transported over Troy dam over a 20-year period and actual transport at Stillwater as calculated by M. Brown *et al.* (1984). Starting year is 1977.

the burden is washed downstream. This model does a fair approximation of the more complicated model's results.

The biological data collected thus far from the estuary do not seem to bear out strongly the ecosystem model's predictions, although this is not entirely true since the number of years elapsed was not specified in the model. Zooplankton data have varied widely over species, seasons, and sampling location, although geographical gradients could be seen (Brown et al., 1985). On the other hand, striped bass have shown a fairly clear trend of declining body concentration, which was not predicted by the Hydroscience model (Table 4.3). This result may call into question the assumption of food chain biomagnification's dominant role invoked by Hydroscience. Biomagnification vs. direct sorption appears to be a major question for which only partial answers are available (e.g., Peters and O'Connor, 1982). Brown et al. (1985) conclude from their analyses that most biotic levels are directly governed by water concentrations, but do not state what the mechanisms for uptake would be.

Flaws notwithstanding, both modelling efforts are to be commended for attempting to link together various parts of the upriver and estuarine systems. The work helped to identify sources of uncertainty and keyed in on possible transport mechanisms that were subsequently investigated. The models, together with the data base, indicate the importance of the Hudson's physical characteristics—flow, sediment transport rates, as

Table 4.3. Comparison of projected striped bass PCB concentrations with measured values from the Hudson estuary.

Measured concentrations[a] (median values, μg/g wet weight)		Model projections[b] (μg/g wet weight)		
		5-yr projection for	(PCB)	
Yr	(PCB)-50% (n)	starting yr-class	Excr. = 0[c]	Excr. = 0.5
1978	10.0 (368)			
1979	5.1 (29)	2 (reaching 7 yr)	5	4.6
1980	4.1 (197)	4 (reaching 9 yr)	7	4.5
1981	3.0 (220)	6 (reaching 11 yr)[d]	14	4.0
1982	2.2 (154)	8 (reaching 12 yr)[e]	16	5.0

[a]From M. Brown et al. (1984). Data read from graph.
[b]From Hydroscience, Inc. (1978). Data are taken from graphs of output of simulations assuming an instantaneous 5-fold reduction in PCB water concentration. Output follows an individual yr-class throughout the simulation period.
[c]Simulations assumed different rates of PCB excretion (excr.) (0 and 0.5 per unit time).
[d]Estimated from graphs; sixth yr-class not specifically simulated.
[e]Oldest age class in model = 12 yr.

well as PCB partitioning—in governing PCB entrance into and residence time in the estuary.

Remedial Actions: Proposed, Rationales, and Controversies

From the battery of scientific investigations, it became clear that PCBs could be found everywhere in the Hudson ecosystem. However, over 60% of the burden remaining in the river was thought to be sequestered in high concentrations ($\geq$ 50 ppm) in various sediments and remnant deposits along the banks downstream of Ft. Edward. Because of this high, localized concentration, it was felt that a cost-effective remedial action plan to remove or secure these deposits could be successfully implemented.

This was discussed at length by the members of the Settlement Advisory Committee. The traditional way of dealing with a sediment-related problem is dredging. However, in this case, it was feared that immediate dredge impacts (such as destruction of existing habitat, extensive resuspension of sediments including PCBs and heavy metals, and excessive traffic noise from dump trucks hauling away the dredge spoil) could outweigh the benefits. An evaluation of alternatives to a remedial dredging plan was done by Horstman (1977). At the time, many potential solutions existed, but few had been tested on any sort of scale approaching the immensity of the Hudson problem. No method appeared to be as efficacious as carefully executed dredging and disposal (MPI, 1980).

The DEC proceeded with feasibility and environmental impact studies of alternative dredging plans. The alternatives included: 1) full-scale (bank-to-bank) dredging of the eight pools of the upper Hudson between Troy and Lock 7 at Ft. Edward; 2) bank-to-bank dredging of only the most heavily contaminated pools, such as the Thompson Island pool (RM 188) about 10 km south of Ft. Edward; 3) a full hot spot dredging program to remove all sediments containing at least 50 μg/g PCBs; 4) a reduced version of hot spot dredging; and 5) a no-action alternative. "No-action" means that no remedial action be taken, except for channel maintenance dredging by the DOT as mandated by the New York State Constitution.

The dredging studies were carried out by Malcolm Pirnie, Inc. for the DEC (MPI 1978a, 1978b, 1979, 1980, and others). The scope of the alternatives was summarized by MPI (1980), as shown in Table 4.4. Of these, the most cost-effective plan was to dredge all or most of the hot spots and remnant deposits.

Outline of the Proposed Dredging Project

The program would consist of dredging up to 40 submerged sites, ranging from Rogers Island (RM 194) to just south of Mechanicsville (RM 164), and would remove up to 8% of the riverbed, thereby posing the least amount of damage to benthic habitat. In addition, 2 to 10 non-riverbed sites (remnant deposits) would be excavated, and their contaminated burden would also be removed to a containment area. This would be in addition to the 141,000 m^3 dredged in 1977–1978 by DOT, under DEC supervision, and taken to the New Moreau demonstration site (MPI, 1980).

The effectiveness of the proposed hot spot and remnant deposit dredging has been the subject of much debate. From the LMS model predictions, the proposed project would reduce the level of PCBs entering the estuary by roughly 40% (of which 21% would be from dredging the hot spots alone; see Table 4.4), and would shorten the total transport time by a score of years.

More recent evaluations of that model and of the PCB budget in the river cast doubts on the prediction. Rather, accumulating evidence suggests that the project may have a greater efficacy (see the "Epilogue" section in this chapter).

The DEC proposed that the entire project be conducted as a demonstration of proper cleanup procedure. As such, special mitigating measures were to be present in all aspects of project execution (MPI, 1980), including special dredge equipment, silt curtains and booms for minimizing losses of material as it is brought up, bank stabilization, etc. In cases where wetlands would be deemed necessary for removal, it was felt that restoration could be accomplished.

The proposed containment site was selected by careful screening from 12 possible candidate sites on the basis of meeting state and federal criteria for disposal sites, accessibility, ease of site preparation, and avoidance of conflict with other land uses. The proposed site, shown in Figure 4.13, is 4 km south of Ft. Edward, is 101.2 ha (250 acres) in size, and is on the average 910 m (3000 ft) from the river's edge. Its soil contains an impermeable clay layer; part of the land parcel is presently in hay production, while the rest is in abandoned fields. Previously dredged material would be brought to the site by truck and placed in the prepared bed, and newly dredged material would be piped in from the river. Materials from remnant deposits 3 and 5 (the most heavily contaminated) would also be placed in the site, as would material from three DOT dredge spoil areas where PCB-containing sediments were placed following channel maintenance dredging. Of the 100 ha, about 68 would be used for containment. The remaining land would serve as a buffer zone from the surrounding farmland, with observation wells and drainage ditches to monitor and capture any PCBs lost from the site via leaching. Provision for anti erosion maintenance, restriction of burrowing animals, and establishment of a suitable vegetation cover were included in the environmental impact statement (MPI, 1980).

Many residents of Washington County felt that the proposed dredge

Table 4.4. Comparison of dredging alternatives.

Action	Dredging[a] volume (cubic yd)	PCB[b] recovered (lb.)	Modeled reduction in annual PCB transport to estuary		Cost in[e] millions of dollars (1981)	Cost per[e] lb. PCB recovered (dollars)
			lb.	%		
Complete dredging (eight pools)	14,510,000	321,300–326,500	n.a.	70[d]	234	720–730
Thompson Island pool, complete	1,720,000	122,200–128,800	n.a.	20[d]	23	180–190
40 Hot Spots	1,452,700	154,300–159,700	1500	21[c]	25.7–25.9	160–165
Thompson Island pool, hot spots 1–20	645,500	96,300–101,400	500	7[c]	11.5–11.7	115–120

From MPI, 1980.
[a]MPI (Jan., Dec. 1978)
[b]Range results from use of factors for dredging system efficiencies and does not imply a minimum and maximum estimate.
[c]LMS (1979).
[d]Thomann and St. John (1979).
[e]Costs in 1981 dollars, although the project would occur in phases. Estimates are higher than previous reports (MPI, Jan., Dec. 1978), reflecting rising fuel costs and inflation.
n.a., not available.

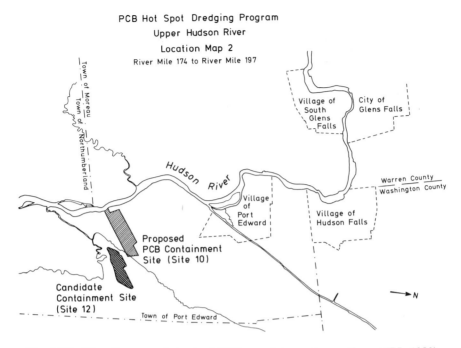

Fig. 4.13. Site of proposed dredged PCBs containment area (from MPI, 1980).

spoil site would seriously conflict with the agricultural uses of surrounding land. Consequently, a citizens' group negotiated with DEC for compensation for residents with land holdings adjacent to the spoil site. Compensation for decreased property values, loss of local taxes, and crop and livestock contamination were among the demands. Additionally, the group wanted the state to pay for monitoring PCB levels in crops and livestock near the site (Pulver and Hafner, 1981).

Federal Involvement

Although New York State took responsibility for all assessment, monitoring, and mitigation studies, federal regulatory agencies also were to become involved. Specifically, the state made a request for federal monies to be allocated to the PCBs cleanup effort under the FWPCA, also known as the Clean Water Act (CWA), and applied to the U.S. Army Corps of Engineers for permits to dredge as mandated under both CWA and the Rivers and Harbors Act (86 Stat. 816, 33 U.S.C. 1344 and 30 Stat. 1151, 33 USC 403) (MPI, 1980). In September 1980, Congress amended the CWA under title I, sections 116(a) and (b). This amendment was entitled the Hudson River PCB Reclamation/Demonstration Project; it authorized (under title II, section 205 [a]) the EPA to spend up to $20 million for the cleanup project (less than half the estimated costs in 1978 for 1980–1981 dredging).

Because of federal involvement with such a large project, a federal environmental impact statement (EIS) was required under the National Environmental Policy Act (NEPA). This was prepared by the Region II office of EPA. A draft EIS (DEIS) appeared in May 1981 (EPA, 1981a), followed by a revised draft in August (EPA, 1981b) and a final version (EPA, 1982) in October 1982. The first draft, prepared by WAPORA, Inc. for EPA, contained summaries of the scientific and engineering findings, and evaluated once again the alternative remedial actions. Supporting documents, as well as reviews of the LMS and Hydroscience models, were included as appendices. The revised versions eliminated most of the technical background information, but included comments and monitoring plans.

The DEIS evaluated eight alternatives, ranging from taking no action to full-scale dredging and containment of hot spots and remnant deposits. In-river detoxification and/or containment were also considered, but were found to be less practicable.

The DEIS appeared rather biased toward its recommended action; that is, the rescoped project that would take out the most heavily contaminated hot spots, cap the two worst remnant deposits, and construct retaining dikes on the channel edge of two wetland hot spots to protect them from lesser floods. The information content varied greatly from one section of the report to the next. Although description of existing conditions was adequate, analysis of environmental consequences was often limited to vague statements, such as "the alternative would have a beneficial effect on fish and aquatic life." The statements of effects were not consistently backed by factual analysis; at least it was not apparent.

Benefits and costs were described in varying detail, but were seldom substantiated. Costs were quantified for the actual dredging/containment project ($40 million for the full-scale; $26.7 million for the rescoped version), but not for long-term maintenance and monitoring, for which further sources of funding would have to be sought and guaranteed. Monetary assessment of benefits accrued by various plans seem limited to a brief mention of the assessed value of fisheries based on a report by Sheppard (1976).

Action. More comprehensive analysis probably should have been performed for the EIS; however, time constraints curtailed the effort. The final EIS was brought out in October 1982 and had the backing of almost all of the involved parties. A final decision was rendered by EPA Administrator Ann Gorsuch in December 1982 (Schafer, 1982).

In a surprise decision, Administrator Gorsuch judged that the project was unsuited to funding under CWA, but would likely be fundable under the Comprehensive Environmental Response, Compensation, and Liability Act of 1980 (CERCLA), also known as Superfund. Under this act, the project would be evaluated along with other problems across the nation. Two steps would be carried out; and at the same time, the project would be assessed for inclusion on the National Priorities List (a list of serious environmentally stressed sites; placement on the list is done by calculation of a score meant to indicate the severity and time constraints of the problem). The first step is called the Remedial Action Master Plan (RAMP), and is an evaluative process by which EPA examines existing information and judges whether or not further action, further data gathering, or nothing at all is required. A second step is to identify responsible parties, presumably to assess a liability fee to be applied to the cleanup. If the project were placed on the National Priorities List, and were given a high enough ranking, then further steps would be carried out. This would include a feasibility study to select the best method for mitigation; and finally, if shown to be cost-effective, the mitigative action would at last be taken.

Reaction. In making this decision, the EPA took on an adversarial role; suddenly EPA was telling all concerned parties that the old rules did not apply in what amounted to a new game. RAMP and all other studies would redo what had already taken years and millions of dollars to do. To many, this appeared to be a stalling tactic, perhaps used to choke off the project in red tape. Reaction was inevitable; on May 19, 1983, a suit was filed against EPA by the Hudson River Sloop Clearwater, Scenic Hudson, Inc., Natural Resources Defense Council, Hudson River Fishermen's Association (HRFA), and Representative Richard Ottinger. The groups sued to release the $20 million in CWA funds for the cleanup project and charged that EPA was ignoring a congressional mandate by changing its plans. The DEC joined the suit in July 1983.

The EPA responded to pressure by beginning to carry out the steps designated in the CERCLA analysis. In May 1983, a consulting firm, NUS Corp., initiated the RAMP study. On September 1, EPA at last placed the Hudson River on the National Priorities List. Meanwhile, time was running out for cleanup funding under CWA; the deadline for use of the funds had been set at September 30, 1983. The plaintiffs requested and were granted by presiding Judge Brient of the Federal Southern District Court of New York a preliminary injunction that overrode the deadline and allowed the judge time to read through the evidence and narrow down the issues.

Also, as promised, the RAMP report was released by EPA in early October 1983 (NUS, 1983). The evaluation covered all of the major studies and pointed out several deficiencies. One major deficiency was the age of the data upon which decisions are to be based; most of the sediment sampling had taken place in 1977 and 1978. Another problem was discovered in the LMS model; not only were fine-grained sediments given a less than significant role, but scour and deposition were predicted in places where it had been shown not to occur. According to NUS (1983), sediment load characteristics of the upper Hudson are such that once in suspension, they tend not to settle out. This means that sediments transported from the PCB-contaminated Thompson Island pool are transported all the way to the estuary (J.E. Sanders, pers. comm.). The RAMP also carried out a formal cost-effectiveness analysis for each alternative cleanup method. This analysis used matrices and weighting factors that were never clearly explained or motivated.

In the executive summary, the report makes the following statement (p. ES-9):

> The matrix evaluation process was used to determine the cost-effective solution as provided by . . . (CERCLA). Based on the current data available on the PCB problem in the Hudson River, the result of a matrix analyses (sic) evaluation with respect to contaminated sediments in the Hudson River is "no remedial action." The limited threat to the public health does not justify the large expenditure of money required to remove a portion of the contaminated sediments.

The summary notes that under the proposed plan, it was estimated that the river would take up to 46 years to be totally cleansed v. 64 years with only routine maintenance dredging. With an estimated project cost of $34 million (including disposal), the project lost on cost-effective grounds, at least for the moment. However, the NUS report recommended capping and fencing of certain remnant deposits and continued river monitoring to protect public health.

Conclusions

The story of PCBs in the Hudson River and decisions on what to do with them is long and complicated. History may fault New York State for its initially poor handling of the problem, but it is to be largely commended for its settlement agreement with General Electric Co., and subsequent assessment work. And yet, even with massive amounts of data, even with guidance of a special committee comprised of scientists, managers, and laypeople, a decision on management has not been reached. One may question whether the key decision variables are still scientific ones in light of EPA's final decision on the reclamation project, or whether politics and dollars now prevail.

Part of the problem may be the issue of future uncertainty. The single clearest trend in the past five years of monitoring is that PCB concentrations declined rapidly after G.E.'s releases halted, and they appear to have more or less stabilized both in the water column and the biota. Against this trend is juxtaposed the prediction that Hudson River flows

Conclusions

are likely to rise again soon as part of a natural long-term periodicity. High flow rates are liable to scour more PCBs out of the sediments, thereby endangering the river's "recovery." Because risks have not really been estimated (such as: What is the risk of increased resuspension of PCBs in the future? or What are possible ranges of PCBs that one might expect to see in fish?), people tend to be divided on the issue. For example, the politics of dredging and reclamation have become geographically split. Many upriver residents see the decreasing trend and note the present improvement in water quality; they would prefer to believe that PCBs will continue to be covered over and think that a major reclamation project would be far more harmful to their region (Hayslip Associates, 1982). The downriver population (i.e., residents of the Mid-Hudson Valley and south) see the problem of PCBs in their fish continuing indefinitely into the future unless something is done now. Certainly, the upriver population has other causes of concern than fish, since the immediate impacts of a dredging program are likely to be severe and might affect agricultural productivity and real estate values.

Another debated issue has been whether or not PCBs from upriver contribute significantly to estuarine concentrations or whether the latter mainly result from municipal discharges. According to Bopp *et al.* (1981), approximately 75% of all PCBs in the lower estuary originated north of the Troy Dam. However, in reports and public statements, the EPA had been quoting Mueller *et al.* (1982), which attributes only up to 40% of the Hudson-Raritan PCB inputs as coming from the river. (It should be noted that this number is not substantiated nor is its derivation explained in Mueller *et al.*, 1982.) By assigning a lesser importance to the upper Hudson's contribution, the EPA may have been building a case against the reclamation project. Again, there appears to be uncertainty as to what the actual situation might be.

In retrospect, several things could have been done to improve the strength of the scientific studies as bases for a decision. For example, one major shortcoming of the biological studies is the virtual absence of discussion of effects. With few exceptions (Smith *et al.*, 1979; Klauda *et al.*, 1981; Merhle *et al.*, 1982; and Powers *et al.*, 1982), not one paper identified toxic effects in Hudson River biota. No mention was made, even in the state toxic substances monitoring reports, of increased mortality or morbidity, declines in reproductive rates, or changes in ecosystem-level processes. Had it been part of the state's goal to look for effects, and had effects been observed, pro-dredging arguments would have had a stronger foundation. The above-mentioned papers do suggest that toxic effects of PCBs can probably be found in Hudson River striped bass and tomcod.

If a "systems approach" is to be truly useful, all of the major environmental variables (forcing functions) should be identified and their impacts assessed. Such topics as synergistic effects of PCBs with other toxicants (e.g., cadmium, lead, and daughter products or contaminants such as dioxins), as well as with power plant operation and sewage releases, are untouched areas in research. As a first step, a conceptual Hudson River model could be helpful. Such a model should contain just enough complexity to represent the main characteristics of the ecosystems, including the human community, but should avoid unintelligibly complicated detail. Enough data are available to be able to assign order-of-magnitude values to various influences; comparisons of scale of different impacts are probably feasible. Development of such a model would be an art and could help in gaining a broader perspective of PCBs' impact.

The effectiveness of simulation studies done in the assessment period have already been discussed. Better coordination between the physical transport and ecosystem efforts would have been highly desirable. The physical model would also have benefited greatly from information on sediment transport characteristics that only became clear after the model had been completed. Perhaps the order of some of the research tasks could have been rearranged so that certain paths of investigation followed others; but it should be remembered that the DEC tried to deal with the problem as quickly as possible, given the potential toxic threat of PCBs. Thus, the fact that the issue has dragged on for so many years through so many assessments and bureaucratic delays is terribly ironic.

Just how the PCB issue will be resolved is moot at present. The politics of decision-making may proceed too slowly for this issue. Certainly, EPA's CERCLA program, with only five completed cleanups in over two years (out of a list of over 300), appears to be a slow and cumbersome vehicle at best (Schabecoff, 1983). Time may, in fact, be running out for anyone's definition of cost-effective remedial action.

Epilogue

Because of the impasse generated by EPA's maneuvering and the response by DEC and environmental groups, decision-making on the PCB issue was at a standstill at the end of 1983. At that point, however, the Hudson River Foundation offered help in the form of sponsoring a workshop held in January 1984. Scientists, technical experts, regulators, and concerned laypeople were invited to participate in an atmosphere free of tension, and to work through many of the technical and scientific issues surrounding the proposed dredging project. Participants were asked to raise and answer as many questions as possible. They were further asked to reach some sort of consensus of agreement on what information needs still remained, as well as agreement on where research had produced good answers to date (HRF, 1984). The main success was in the dissemination of information and the fostering of a cooperative spirit among the various parties—a spirit that, at times, had been dimmed. The largest disappointment was the absence of EPA and NUS, the contractors who had prepared the RAMP study. Many of the points raised in the RAMP were discussed, but, of course, the group could get no official response from the EPA.

The discussions may be grouped into three categories: scientific, technical, and social. The main points are summarized in Table 4.5. One of the more striking conclusions was derived from evidence, presented by several different groups of scientists, of a significant loss of lesser-chlorinated PCBs over time as a result of solubilization and volatilization. At the same

Table 4.5. Consensus issues raised at PCB Workshop, Jan. 1984.

I. Scientific
   A. Trend analysis
      1. Declines in PCB concentrations in the water column and in fish over the period 1977–1983.
      2. Loss of lower-chlorinated PCB congeners from the upper river and estuary; persistence of higher-chlorinated congeners into New York Harbor.
      3. Selective accumulation of higher-chlorinated congeners in food chain.
      4. Identification of Thompson Island pool near Ft. Edward as most likely source of PCBs to estuary.
   B. Questions and issues raised:
      1. Have the hot spot sediments moved or been reworked since the 1977 survey? New survey needed.
      2. Leaching/volatilization from remnant deposits? From dump sites? From river?
      3. What are the major mechanisms of PCB air transport?
      4. What are the mechanisms of sediment (and hence PCB) transport in the upper river? Research on cohesive particle transport, sediment transport studies needed to reevaluate models.
      5. What are the physicochemical properties of different PCB congeners? How do they behave in the environment? In biological systems? What are their health/environmental effects?

II. Technology
   A. Currently proposed dredging project
      1. Near agreement that removal is a necessary first step.
      2. Coupled to resurvey of sediments in Thompson Island pool.
      3. Question of how to manage remnant deposits: removal, fencing, in-place capping of some.
      4. Design of landfill: planning for uncertainties (volatilization risks, water treatment, monitoring, maintenance in perpetuity).
      5. Design of actual implementation schedule: temporal and meteorological considerations, worker protection, monitoring.
   B. Alternative technology
      1. Wright-Malta process: steam gasification to destroy PCBs, coupled to electrical generation.
      2. (Other proposals not discussed.)

Table 4.5. *Continued*

III. Social issues (mainly associated with upriver community)
    A. Health-related
        1. What is known about PCB epidemiology? Risks? What about psychological burden?
        2. What constitutes a "human health hazard?"
        3. What about carcinogenic risks?
    B. Farm-related
        1. What are the risks to crop contamination associated with low levels of PCBs in the air, such as would be emitted by a landfill site?
        2. Contamination of milk
        3. Compensation for contaminated crops, animals, water; compensation for devalued land; easements
    C. General
        1. Concern for groundwater contamination
        2. Potential decline in property values due to proximity of hazardous waste site.
        3. Assurances by the state for monitoring, informing, and cooperation with residents.

time, the more heavily chlorinated PCBs persist in the system, remaining in the food web and in the river system all the way to the mouth. Another scientific issue considered crucial to understanding the magnitude of the present PCB problem is that of PCB transport mechanisms in the upper Hudson. Two alternative hypotheses for sediment (and hence PCB) transport had been highlighted in the RAMP study, and were discussed at length in the meeting. One hypothesis holds that sediments are transported mainly as graded suspensions; that is, they are transported from one pool to the next via scour and deposition, so that the PCBs ultimately enter the estuary from the pool directly upstream from the Troy Dam. The alternative hypothesis is that PCB-laden sediment behaves as a uniform suspension derived mainly from the Thompson Island pool (where the hot spots are located), and is transported more or less directly to the estuary. The general consensus was that more research is needed to ascertain which of the competing hypotheses is more correct. If the latter hypothesis is correct, then cleaning up the hot spots might eliminate nearly the entire riverine source of PCBs to the estuary.

    Another positive outcome of the meeting was the discussion of social issues, and how these interfaced with scientific and technical issues. A contingent of upriver residents was able to voice its doubts and ask questions directly of the scientists and engineers. A feeling of true concern for the safe management of the PCB mitigation project was evident. Also, these residents had an opportunity to listen to a representative from a small firm of former G.E. engineers who have developed an unusual technique for destroying PCBs by steam gasification rather than incineration. The process holds promise for providing multiple benefits, namely, destruction of PCBs, handling of organic wastes and sludge, and power generation from the organic materials. However, more pilot testing is required.

During the remainder of 1984, further studies were made. More attention was given to higher-resolution PCB analyses in ongoing monitoring work. Also, the DEC carried out an extensive resurvey of sediments in the Thompson Island pool to determine whether or not, and to what degree, the hot spots had shifted or dispersed since the 1977 Normandeau study. Various alternatives were discussed for PCB destruction, including biological, chemical, and physical techniques (Hudson River PCB Settlement Committee, minutes of Committee meetings). The proposed site for dredge spoil was declared illegal in the spring of 1984, and the siting issue remained unresolved into 1985.

References

Ahmed, A.K. 1976. PCBs in the environment. Environment 18(2):6–11.

Apicella, G.A. and T.F. Zimmie. 1978. Sediment and PCB transport model of the Hudson River, pp. 645–653 In Proc. 26th Ann. Hydraulics Division Specialty Conference, ASCE., Aug. 9–11, 1978. University of Maryland, College Park, MD.

Armstrong, R.W. and R.J. Sloan. 1980. Trends in levels of several known chemical contaminants in fish from New York State waters. Technical Report 80-2. N.Y.S. Dept. of Environmental Conservation, Albany, NY. 77 pp.

Armstrong, R.W. and R.J. Sloan. 1981. PCB patterns in Hudson River fish. I. Resident/freshwater species. Proc. Hudson River Fisheries Conf. (C.L. Smith and G. Tauber, eds.). Hudson R. Environ. Soc. Hyde Park, NY.

Bickel, M.H. Polychlorinated persistent compounds. Experientia 38(8):879–882.

Bopp, R.F. 1979. The geochemistry of polychlorinated biphenyls in the Hudson River. Ph.D. Dissertation. Columbia University, New York, NY.

Bopp, R.F., H.J. Simpson, C.R. Olsen, and N. Kostyk. 1981. Polychlorinated biphenyls in sediments of the tidal Hudson River, New York. Environ. Sci. Technol. 15(2):210–216.

Bopp, R.F., H.J. Simpson, C.R. Olsen, R.M. Trier, and N. Kostyk. 1982. Chlorinated hydrocarbons and radionuclide chronologies in sediments of the Hudson River and estuary, New York. Environ. Sci. Technol. 16(10):666–676.

Bopp, R.F., H.J. Simpson, B.L. Deck, and N. Kostyk. 1984. The persistence of PCB components in sediments of the lower Hudson. Northeastern Environ. Sci. (3/4):180–184.

Boyle, R.H. and J.H. Highland. 1979. The persistence of PCBs. Environment 21(5):6–37.

Brown, J.F., Jr., R.E. Wagner, D.L. Bedard, M.J. Brennan, J.C. Carnahan, and R.J. May. 1984. PCB transformations in upper Hudson sediments. Northeastern Environ. Sci. 3(3/4):167–179.

Brown, M.P., J.J.A. McLaughlin, J.M. O'Connor, and K. Wyman. 1982. A mathematical model of PCB bioaccumulation in plankton. Ecol. Modelling 15:29–47.

Brown, M.P., M.B. Werner, R.J. Sloan, and K.W. Simpson. 1984. Recent trends in the distribution of polychlorinated biphenyls in the Hudson River system. (Mss.)

Brown, M.P., M.B. Werner, R.J. Sloan, and K.W. Simpson. 1985. Polychlorinated biphenyls in the Hudson River. Environ. Sci. Technol. 19(8):656–661.

Buckley, E.H. 1982. Accumulation of airborne polychlorinated biphenyls in foliage. Science 216:520–522.

Buckley, E.H. and T.J. Tofflemire. 1983. Uptake of airborne PCBs by terrestrial plants near the tailwater of a dam, pp. 662–669 In Proc. 1983 National Conference on Environ. Eng., ASCE Specialty Conference, July 6–8, 1983. Boulder, CO.

Califano, R.J. 1981. Accumulation and tissue distribution of polychlorinated biphenyls (PCBs) in early life stages of the striped bass, *Morone saxatilis*. Ph.D. Dissertation. New York University, New York, NY. 178 pp.

DHEW (Subcommittee on Health Effects of PCBs and PBBs). 1978. General summary and conclusions. Environ. Health Pers. 24:191–198.

DiNardi, S.R. and A.M. Desmarais. 1976. Polychlorinated biphenyls in the environment. Chemistry 49(4):14–17.

EPA. 1981a. Environmental impact statement on the Hudson River PCB Reclamation Demonstration Project. Draft, May 1981. U.S. Environmental Protection Agency, Region II Office, NY.

EPA. 1981b. Environmental impact statement on the Hudson River PCB Reclamation Demonstration Project. Supplemental draft, August 1981. U.S. Environmental Protection Agency, Region II Office, NY.

EPA. 1982. Environmental impact statement on the Hudson River PCB Reclamation Demonstration Project. Final, October 1982. U.S. Environmental Protection Agency, Region II Office, NY.

Hayslip Associates. 1982. Washington County residents' opinions of the Hudson River PCB clean-up program. A public opinion survey conducted for NYSDEC. R.A. Hayslip Associates. (No address.)

Hetling, L., E. Horn, and J. Tofflemire. 1978. Summary of Hudson River PCB study results. Technical Paper 51, N.Y.S. Dept. of Environmental Conservation, Albany, NY. 88 pp.

Hetling, L.J., T.J. Tofflemire, E.G. Horn, R. Thomas, and R. Mt. Pleasant. 1979. The Hudson River PCB problem: management alternatives, pp. 630–650 In Health Effects of Halogenated Aromatic Hydrocarbons (W.J. Nicholson and J.A. Moore, eds.). Annals N.Y. Acad. Sci., Vol. 230.

Horn, E.G., L.J. Hetling, and T.J. Tofflemire. 1979. The problem of PCBs in the Hudson River system. Annals N.Y. Acad. Sci. 320:591–609.

Horn, E., R. Sloan, and M. Brown. 1983. PCB in Hudson River striped bass, 1983. N.Y.S. Dept. of Environmental Conservation, Albany, NY. 8 pp. (Mss.)

Horstman, K.H. 1977. Evaluation of non-dredging alternatives to the removal of PCB contamination in the Hudson River. Comp. Ed. 18 Thesis. Union College, Schenectady, NY. 51 pp.

HRF. 1984. Proceedings of a PCB workshop held January 19, 1984. Hudson River Foundation for Science and Environmental Research, Inc., NY.

Hullar, T., R. Mt. Pleasant, S. Pagano, J. Spagnoli, and W. Stasiuk. 1976. PCB data in Hudson River fish, sediments, water and wastewater. Prepared for PCB Task Force, N.Y.S. Health Advisory Council. N.Y.S. Dept. of Environmental Conservation, Albany, NY. 24 pp.

Hutzinger, O., S. Safe, and V. Zitko. 1974. The chemistry of PCBs. CRC Press, Cleveland, OH. 269 pp.

Hydroscience. 1979. Analysis of the fate of PCB's in the ecosystem of the Hudson estuary. Prepared by Hydroscience, Inc., Westwood, NY for N.Y.S. Dept. of Environmental Conservation, Albany, NY.

Jensen, S. 1966. A new chemical hazard. New Scientist 32:612.

Jensen, S., A.G. Johnels, M. Olsson, and G. Otterlind. 1969. DDT and PCB in marine animals from Swedish waters. Nature 224:247–250.

Johnson, B.J. 1981. PCB's in Hudson River sediments. 10/7/81–10/22/81. U.S. Environmental Protection Agency Surveillance and Monitoring Branch, Region II, Edison, NJ.

Kimbrough, R. 1974. The toxicity of polychlorinated polycyclic compounds and related chemicals. CRC Crit. Rev. Toxicol. Jan. 1974:445–498.

Kimbrough, R., J. Buckley, L. Fishbein, G. Flamm, L. Kasza, W. Marcus, S. Shibko, and R. Teske. 1978. Animal toxicology. Environ. Health Pers. 24:173–184.

Klauda, R.J., T.H. Peck, and G.K. Rice. 1981. Accumulation of polychlorinated biphenyls in Atlantic tomcod (Microgadus tomcod) collected from the Hudson River estuary, New York. Bull. Environ. Contam. Toxicol. 27:829–835.

Kramer, K.E. 1975. State of concerns of the Lake Michigan Toxic Substances Committee related to polychlorinated biphenyls. Prepared for the U.S. Environmental Protection Agency, Region V, Chicago, IL.

Letz, G. 1983. The toxicology of PCB's — an overview for clinicians. West. J. Med. 138:534–540.

Limburg, K.E. 1984. Environmental impact assessment of the PCB problem: a review. Northeastern Environ. Sci. 3(3/4):124–137.

LMS. 1978. Upper Hudson River PCB no action alternative study: final report. Report to New York State DEC. Lawler, Matusky, and Skelly, Engineers, Pearl River, NY. 190 pp.

LMS. 1979. Upper Hudson River PCB transport modeling study. Final report to New York State DEC. Lawler, Matusky, and Skelly, Engineers, Pearl River, NY. 118 pp.

Merhle, P.M., T.A. Haines, S. Hamilton, J.L. Ludke, F.L. Mayer, and M.A. Ribick. 1982. Relationship between body contaminants and bone development in east-coast striped bass. Trans. Am. Fish. Soc. 111:231–241.

MPI. 1978a. Feasibility report, dredging of PCB contaminated river bed materials, Upper Hudson River, New York. Prepared by Malcolm Pirnie, Inc. for N.Y.S. Dept. of Environmental Conservation, Albany, NY.

MPI. 1978b. Environmental assessment, remedial measures remnant deposits, former Fort Edward Pool, Fort Edward, NY. Prepared by Malcolm Pirnie, Inc. for N.Y.S. Dept. of Environmental Conservation, Albany, NY.

MPI. 1979. Removal and encapsulation of PCB-contaminated Hudson River bed materials. Prepared by Malcolm Pirnie, Inc. for NYSDOT and NYSDEC.

MPI. 1980. PCB hot spot dredging program, Upper Hudson River, New York.

Draft Environmental Impact Statement. Prepared by Malcolm Pirnie, Inc. for N.Y.S. Environmental Quality Review, N.Y.S. Dept. of Environmental Conservation, Albany, NY.

Mueller, J.A., T.A. Gerrish, and M.C. Casey. 1982. Contaminant inputs to the Hudson-Raritan estuary. NOAA Technical Memorandum OMPA-21. NOAA, Office of Marine Pollution Assessment, Boulder, CO.

Nadeau, R.J. and R. Davis. 1974. Investigation of polychlorinated biphenyls in the Hudson River: Hudson Falls-Fort Edward Area. U.S. Environmental Protection Agency Region II Report.

National Research Council. 1979. Polychlorinated biphenyls. A report prepared by the Committee on the Assessment of Polychlorinated Biphenyls in the Environment. National Academy of Sciences, Washington, D.C. 182 pp.

Nebeker, A.V. and F.A. Puglisi. 1974. Effect of polychlorinated biphenyls (PCBs) on survival and reproduction of *Daphnia, Gammarus,* and *Tanytarsus.* Trans. Am. Fish. Soc. 103(4):722–728.

NUS. 1983. Remedial action master plan for the Hudson River. Prepared for U.S. Environmental Protection Agency. NUS Corp., Pittsburgh, PA. (Draft.)

NYSDEC. 1976. In the matter of alleged violations of Sections 17-0501, 17-0511, and 11-0503 of the Environmental Conservation Law of the State of New York by General Electric Co., Respondent. File No. 2833. September 1976.

NYSDEC. 1979. Hudson River PCB study description and detailed work plan: implementation of PCB settlement. Technical Paper No. 58 (revised 1979). N.Y.S. Dept. of Environmental Conservation, Bureau of Water Research, Albany, NY.

NYSDEC. 1981a. Toxic substances in fish and wildlife. 1979 and 1980 Annual Reports. Vol. 4, No. 1, Technical Report 81-1 (BEP). N.Y.S. Dept. of Environmental Conservation, Division of Fish and Wildlife, Albany, NY. 138 pp.

NYSDEC. 1981b. Toxic substances in fish and wildlife: May 1 to Nov. 1, 1981. Vol. 4, No. 2. Technical Report 82-1 (BEP). N.Y.S. Dept. of Environmental Conservation, Division of Fish and Wildlife, Albany, NY. 45 pp.

NYSDEC. 1982a. Toxic substances in fish and wildlife. November 1, 1981 to April 30, 1982. Vol. 5, No. 1., Technical Report 82-2 (BEP). N.Y.S. Dept. of Environmental Conservation, Division of Fish and Wildlife, Albany, NY. 25 pp.

NYSDEC. 1982b. Environmental monitoring program: Hudson River PCB Reclamation Demonstration Project. N.Y.S. Dept. of Environmental Conservation, Bureau of Water Research, Albany, NY.

NYU Medical Center. 1982. The biology of PCBs in Hudson River zooplankton. Final Report by New York University Medical Center to N.Y.S. Dept. of Environmental Conservation, Albany, NY. (Cited in Brown *et al.,* 1985.)

Odum, H.T. 1971. Environment, power, and society. John Wiley & Sons, New York, NY. 331 pp.

Peters, L.S. and J.M. O'Connor. 1982. Factors affecting PCB and DDT uptake by zooplankton and fish from the Hudson estuary, pp. 451–465 *In* Environmental Stress and the New York Bight: Science and Management (G. Mayer, ed.). Estuarine Research Federation, Charleston, SC.

Pizza, J.C. and J.M. O'Connor. 1983. Polychlorinated biphenyl dynamics in Hudson River, New York (USA) striped bass *Morone saxatilis.* II. Accumulation from dietary sources. Aquatic Toxicol. 3(4):313–328.

Powers, C.D., G.M. Nau-Ritter, R.G. Rowland, and C.F. Wurster. 1982. Field and laboratory studies of the toxicity to phytoplankton of polychlorinated biphenyls (PCBs) desorbed from fine clays and natural suspended particulates. J. Great Lakes Res. 8(2):350–357.

Pulver, D.A. and R. Hafner. 1981. Letter to James DeZolt, NYSDEC Project Manager of the Hudson River Reclamation Project, Aug. 31, 1981.

Roscigno, P.F. and L.L. Punto. 1982. Physical, chemical, and biological parameters associated with phytoplankton in the Hudson River ecosystem with studies of ^{14}C-PCB interaction with phytoplankton and sediment. Louis Calder Conservation and Ecology Study Center of Fordham University, Armonk, NY.

Sanders, J.E. 1982. The PCB-pollution problem of the upper Hudson River from the perspective of the Hudson River PCB Settlement Advisory Committee. Northeastern Environ. Sci. 1(1):7–18.

Sanders, J.E. 1983. Affadivit of John E. Sanders, Ph.D. United States District Court. Southern District of New York. 83 Civ. 3861 (CLB) and 83 Civ. 4890 (CLB).

Sanders, J.E. 1984. Some bits of history and accomplishments of the Advisory Committee. Memo to participants in the Hudson River Foundation's PCB Workshop, Jan. 19, 1984.

Schafer, J.E. 1982. Record of decision for the environmental impact statement on the Hudson River reclamation/demonstration project. U.S. Environmental Protection Agency, Region II, NY.

Schmidt, H. and G. Schultz. 1881. Uber Diphenyl-basen. Annalen 207:338.

Severo, R. 1975a. "Reports of chemical in fish initially withheld." New York Times, Aug. 17, 1975, p. 44.

Severo, R. 1975b. "State says some striped bass and salmon pose a toxic peril." New York Times, Aug. 8, 1975, p. 1.

Severo, R. 1975c. "Warning ignored on striped bass." New York Times, Aug. 9, p. 21.

Shabecoff, P. 1983. "EPA adds 133 sites to hazardous waste list." New York Times, Sept. 2, 1983.

Sheppard, J.D. 1976. Valuation of the Hudson River fishery resources: past, present and future. N.Y.S. Dept. of Environmental Conservation, Bureau of Fisheries, Albany, NY. 51 pp. (Mss.)

Simpson, K.W. 1982. PCBs in multiplate residues and caddisfly larvae from the Hudson River, 1977–1980. Technical Paper (in prep.) N.Y.S. Dept. of Environmental Conservation, Albany, NY.

Sloan, R.J. and R.W. Armstrong. 1981. PCB patterns in Hudson River fish. II. Migrant/marine species. Proc. Hudson R. Environ. Soc. Hyde Park, NY.

Sloan, R.J., K.W. Simpson, R.A. Schroeder, and C.R. Barnes. 1983. Temporal trends toward stability of Hudson River PCB contamination. Bull. Environ. Contam. Toxicol. 31:377–385.

Smith, C.E., T.H. Peck, R.J. Klauda, and J.B. McLaren. 1979. Hepatomas in Atlantic tomcod *Microgadus tomcod (Walbaum)* collected in the Hudson River estuary in New York. J. Fish Diseases 2:313–319.

Sofaer, A.D. 1976. Interim opinion and order. Unpublished opinion in the matter of violations of the Environmental Conservation Law of the State of New York by General Electric Co. NYSDEC File No. 2833. Feb. 9, 1976. 77 pp.

Stalling, D.L. and F.L. Mayer, Jr. 1972. Toxicities of PCBs to fish and environmental residues. Environ. Health Perspectives 1:159–164.

Stone, W.B., E. Kiviat, and S.A. Butkas. 1980. Toxicants in snapping turtles. N.Y. Fish & Game J. 27(1):39–50.

Thomann, R.V. and J.P. St. John. 1979. The fate of PCBs in the Hudson River ecosystem. Annals N.Y. Acad. Sci. 230:610–629.

Tofflemire, T.J. and S.O. Quinn. 1979. PCB in the upper Hudson River: Mapping and sediment relationships. NYSDEC Technical Paper No. 56. Albany, NY. 140 pp.

Tofflemire, T.J., L.J. Hetling, and S.O. Quinn. 1979. PCB in the upper Hudson River: Sediment distributions, water interactions and dredging. NYSDEC Technical Paper No. 55. Albany, NY.

Tofflemire, J.T., T.T. Shen, and E.H. Buckley. 1981. Volatilization of PCB from sediment and water: experimental and field data. Technical Paper No. 63. N.Y.S. Dept. of Environmental Conservation, Albany, NY. 37 pp.

Turk, J.T. 1980. Applications of Hudson River Basin PCB-transport studies, pp. 171–183 *In* Contaminants and Sediments Vol. I (R.A. Baker, ed.). Ann Arbor Science Pub., Inc., Ann Arbor, MI.

Turk, J.T. and D.E. Troutman. 1981. Polychlorinated biphenyl transport in the Hudson River, NY. Water-Resources Investigations 81-9. U.S. Geological Survey, Albany, NY. 11 pp.

USGS. 1977. Water resources data for New York. Water Year 1976. U.S. Geological Survey, Water-Data Report NY-76-1, Albany, NY.

Werner, M.B. 1981. The use of a freshwater mollusc (*Elliptio complanatus*) in biological monitoring programs. Toxic Substances Control Act Cooperative Agreement No. 3. N.Y.S. Dept. of Environmental Conservation, Division of Water, 47 pp.

Wolff, M.S., A. Fischbein, K.D. Rosenman, and I.J. Selikoff. 1981. Comparison of polychlorinated biphenyl residues in humans with varying exposures. Paper presented before the Division of Environmental Chemistry, American Chemistry Soc., New York, NY. Aug. 1981 (abstract).

5
The Westway

MARY ANN MORAN and DOOLEY S. KIEFER

Manhattan Island has been increasing in area since the late 1700s. Shoreline expansion into the Hudson and East Rivers and into New York Bay has been used for creating more coveted real estate and for convenient waste and construction debris disposal. In past years, expansion of the southern end of Manhattan through encroachment into the Hudson and East Rivers has occurred with regularity. Most recently, Battery Park City, a commercial and residential development currently under construction, involved filling in almost 100 acres of the Hudson River at the southwestern tip of the island.

Considering Manhattan's history, it was no surprise when, in 1974, plans for construction of the Westway Highway included the creation of a 200-acre landfill in the Hudson River. Yet, as the routine environmental permitting procedure was set in motion, concern over the highway's impacts surfaced. Eventually, potential environmental effects of the landfill on the Hudson River ecosystem stopped all work on the Westway project. Disruption of the project for environmental reasons surprised New Yorkers. Many people had believed this region of the Hudson to be totally polluted and incapable of supporting life, and they saw the lower Hudson as an unlikely focus of a major environmental debate.

Although the Hudson River has a long history of environmental controversies, the Westway conflict has several unique aspects. Unlike the thermal pollution and destruction of organisms associated with power plant cooling, or the fate and toxicity of PCBs in the river sediments, Westway involves irreversible loss of estuarine habitat. In addition, Westway would contribute to the cumulative effects of all the previous landfill encroachments into the river. At stake is a recreational and commercial fishery placed in jeopardy not through cropping or toxification of fish, but through destruction of the system on which the fish depend.

The environmental assessment process for the Westway landfill began in 1977 and was still being contested in late 1985. The landfill permit decision had been passed through several administrative levels in the U.S. Army Corps of Engineers, and eventually to the courts. The lack of agreement by scientists on potential landfill impacts was the most substantial obstacle to project approval. The following sections examine the chronology of the Westway project and the legal and scientific controversies of impact assessment.

History

Westway originated in 1956 with the realization that it was necessary to rebuild New York City's decaying West Side Highway. The road needing replacement runs from the Battery (southern tip of Manhattan Island) to 72nd Street; its deterioration was dramatically underlined when in December 1973, a portion of the elevated road near West 12th Street collapsed and two vehicles fell through to the pavement below. The highway was then closed to vehicles south of 46th Street (the portion south of 42nd Street is now demolished). By 1966, a modest proposal to refurbish the old highway had developed into a far more ambitious idea of coupling highway renewal with industrial and commercial development. In 1972, New York State Governor Nelson A. Rockefeller and New York City Mayor John V. Lindsay established the West Side Highway Project (the Project) as part of the New York State Department of Transportation (DOT) for the purpose of developing proposals for improvement of the west side waterfront.

By 1974, the Project had produced a final Westway design. It was proposed that a 4.2-mile highway be constructed along the west side of Manhattan in New York City (Figure 5.1), from the Battery (southern tip of the island) to 42nd Street. Westway would include a six-lane interstate highway (I-478) with two shoulder lanes, a four-lane service road, and a rebuilt six-lane West Street (the road under the old elevated West Side Highway) to serve local traffic.

Westway would also encompass more than highway construction. The New York State DOT proposed to dredge, construct an embankment, and place fill in the Hudson River to cover 200 acres of the river bottom and create 165 acres of land. An additional 10 acres of land would be created by construction of a platform. Thirty-five acres of the proposed project area would be used for the highway (buried in the landfill for 2.6 miles), interchanges, and ramps. The rest would be used for parkland and commercial, residential, and industrial development. The fill and the platform

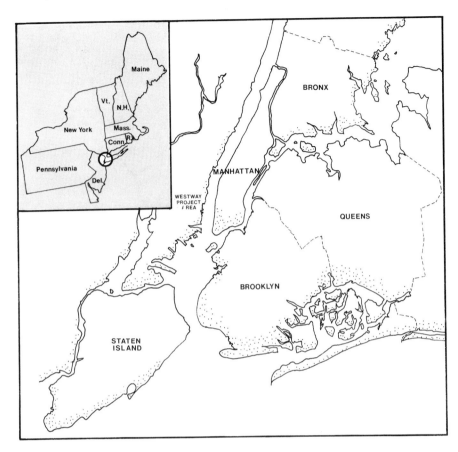

Figure 5.1. New York City Metropolitan Area.

would encroach into the Hudson (Figure 5.2) in an area currently occupied by obsolete piers (referred to as the "interpier area") and would block approximately 10% of the cross-sectional area of the river (Figure 5.3).

Construction of Westway was estimated to take at least 10 years, with an original (1977) cost estimate of $1.16 billion and a 1983 estimate of $2.3 billion (in 1979 dollars); unofficial estimates now range up to $4 billion. However, since Westway was planned as part of the interstate highway system, 90% of the construction cost would be paid by federal dollars and only 10% by New York State funds. This funding arrangement made the large development scheme financially feasible for New York City.

Opponents to Westway considered the billion dollar project "a gargantuan plan to develop housing along the river . . . (and) to create inexpensive new land near Wall Street and the World Trade Center" (Goldstein, 1978) at the expense of New York City's mass transit system. Since the

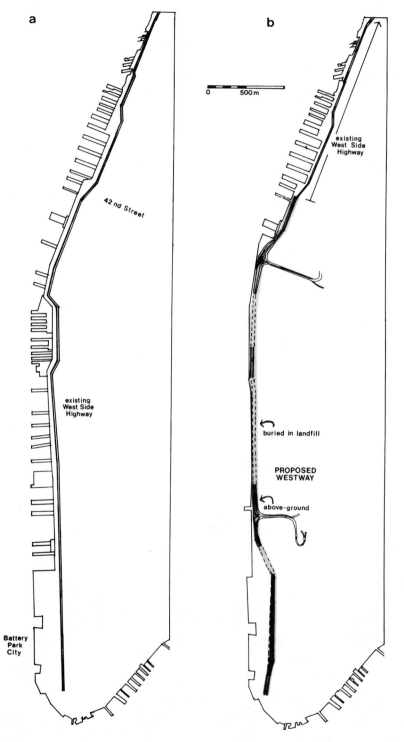

Figure 5.2. The southwestern tip of Manhattan in present form (a) and with proposed landfill and highway (b) (Federal Highway Administration, 1977).

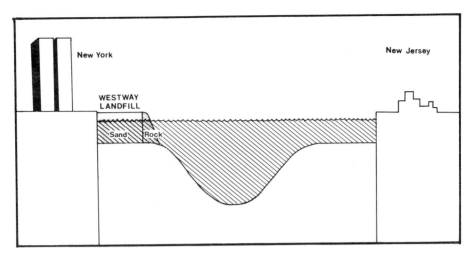

Figure 5.3. Cross-section of the Hudson River with proposed Westway landfill (from Wilder, 1981).

Federal Highway Trust Fund allows funds for "non-essential" interstate highways in urban areas to be transferred for use on mass transit and local road development, construction of a modest replacement highway would permit reallocation of Westway monies for improving mass transportation on Manhattan Island. Although Congress set a trade-in deadline of September 30, 1983 for all unfinished portions of the interstate highway system, Westway was exempted because of litigation and the deadline was extended to September 1985.

The Westway project endured more than ten years of uncertainty since the West Side Highway Project released the final design for the roadway. The legal obstacle to Westway construction lay in questionable assessment of environmental impacts of the 200-acre landfill in which the proposed highway is to be buried. At issue in the dispute were the compliance with federal environmental legislation by state and federal agencies and the objectivity of those scientists involved in assessing landfill impacts.

Impact Assessment Process

Under section 404 of the Clean Water Act (CWA) and section 10 of the Rivers and Harbors Appropriations Act, the New York State DOT was required to obtain a permit to place landfill in the Hudson River. In April 1977, a permit application was made to the U.S. Army Corps of Engineers. An environmental impact statement (EIS) for the proposed landfill had been released previously by the Federal Highway Administration

(FHWA) and the New York State DOT. The Corps deemed this earlier EIS to be sufficient for its evaluation procedures.

In the EIS, potential disturbance of the Hudson River estuarine ecosystem by the Westway project was discussed. Five areas of potential change were identified: 1) alterations in shoreline configuration (including a 10% decrease in cross-sectional area of the river in the vicinity of Westway); 2) relocation of sewer outfalls and the stirring up of many years' accumulation of untreated waste materials and polluted sediment; 3) covering of bottom materials with fill; 4) loss of estuarine habitat; and 5) changes in pollution from highway runoff. Loss of estuarine habitat, discussed in a four-paragraph section, was considered to be of no concern. The Westway area was described as "biologically impoverished" and "almost devoid of macroorganisms" (Federal Highway Administration, 1977). In the opinion of judge Thomas Griesa in the Westway trial that was eventually to follow, the conclusion to be drawn from the EIS was that "the proposed Westway landfill will have no significant impact on fish resources, because there is no fish life in the proposed landfill area even worth mentioning" (*ART v. West Side Highway Project*, 536 F. Supp. 1225 [S.D.N.Y. 1982]).

The findings of the EIS were based on a two-volume report, "Technical Report on Water Quality" (TRWQ), prepared by the consulting firms of Quirk, Lawler, and Matusky, Engineers and Alpine Geophysical Associates. The biological field studies that provided the factual basis for the report were conducted only during May and June 1973. Fish were sampled in baited traps with 1.75-inch diameter entrances, and sediment cores, bottom grabs, and plankton tows were also made. Reporting of biological data was scanty, but authors of this section of the TRWQ recorded that the bottom areas "contained many macroorganisms including snails, crabs, anemones, worms, tunicates, and barnacles," that the upper 1 to 2 inches of bottom muds contained "rich microscopic biota including bacteria, fungi, protozoans, nematodes and polychaete worms," and that an "unexpected" benthic community existed (Alpine Geophysical Associates, 1974). Fish (mainly tomcod) were caught at midwater levels, and it was found that a "full cycle of biological food chain exists," with final steps being "fish and subsequently the birds or larger fish which prey on these small fish" (Alpine Geophysical Associates, 1974).

In a supplement to the TRWQ (Parsons *et al.,* supplement to Alpine Geophysical Associates, 1974), the proposed project area is characterized quite differently from the main report as "almost devoid of organisms," despite the earlier-quoted sections and lack of further data collection. Much of the wording of the Westway EIS discussion of estuarine habitat loss and its minor impact is taken directly from this supplement.

During the U.S. Army Corps of Engineers' deliberation over the

Westway landfill permit, three federal agencies submitted objections to issuance of the permit: 1) the National Marine Fisheries Service (NMFS) recommended denial of the fill permit on grounds that the landfill would have negative effects on the marine habitat (*Sierra Club v. U.S. Army Corps of Engineers,* 701 F. 2d 1011 [2d Cir. 1983]); 2) The Environmental Protection Agency (EPA) asserted that the biological sampling forming the basis of the EIS was superficial and insufficient for determining impacts on marine life; and 3) the Fish and Wildlife Service (FWS) believed that the landfill would destroy a habitat with the potential to recover from pollution impacts in light of recent pollution control regulations (Choy, 1978).

Under CWA provisions, the comments of these specific federal agencies were to be accorded great weight by the Corps during the permitting process. The U.S. Army Corps of Engineers' district engineer, however, simply forwarded the objections to the Project. The West Side Highway Project responded, but all three federal agencies found the responses unsatisfactory and restated their positions to the Corps. By late December 1978, EPA persuaded the New York State DOT to fund further biological study. The consulting firm Lawler, Matusky, and Skelly (LMS; formerly Quirk, Lawler, and Matusky) was retained, and work on a Westway fish survey began in April 1979.

The LMS study spanned 1.5 years, concluding with a report published in two volumes in September and October 1980. The sampling scheme was based on eight primary stations in the river region of the Westway—two of these within the actual project (interpier) area. Samples taken monthly or semimonthly for one year (April 1979 to April 1980) at the eight stations included quantification of fish, ichthyoplankton, macrozooplankton, microzooplankton, phytoplankton, and benthos. Concurrent water quality sampling consisted of measuring temperature, salinity, dissolved oxygen, ammonia, pH, turbidity, and bacterial numbers. Early in 1980, additional sampling was conducted outside the study area (RM − 11.5 to RM 37.5) in order to determine the winter and early spring distribution and abundance of major fish species.

Results of the LMS study showed a diverse fish community. Tomcod was found to be the most abundant fish at the primary stations (69.7% of the total catch). Other important fish included hogchoker (13.4%), weakfish (1.5%), bay anchovy (1.6%), winter flounder (5.9%), striped bass (3.6%), and white perch (2.4%). The latter three species showed a preference for the interpier area over the channel zone. Winter flounder made up 23.3% of all fish caught in the interpier area, striped bass accounted for 21.4%, and white perch for 12.6%. An important result of the sampling was that the striped bass present in the project area were primarily young-of-the-year (y-o-y) and yearling fish.

With some exceptions and seasonal differences, water quality, phytoplankton, microzooplankton, macrozooplankton, and benthos were found to be comparable between interpier and river channel stations. LMS concluded that the attractiveness of the interpier area to striped bass, winter flounder, and white perch did not originate with a particularly unusual or abundant food source. The LMS report suggested that it was the combination of adequate food with structural shelter provided by the piers that created a suitable winter habitat.

The additional fish sampling effort encompassing RM -11.5 to RM 37.5 revealed that February 1980 tomcod populations were centered north of the project area; winter flounder, in contrast, were found in highest numbers to the south of the proposed Westway site. Although not directly stated in the LMS report, but deducible from the data, the most important finding was that the portion of the river slated for occupation by the Westway facility harbored the most extensive population of striped bass of any areas sampled. This finding would eventually become the key legal issue.

In May 1979, the U.S. Army Corps of Engineers' district engineer recommended issuance of the landfill permit before completion of the LMS study. Because of unresolved objections from federal agencies, the decision was elevated to the division level of the Corps. During this time period, the Project did not release interim study results to the Corps because they wished to prevent the biological sampling information from becoming public (*ART v. West Side Highway Project*, 536 F. Supp. 1225 [S.D.N.Y. 1982]). It was in June 1980, 15 months after the LMS study began, that the Corps was finally given a progress report. The report included data collected only through November 1979, even though data from winter samples, showing the greatest abundance of fish, were available at the time (*Sierra Club v. U.S. Army Corps of Engineers*, 701 F. 2d 1011 [2d Cir. 1983]).

By September 1980, a draft of volume I of the LMS report was released to federal agencies. The EPA and the Sierra Club raised questions about the need for a new EIS in light of the LMS findings, but the FHWA and the Project assured the Corps that a new or supplemental EIS was not necessary (*ART v. West Side Highway Project*, 536 F. Supp. 1225 [S.D.N.Y. 1982]). In November, the division engineer recommended issuance of the landfill permit, but because the three federal agencies still objected, the controversy was further elevated within the Corps hierarchy. In February 1981, the Chief of Engineers concurred in the issuance of the permit.

Meanwhile, lawsuits questioning the adequacy and legality of the Westway environmental review process had begun. As early as 1974, Action for Rational Transit (ART), a Manhattan citizen group, brought suit against the West Side Highway Project for inadequate evaluation of air and noise pollution. In 1979, the Sierra Club filed a suit directed against the U.S. Army Corps of Engineers' landfill permitting process.

Legal Controversies

Air Pollution Issues

As early as February 1977, the EPA issued a detailed report describing Westway as "environmentally unsatisfactory," saying that, among other things, it would promote air pollution and inefficient use of energy resources; EPA called for trading in Westway for mass transit. These were the issues forming the basis of the first court cases to be filed. In 1978 and 1979, the air pollution level question postponed issuance of the necessary Indirect Source permit.

When ART and the Sierra Club cases were considered by the U.S. District Court in late 1981, air pollution and traffic impacts were of primary concern. ART claims against the federal and state agencies directing the Westway project included: 1) potential violations of emission standards set by New York State under the Clean Air Act; 2) unwarranted use of federal funds to encourage vehicular traffic and accompanying air pollution on Manhattan; and 3) violation of the National Environmental Policy Act (NEPA) requirements through inadequate consideration of mass transit alternatives, potential funding problems, and loss of parkland in a possible future extension of the highway.

Sierra Club actions against the Westway project were based on allegations of NEPA violations by the U.S. Army Corps of Engineers. The Sierra Club claimed that the EIS was inadequate with regard to questions of air quality impacts, development alternatives, traffic impacts, and aquatic impacts. The Sierra Club also claimed violations of the CWA and the Rivers and Harbors Appropriations Act of 1899.

Somewhat ironically for a proposed highway project, the clean air issues lost in court. All of the ART claims were dismissed in November 1981 by Judge Thomas P. Griesa of the U.S. District Court. All Sierra Club claims involving allegations of inadequate consideration of air and traffic impacts were also dismissed. Only the Sierra Club claims based on potential impacts on the Hudson fishery were deemed to be of sufficient merit for litigation. Thus, the focus of the Westway controversy increasingly narrowed to a point where, as with previous cases involving power generation, Westway's potential impact on the Hudson River's striped bass fishery became the main environmental issue.

Fisheries Issues

The Sierra Club challenge of the U.S. Army Corps of Engineers' permitting process was decided by Judge Griesa in March 1982 (*ART v. West Side Highway Project,* 536 F. Supp. 1225 [S.D.N.Y. 1982]), when the Corps was found to be in violation of NEPA. The purpose of NEPA is to provide the public with information about significant environmental effects associated with a proposed project and with potential alternative actions. NEPA also mandates that government agencies fully consider the potential adverse effects during the decision-making process. In March 1982, Judge Griesa concluded that the U.S. Army Corps of Engineers (as permitting agency) was directed under NEPA to assay the impacts of the proposed Westway landfill on fisheries resources and to give adequate consideration to these impacts during its review. Additionally, NEPA required that the Corps make full disclosure of fisheries information to the public and allow opportunity for public comment. The views of the EPA, FWS, and NMFS, as federal agencies with jurisdiction in fisheries, should have been made public by the Corps and given great weight in the decision-making process. Judge Griesa was persuaded by the evidence that the Corps had failed to comply with these obligations, and the landfill permit was set aside.

In a subsequent trial commencing in June 1982, Judge Griesa also found the FHWA to be in violation of NEPA (*Sierra Club et al. v. U.S. Army Corps of Engineers*, 541 F. Supp. 1367 [S.D.N.Y. 1982]). The FHWA had failed to develop a supplement to the 1977 Westway EIS after significant new information relative to environmental impacts (i.e., the LMS 1980 study) came to light. "The FHWA, in collaboration with the New York State DOT, acted in willful derogation of the requirements of law in failing to issue a corrective supplemental environmental impact statement. The FHWA fully recognized the serious nature of the environmental impact which had been revealed by the new fisheries data, but refrained from making the required public disclosure" (*Sierra Club et al. v. U.S. Army Corps of Engineers*, 541 F. Supp. 1367 [S.D.N.Y. 1982]).

The second environmental law at issue in the Westway case was the CWA (Federal Water Pollution Control Act, 33 U.S.C., sections 1251 *et seq.*). The CWA states that following public hearings, the Secretary of the Army may issue dredge or fill permits in accordance with regulations developed by the Corps of Engineers. These regulations provide for public review of the project and consideration of potential impacts on human interests. One specific issue that must be considered in the Corps review process is the value of fish and wildlife, which includes consideration of views of the NMFS, FWS, and EPA.

EPA regulations, also promulgated under the CWA, require that the Corps district engineer should consider the "biological integrity of the aquatic ecosystem" in acting on any dredge-and-fill application. An important objective is the avoidance of disturbing fauna moving in or out of feeding, spawning, breeding, or nursery areas. In regard to CWA obligations, Judge Griesa held in his March 1982 decision that the U.S. Army Corps of Engineers had issued the Westway landfill permit without a legally sufficient basis.

An important distinction exists between the U.S. Army Corps of Engineers' violations of NEPA and its violations of CWA. Although NEPA has established substantive environmental goals for the U.S., the balancing of the substantive issues is left to the judgment of the federal agencies involved. Thus, court review of NEPA duties essentially covers only procedural aspects of the law (*Sierra Club v. U.S. Army Corps of Engineers*, 701 F. 2d 1011 [2d Circ. 1983]), and final balancing of environmental issues with economic and social values is up to the agency. Nonetheless, the Corps failed to carry out its NEPA procedural obligations. Since the EIS fisheries conclusions lacked "substantial basis in fact," and since a decision-maker relying on the EIS "could not have fully considered and balanced the environmental factors," the Corps violated NEPA (*Sierra Club v. U.S. Army Corps of Engineers*, 701 F. 2d 1011 [2nd Circ. 1983]).

The CWA places somewhat different obligations. CWA section 404, although requiring public hearings and consideration of "public interest" before a permit decision is made, does not require the same type of environmental disclosures as required by NEPA. Rather, the Corps must give notice, conduct hearings, make its own assessment of the impacts of the proposed project, and create a reasoned administrative record for its decision (*Sierra Club v. U.S. Army Corps of Engineers*, 701 F. 2d 1011 [2d Circ. 1983]). The court found the Corps in violation of the CWA because "the Corps simply ignored the views of sister agencies that were, by law, to be accorded great weight" and because "the Corps' unquestioning reliance on the FEIS must be regarded as arbitrary and capricious" (*Sierra Club v. U.S. Army Corps of Engineers*, 701 F. 2d 1011 [2d Circ. 1983]). In addition, the dredge-and-fill portion of the CWA deals in part with protection of ecosystem integrity. An agency must show reasonable deliberation based on available scientific information and must act in a manner necessary to protect the biological integrity of the ecosystem.

Scientific Controversies

The earliest discord among scientists involved in the Westway project arose as the two initial documents assessing fisheries impacts, the 1977 EIS and the 1980 LMS study, were evaluated. EPA scientists objected to the EIS conclusion that there were no significant impacts from estuarine habitat loss. In a 1977 communication to the U.S. Army Corps district engineer, EPA labelled the Westway EIS and the supporting TRWQ as superficial, and asserted that information based on only a few weeks of sampling during the early summer (May and June 1973) did not provide sufficient knowledge about the water resources; therefore; it was insufficient to support the Corps' conclusion that the landfill would have no adverse effects on Hudson fisheries (*ART v. West Side Highway Project*, 536 F. Supp. 1225 [S.D.N.Y. 1982]). In his affidavit as an expert witness at the Westway trial, Fletcher (1981) attributed the conclusions of the EIS to the prevalent layman's sense that an area so polluted with sewage discharges was unlikely to support marine life. In addition, the sampling was undertaken at the time of year when biological activity in the interpier area would be low because of increasing water temperatures and decreasing dissolved oxygen levels (Fletcher, 1981).

LMS, as consultant to the New York State DOT, duplicated the minnow trap sampling method relied on in the EIS for comparison with bottom trawl sampling (LMS, 1980). Results showed the use of minnow traps to be inappropriate for determining abundance or composition of the fish community of the Westway area (LMS, 1980). Only six Atlantic tomcod were collected in the minnow traps in the May-June 1973 EIS samples, and only three American eels were collected by LMS in 1979. This contrasts strongly with the LMS bottom trawls, which showed a viable and diverse fish community in the interpier region.

At the commencement of litigation, the inadequacy of biological sampling for the Westway EIS was largely recognized by both sides. During the trial, officials in the FHWA and the Project "admitted knowing that the reason the earlier study revealed virtually no fish in the interpier area was that the study was made at a time of year which was not representative and the sampling techniques were faulty" (*Sierra Club et al. v. U.S. Army Corps of Engineers*, 541 F. Supp. 1367 [S.D.N.Y. 1982]). However, in contrast to the general recognition of the Westway EIS and TRWQ as inadequate scientific evaluations, the LMS 1980 study was the focus of much more controversy. The substance of the dispute was a question of bias.

Scientific Objectivity

The objectivity of the LMS study (1980) was challenged several times during the Westway trials. At issue were the analysis and presentation of data as well as conclusions drawn in the report. The most critical example involves the evaluation of data on striped bass abundance in the project area and comparison of striped bass data from the Westway area to data collected at stations to the north and south.

When presenting the data on striped bass abundance in graphic form, authors of the LMS report averaged data from two interpier stations, where substantial numbers of striped bass were found, with data from the pierhead and midchannel stations at the same mile point (Figure 5.4); these stations generally contained very few striped bass. The resulting graph did not show abundance in the proposed landfill area, but instead presented river-region abundance of striped bass as compared to other river regions; thus, the high numbers of striped bass found precisely in the project area (Figure 5.5) were not revealed (Fletcher, 1981). By contrast, the catch data at the New Jersey interpier zone, across the river from the Westway site, were not analyzed in a similar manner. The high numbers of fish in the New Jersey interpier zone were presented without averaging

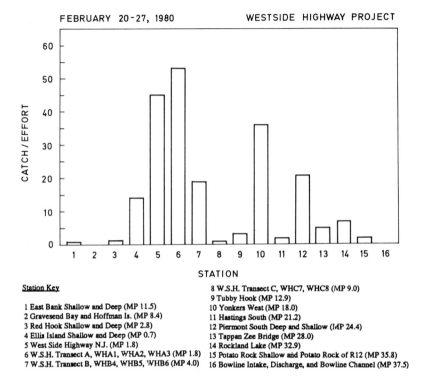

Figure 5.4. Figure 4.0-3 from the LMS (1980) report showing striped bass abundance in the lower estuary during February 1980. Stations 6 and 7 represent a project (interpier) area sample averaged with two river samples at the same mile point. Station 5 is the non-averaged New Jersey interpier sample (from LMS, 1980).

with midriver collections. About this section of the LMS study, Judge Griesa stated in his trial opinion, "a bar graph purporting to deal with this (sic) data is misleading, since it averages the data in such a way as to show the numbers of striped bass in the interpier area as only about one-third of what they actually were. Only by an analysis of this graph with other materials in the report can the facts about the relative magnitude of striped bass found at the different sites be pieced together."

Scientific objectivity was also questioned when LMS invoked a "mild winter" hypothesis in interpreting results of fish sampling. In accounting for the abundant striped bass in the interpier area, LMS (1980) concluded that the mild temperatures during the 1979–1980 winter made the Westway region unusually hospitable to overwintering juveniles. Because of the physiological constraints at water temperatures below 1.5°C, LMS

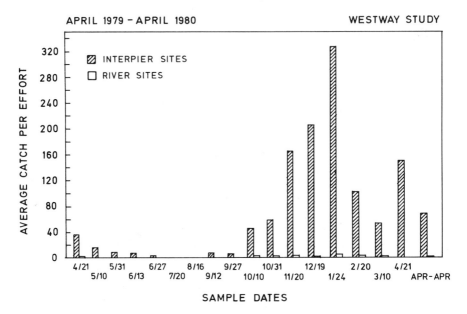

Figure 5.5. Figure 3.0-11 from the LMS (1980) report, showing the average number of striped bass at interpier and river stations (from LMS, 1980).

asserted that it is only during mild winters that striped bass would be expected to overwinter in large numbers in the interpier zone (LMS, 1980). The temperature data given in Appendix C of the LMS report, however, do not support the claim that the 1979–1980 winter was milder than normal (*ART v. West Side Highway Project,* 536 F. Supp. 1225 [S.D.N.Y. 1982]).

The possibility of an atypical winter was also examined following the March 1982 decision by the U.S. District Court to set aside the Westway landfill permit. The U.S. Army Corps of Engineers hired a consulting firm, Malcolm Pirnie, Inc., to reexamine the existing Westway data base and compile a thorough review of life histories of major Hudson River fish species. As part of the Malcolm Pirnie, Inc. review, temperature and salinity data during the 1979–1980 winter were compared to historical data to investigate the LMS claim of an atypically mild winter. Their conclusion was that "temperature and salinity values appeared to be within normally expected ranges for November through February" (Malcolm Pirnie, Inc., 1982a). Judge Griesa, in his findings of fact at the close of the Westway trial, stated: ". . . some phases of the LMS report are misleading and represent rather obvious attempts to avoid the full impacts of the facts revealed by the study. These facts were obviously unpalatable to LMS's employer, the Westway Project'' (*ART v. West Side Highway Project,* 536 F. Supp. 1225 [S.D.N.Y. 1982]).

Lack of objectivity in the analysis of environmental perturbations is a familiar issue in current environmental analysis procedures. There is obvious benefit to a hired consultant and its client in underplaying or concealing environmental damage, and this can encourage the misuse of science to advocate a prodevelopment position. Albert Butzel, counsel for the Sierra Club in the Westway litigation, emphasized the importance of distinguishing between position advocacy and scientific objectivity based on his familiarity with the Westway predicament (Butzel, 1983). Although data collection may be objective, interpretation in the impact assessment arena can occur in a judgmental fashion that needs to be recognized as advocacy.

Rosenburg *et al.* (1981) and Hanson (1976) considered a further complication to scientific objectivity with regard to the fate of scientific research conducted as the property of a client. In the current environmental assessment process, clients often retain the right to approve all study results before publication, and thereby can control the selective release of impact information.

Tokenism

In a review of some of the highest quality environmental impact assessments (EIAs) conducted in recent years, Rosenburg *et al.* (1981) identified tokenism as one of the most consistent problem areas. Many impact assessments played no serious role in the decision of whether or not to go ahead with a project, since adverse environmental impacts were not regarded as sufficient reason to stop the project or because the proposed project was already locked into specific design or contract obligations. Superficial, non-specific, or unsubstantiated findings appeared in those assessments written purely to satisfy a necessary legal hurdle before beginning a project (Rosenburg *et al.,* 1981; see also Kreith, 1973 and Schindler, 1976).

In the Westway trial, Judge Griesa concluded that "the actions of the (Army Corps of Engineers) Division office following receipt of the LMS material can only be explained as resulting from an almost fixed predetermination to grant the Westway landfill permit" (*ART v. West Side Highway Project,* 536 F. Supp. 1225 [S.D.N.Y. 1982]). When faced with data that indicated the landfill would eliminate a significant fishery habitat, Griesa found that "(it) was agreed among the Project, the FHWA and the Corps that, in order to make issuance of the landfill permit more defensible, and to defuse the opposition as much as possible, mitigation concepts would be worked out to attempt to compensate for the loss of fish habitat. There would then be an attempt to make mitigation the 'focal point' of the discussion with the other federal agencies, rather than the LMS report itself" (*Sierra Club et al. v. U.S. Army Corps of Engineers*, 541 F. Supp. 1367 [S.D.N.Y. 1982]).

Estimating and Interpreting Uncertainty

In the Westway impact assessment process, biological sampling for the 1977 EIS and subsequent LMS study was intended to characterize the benthic, planktonic, and fish populations residing in the proposed landfill area. Knowledge of predevelopment conditions would provide a baseline against which potential alterations from the Westway project could be measured. For the lower Hudson River, however, characterizing the "baseline community" was no easy task.

Component population densities are expected to vary within an ecosystem; this applies both to human-disturbed systems and to undisturbed or pristine systems (Ward, 1978). Thus, the environmental impact assessment process is forced to operate within the context of daily, seasonal, and yearly fluctuations and spatial heterogeneity of populations. Although the two-week minnow trap sampling during the early summer of 1973 was quite clearly an inadequate evaluation of the Westway area fish populations, the 13-month sampling undertaken by LMS in 1979–1980 still did not provide an adequate characterization of population levels. Large yearly fluctuations in abundance and distribution are common for many fish species of the Hudson. The winter of 1979 could well represent a year of particularly high fish densities in the Westway project area or of particularly low densities. It was because of this large natural variance that LMS was able to propose a "mild winter" hypothesis and cast doubt on the significance of the high fish abundance found in the project area. Scientists advising the Sierra Club and other claimants disagreed that juvenile striped bass populations were unusually high during that particular winter, arguing that 1979 was a typical year and that striped bass abundance could be considered typical.

Butzel (1983) felt that uncertainty in measuring fish abundance was a major obstacle in resolution of Westway conflicts. Following the Westway trials, Malcolm Pirnie, Inc. (1982a) and federal agencies advised the U.S. Army Corps of Engineers that additional studies were needed to evaluate the landfill impacts. The Corps then asked Malcolm Pirnie, Inc. to design a sampling program adequate for quantifying impacts on overwintering fish; during October 1982, Malcolm Pirnie, Inc. assembled a panel of fisheries biologists. A consensus was reached among panel members to endorse a generalized study plan extending over a 17-month period and encompassing a sampling area from south of Manhattan to the Haverstraw-Peekskill area (RM 35) (Malcolm Pirnie, Inc., 1982b). It was made clear to the Corps, however, that although a minimum of two years of additional studies could help to characterize interpier fish populations, and would allow the Corps to make a decision with greater confidence, scientific uncertainty would still exist and the fate of fish displaced by the proposed landfill would not be known. Faced with the realization that questions would still linger after the new study, the Army Corps district engineer decided to do no additional investigations and to evaluate the permit application based only on existing data (Butzel, 1983).

According to Butzel (1983), this interface of science and the regulatory arena was not successful. The problems with the existing scientific data for evaluation of Westway landfill impacts were not clear to decision-makers at the U.S. Army Corps of Engineers. The benefits to be gained from further, admittedly limited, studies were also not clear to the Corps. Although still not providing certainty, these studies could have substantially increased the confidence of scientific evaluations and predictions; they would have allowed the Corps to make their permit decision in the context of reality rather than pure theory (Butzel, 1983). In speaking of the certainty associated with an impact assessment or prediction, Butzel concludes that "of major importance . . . is the need for scientists to begin to provide input in terms of confidence levels, and to educate decision-makers as to varying levels of confidence. Also important is the need to educate regulators as to how, if at all, confidence levels can be increased within the finite time periods that such regulators have to act."

The use of one version of confidence limits was actually attempted for the Westway EIA. In a controversial section of the August 1982 Malcolm Pirnie, Inc. report to the U.S. Army Corps of Engineers, an effort was made to obtain a range of striped bass population estimates for the Westway site. Results of the analysis were expressed as three scenarios labelled "worst" case, "reasonable" case, and "best" case. The worst case analysis estimated than 4.3% of the Hudson y-o-y and 21.1% of the yearlings use the interpier area for overwintering. Reasonable case analysis estimated 0.4% of the y-o-y and 0.9% of the yearlings overwinter within the interpier area; and best case analysis estimated that 0.2% of the y-o-y and 0.5% of the yearlings overwinter in the project area. For both the reasonable and best cases, more than 99% of the striped bass juveniles were calculated to be unaffected by Westway.

Two other attempts at predicting the percent of Hudson River striped bass overwintering in the interpier region are discussed in the Malcolm Pirnie, Inc. report (Malcolm Pirnie, Inc., 1982a). In a June 1982 meeting with the U.S. Army Corps of Engineers, LMS presented "worst" and "realistic" case analyses of striped bass levels. Worst case calculations indicated that 0.46% of the y-o-y striped bass and 0.67% of the yearlings were winter residents of the interpier area, while for the realistic analysis, 0.09% of the y-o-y and 0.12% of the yearlings were expected to overwinter within the Westway area. Dr. Robert Pierce, a biologist for the U.S. Army Corps of Engineers, also performed a preliminary worst case analysis of interpier use by striped bass. In contrast to the preceding scenarios, Pierce estimated that up to 70% of both juvenile age classes might be found overwintering within the Westway site, leaving only 30% of the young striped bass unaffected by the project. Thus, estimates of the range of Westway impacts differ by two orders of magnitude. The large discrepancy between estimates stemmed from an assumption that striped bass occur only in the bottom 1 to 2 m of the water column (LMS; Malcolm Pirnie, Inc.), as opposed to being found up to 4 m above the river bottom (Pierce); and, there was a substantial difference in calculations of Hudson estuary standing crop of striped bass.

The connotations of objectivity associated with employing a range of assumptions in the analysis makes this the most frustrating of the Westway controversies. Although the "best," "realistic," and "worst" case analyses used by the scientific consultants superficially appeared as an attempt to delimit a range of potential fish losses, the scanty data base provided too much reliance on "estimating" parameters. These confidence limits reflected an underlying bias, and thus Westway regulators were given a range of potential striped bass losses that varied between 0.2 and 70%. For those hoping that probabilities and ranges in scientific evaluations can be an improvement to the handling of scientific uncertainties in EISs, the Westway experience has been a disappointment. Under a guise of objectivity, ranges and probabilities may actually be no closer to the unbiased scientific evaluations called for by regulators.

Final Phases

In September 1982, the U.S. Army Corps of Engineers, FHWA, and New York State DOT appealed their case to the U.S. Court of Appeals. In a February 1983 decision, the U.S. Court of Appeals agreed with Judge Griesa that the Corps had violated NEPA and CWA, and that FHWA had violated NEPA (*Sierra Club et al. v. U.S. Army Corps of Engineers*, 701 F. 2d 1011 [2d Circ. 1983]). This decision meant that the injunction against the landfill permit was reaffirmed.

In April 1982 (one month after Judge Greisa's decision to void the landfill permit), permit proceedings were reinitiated when the New York State DOT renewed its application to the Corps for placing landfill in the Hudson River. Between April 1982 and February 1984, the U.S. Army Corps of Engineers announced two decisions on the fate of Westway, followed by a reevaluation and rescinding of each decision. In December 1982, Colonel W.M. Smith, Jr. announced that there would be no new fish sampling program in the Hudson to determine landfill impacts and, furthermore, that he intended to reissue the landfill permit. When it came to light that Colonel Smith was seeking employment with the main engineering consultant for Westway during the time he was deliberating on whether to conduct new studies (New York Times, 1983a), the Corps rescinded Colonel Smith's decision against further study; Colonel F.H.

Griffis, appointed to replace the retiring Colonel Smith as New York district engineer, was given the responsibility to review the question of further fish surveys.

In his review of the Westway case, Colonel Griffis convened a scientific workshop to reexamine the need for further fishery studies in the Westway project area. In the July 1983 workshop, scientists debated both the importance of the Westway area as a fishery habitat and the impact of the loss of this habitat on the Hudson River and coastal fishery. Following the majority recommendations of the workshop participants, Colonel Griffis decided to sponsor a two-year study before reaching a decision on the Westway landfill permit (Village Voice, 1983a).

In the final months of 1983, the Westway issue moved into new political arenas. New York Governor Mario Cuomo took his support of the Westway project to the U.S. Congress, where amendments to exempt Westway from the court-ordered environmental review were introduced in both the House and Senate (Village Voice, 1983b). Congress took no action on granting a Westway loophole to national environmental legislation. Indeed, the New York State attorney general telegrammed members of the N.Y.S. congressional delegation that "whatever your position on the merits of Westway, exempting any particular project from the impartial application of environmental laws could do great and lasting harm to the national effort to clean our rivers and streams and preserve our natural resources." Governor Cuomo did gain a different pro-Westway victory; he requested that Colonel Griffis' decision to undertake a two-year study be reviewed by Corps superiors. The review resulted in the overturning of that decision in December 1983 by Chief of Engineers Lieutenant General J. Bratton (New York Times, 1983b; Village Voice, 1983c).

Nevertheless, an additional winter of sampling was funded by the New York State DOT and the Corps, and two concurrent studies were carried out in the winter of 1983–1984 (LMS, 1984; Malcolm Pirnie, Inc., 1984; NJMSC, 1984). These additional studies formed the basis of a supplemental environmental impact statement (SEIS) released at the close of 1984 by the U.S. Army Corps of Engineers and the FHWA (COE and FHWA, 1984).

The draft SEIS (DSEIS) appeared in June 1984 and drew enormous public comment—most of it negative. Federal agencies charged with the duty of protecting fish and wildlife (EPA, FWS, NMFS, and National Oceanic and Atmospheric Administration) objected particularly strongly: "The basic conclusion presented in the DSEIS is that the proposed 'Westway' project site represents a critical portion of the striped bass fishery habitat, the loss of which would pose a significant adverse impact on the Hudson River stock of that species . . . In reviewing the adequacy of the DSEIS, we have determined that it does not contain sufficient information to assess fully the environmental impact of the proposed project. As the DSEIS points out, many of the fishery studies conducted to date have flaws regarding methodology, duration or scope" (EPA, 1984). "In summary, the NMFS concludes that construction of Westway would contribute to a significant degradation of the waters of the United States" (NMFS, 1984).

The grounds for these and other similar comments was the presentation, in the DSEIS, of the latest efforts to determine the relative use of the Westway project area by striped bass. Fish were sampled by bottom trawl at a large number of sites extending from the Haverstraw Bay area down to upper New York Harbor, Newark Bay, and Jamaica Bay (Figure 5.6) in an effort to determine patterns of striped bass distribution throughout the region during the winter months (COE and FHWA, 1984). Some limited sampling was done to determine striped bass' vertical distribution in the water column, to evaluate effectiveness of various sampling gear, and also to investigate diurnal preferences for the pier-interpier areas in the Westway zone.

The 1983–1984 surveys did demonstrate that striped bass y-o-y and yearling fish utilized the Westway project area heavily at times during the course of the sampling programs. Other areas adjacent to Westway or across the river on the New Jersey shore also showed occasional heavy use, as did areas in Haverstraw Bay (NJMSC, 1984; LMS, 1984). High sample variance characterized most of the areas, making statistical analysis difficult. As is almost inevitable, bias was introduced by sampling gear, although bottom trawls were the best of the available methods. A hydroacoustic survey made in the Westway area by Biosonics, Inc. suggested that bottom trawl samples favored smaller fish (NJMSC, 1984).

A statistical model was employed to estimate P, the percentage of y-o-y and yearling striped bass that used the Westway area (MMES, 1984). The highest percentages obtained from that model were then to be used in a "worst case" analysis of impact. The model lumped Westway data in with data from other sampling zones into a "superzone," according to an algorithm that minimized the overall variance on the estimate of P. However, the Corps determined that the statistical variances resulting from the model projections were too high and, therefore, that the worst case projection usages (78.9% for the 1983 year-class y-o-y and 92% for the 1982 year-class, respectively) were unreliable (COE and FHWA, 1984). The Corps also felt that the use of superzones created an inflated estimate of P. For these reasons, the model projections were downplayed. Instead, the SEIS finally projected a possible loss of 20 to 33% of juvenile striped bass based upon the direct geographic proportion of pier habitat that Westway represents in the study region. This was done because "nothing learned . . . demonstrated any unique quality of the WW area that would distinguish it from any of the other pier complexes . . ." (COE and FHWA, 1984, p. 38).

Despite the intensive sampling and analytical efforts that went into development of the SEIS, the level of information gained was still consid-

Final Phases

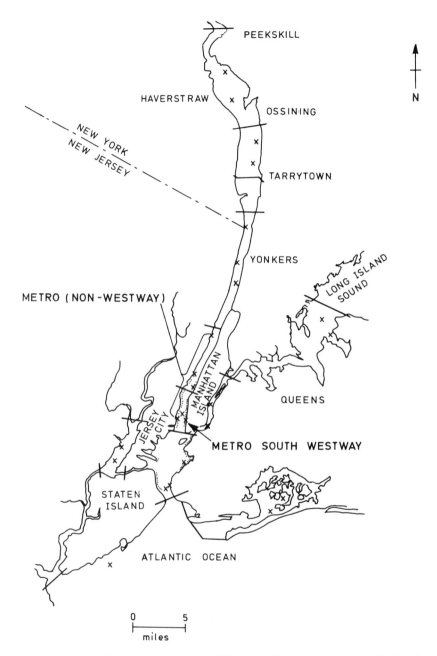

Figure 5.6. Sampling zone locations, 1983-1984 Westway Fisheries Study (from COE and FHWA, 1984).

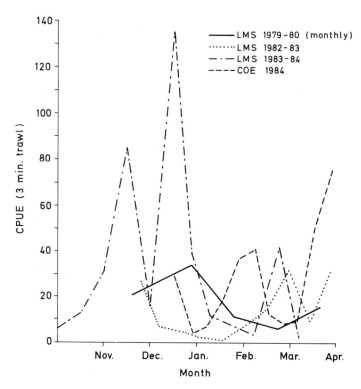

Figure 5.7. Striped bass catch, all ages, per unit effort (three-minute tow) in Westway by sampling date. LMS = Lawler, Matusky and Skelly studies; COE = Corps of Engineers (New Jersey Marine Sciences Consortium) study. Note that COE and LMS are not completely comparable due to differences in vessels and gear (from COE and FHWA, 1984).

ered by many (including the federal resource agencies) to be unsatisfactory. Major conclusions by reviewers of the DSEIS (e.g., Fletcher, 1984) were that the sampling period chosen was too short and that a potentially critical time period had been missed by not beginning sampling earlier. As can be seen in Figure 5.7, the catch-per-effort data collected by LMS in 1983 (but not officially used in the estimate of P) indicated major peaks in striped bass abundance in late November and December, during which time the NJMSC project did not take any usable data (COE and FHWA, 1984). Also, the SEIS failed to elucidate the importance of the Westway zone as a habitat, stating that it was unlikely to exceed the zone's importance as a physical shelter for bass. Nor did the fact that the Hudson stock had recently become more important to the maintenance of the entire east coast fishery (because of the major decline in the Chesapeake stocks) receive adequate weight.

Nevertheless, in late January 1985, Colonel Griffis announced his intent to approve Westway's remaining permit and clear the way for construction to begin (New York Times, 1985). His decision was based on a number of issues, and one of his conclusions was that the project would present "potentially minor impacts to striped bass." Opponents of Westway returned to the courts, and in August 1985, Judge Thomas Griesa once again barred construction of the highway and development project based on the improper issuing of the landfill permit. Finally, on September 20, 1985, after losing an appeal and facing mounting pressure from Congress, both New York's Mayor Koch and New York State Governor Cuomo announced the abandonment of the full Westway project in favor of a reduced plan that would minimally affect the Manhattan/Hudson shoreline.

References

Alpine Geophysical Associates. 1974. West Side highway project technical report on water quality. Part II. Water quality sampling program, biological populations and inshore area studies.

Butzel, A.K. 1983. Letter to Simon A. Levin, Ecosystems Research Center. February 7, 1983. 15 pp.

Choy, R. 1978. Westway receives bad marks on jobs, air, water. Sierra Atlantic 5(15):3, 7.

COE and FHWA. 1984. Final supplemental environmental impact statement, Westside Highway project. Prepared jointly by New York District, U.S. Army Corps of Engineers, and by U.S. Dept. of Transportation, Federal Highway Administration, Region One. November 1984. (2 vols.)

EPA. 1984. Review comments on Westway DSEIS by U.S. Environmental Protection Agency. Letter, July 16, 1984 to New York District, U.S. Army Corps of Engineers.

Federal Highway Administration. 1977. West-Side highway project. Final environmental impact statement. U.S. Dept. of Transportation. 300+ pp.

Fletcher, R.I. 1981. Affidavit of R. Ian Fletcher. U.S. District Court. Southern District of New York. 81 Civ. 3000 (TPG).

Fletcher, R.I. 1984. Commentary on Volume II of the draft supplemental environmental impact statement, Westside Highway project. Letter, July 10, 1984, on behalf of the Hudson River Fishermans' Association.

Goldstein, N. 1978. Westway—worst way? Sierra Feb./March:11–14.

Hanson, C.H. 1976. Commentary—Ethics in the business of science. Ecology 57:627–628.

Kreith, F. 1973. Lack of impact. Environment 15:26–33.

LMS. 1980. Biological and water quality data collected in the Hudson River near the proposed Westway project during 1979–1980. Lawler, Matusky, and Skelly, Engineers. Prepared for N.Y.S. Dept. of Transportation and System Design Concepts, Inc. (2 vols.)

LMS. 1984. 1983–1984 Westway winter sampling program—draft trawl data. Prepared by Lawler, Matusky, and Skelly, Engineers, Pearl River, NY for N.Y.S. Dept. of Transportation.

Malcolm Pirnie, Inc. 1982a. Hudson River Estuary fish habitat study. Prepared for U.S. Army Corps of Engineers, New York, NY.

Malcolm Pirnie, Inc. 1982b. Study of sampling requirements for the Westway interpier area and Hudson River estuary. Prepared for New York District, Dept. of the Army. 21+ pp.

Malcolm Pirnie, Inc. 1984. Westway fisheries study data evaluation workshop, April 16–18, 1984. Final summary report. Prepared by Malcolm Pirnie, Inc., White Plains, NY for New York District, U.S. Army Corps of Engineers.

MMES. 1984. Analysis of 1982 and 1983 year-class striped bass utilization of the Westway fisheries study area of the Hudson River estuary. Prepared by Martin Marietta Environmental Systems, Columbia, MD for Malcolm Pirnie, Inc., White Plains, NY. (5 vols.)

New York Times. 1983a. "Judge will hold contempt trial in the Westway to review compliance in study of river fish." New York Times, May 18, 1983, p. B1.

New York Times. 1983b. "Army engineers to end fish study a year early." New York Times, Dec. 16, 1983, p. 83.

New York Times. 1985. "Westway landfill wins the support of army engineer." New York Times, Jan. 25, 1985, p. A1.

NJMSC. 1984. Summary report, Westway fisheries study. Prepared by New Jersey Marine Sciences Consortium for New York District, U.S. Army Corps of Engineers. (7 vols.)

NMFS. 1984. Response to Westway DSEIS by National Marine Fisheries Service, Washington, DC. Letter, July 16, 1984 to Col. F.H. Griffis, New York District Chief Engineer, U.S. Army Corps of Engineers.

Rosenburg, D.M., V.H. Resh, S.S. Balling, M.A. Barnby, J.N. Collins, D.V. Durbin, T.S. Flynn, D.D. Hart, G.A. Lamberti, E.P. McElravy, J.R. Wood, T.E. Blank, D.M. Schultz, D.L. Marrin, and D.G. Price. 1981. Recent trends in environmental impact assessment. Can. J. Fish. Aquat. Sci. 38:591–624.

Schindler, D.W. 1976. The impact statement boondoggle. Science 192:509.

Village Voice. 1983a. "Undermining the law. Cuomo's cynical Westway strategy." Village Voice, Nov. 15, 1983, pp. 7, 96.

Village Voice. 1983b. "Congress drops Westway loophole." Village Voice, Nov. 29, 1983, p. 5.

Village Voice. 1983c. "A 'small victory' for Westway." Village Voice, Dec. 27, 1983, p. 5.

Ward, D.V. 1978. Biological environmental impact studies: Theory and methods. Academic Press, New York, NY. 157 pp.

Wilder, S.F. 1981. Westway: An idea whose time has long gone. Sierra Atlantic 8(1):12.

6
Synthesis and Evaluation

KARIN E. LIMBURG and MARY ANN MORAN

What has been depicted in the preceding chapters is a portrait of the Hudson River under somewhat haphazard management. Three distinct types of threats to the Hudson ecosystem were at issue, representing direct reductions of animal populations (power plant operation), removal of toxic substances (PCB pollution), and habitat destruction (Westway construction). Each situation that we have chosen to study has had the same characteristics: 1) scientific investigations have been used to help gather information, to clarify phenomena, or to explain effects; 2) none of the findings have gone unchallenged; so that 3) aspects of all of these impacts have gone to trial; and 4) action, if any, has proceeded by court edict more often than not.

For all three Hudson case studies, no ultimate legal resolution of the environmental issues occurred. The passage of the National Environmental Policy Act (NEPA) in 1969 and the Clean Water Act (CWA) in 1972 provided the legislative basis for litigation over power plant impact on Hudson River fisheries. Today, although 15 years have passed and a temporary truce has been called, the power plant controversies legally remain in limbo. In 1990, when the temporary agreement expires, the issue of cooling towers in the Hudson estuary may once again become the subject of a major legal contest. Also, the PCB case is legally unresolved, even though PCBs were recognized as a major problem in 1975.

Parallel to the legal issues, none of the major scientific disputes have ever been definitively laid to rest. In our Hudson River case studies, we found that the inability of science to contribute efficiently to major regulatory decisions was due to two aspects of the impact assessment process. First, the limitations of science were not acknowledged by regulatory and judicial bodies, so that scientists were asked to provide precise, unequivocal answers to questions that could not be answered in that fashion. Second, scientists often became trapped in advocacy roles, at times interpreting their analyses with their employers' implicit biases and carrying on exercises in frustration when, as expert witnesses, they contradicted one another in the courtroom.

In Chapter 1, five questions were raised about various aspects of the environmental assessment work done on the Hudson over the past 15 years. These were addressed to some degree in subsequent chapters dealing with different case studies, but we restate and answer them more completely here.

1. *Have appropriate aspects of the Hudson ecosystem been emphasized? Have the data collected been proven adequate for the estimation of the impacts under consideration?*

This double question receives a mixed answer. For each impact, the laws and regulations were interpreted in such a way that the resulting studies were, in fact, appropriately focused. (In each case, fish were the primary object of attention.) Yet other interpretations of the laws could have been made and other ecosystem features could have been carried out more thoroughly. In the final analysis, each of the scientific studies carried out for impact assessment represented compromises between the goal of answering all relevant questions and the availability of two essential resources—money and time.

For example, studies of the actual effects of PCBs in the Hudson ecosystem, complementing the extensive environmental fate studies, would have created a stronger basis for making a decision on what to do with the remaining load of PCBs in the river. However, such studies are costly, and effects may be subtle and require long periods of observation before they become manifest. For this reason, environmental assessment and regulation of PCBs have been carried out on the basis of a concentration in consumable biomass allowed by the U.S. Food and Drug Administration (FDA).

In a second example, taken from the power plant impact assessments, the federal Atomic Safety and Licensing Appeals Board (ASLAB) gave the utilities five years to prove that once-through cooling at Indian Point Unit-Two was an acceptable alternative to cooling towers. That seemingly generous time allowance was sufficient only for obtaining estimates of direct power plant effects on individual year-classes of five fish species (L. Barnthouse, pers. comm.). Again, the information that could be gathered was used to the fullest extent possible to make the final decision agreed upon in the Hudson River Settlement.

Adequacy and quality of data were also major issues in all of the cases we profiled. A parallel issue was that of unethical interpretation or suppression of data. A major mechanism of "quality control" in both the power plant and Westway cases was the scrutiny the data received in the courtroom. Certainly the quality of the data was improved by cross-examination.

Nevertheless, much of the collected data failed to yield clear-cut answers; often, questions had to be narrowed in scope to be tractable, particularly in the power plants case. Asking questions of a biological nature leads to answers, but those answers are associated with considerable uncertainty. Populations comprising a biological system of interest are inherently variable with respect to organismic physiology and behavior. Variation in the physical environment overlays further patterns, which may be reflected in organisms as clear signals, noise, or something in between. When the system of interest is large and complex, as is the Hudson River, variability in each of the individual components makes the job of first understanding and ultimately predicting the outcome of a disturbance to the system a difficult one. For instance, Barnthouse *et al.* (1984) explain that they were unable to predict long-term effects of once-through cooling on the Hudson fisheries not because of lack of effort, incompetence of the scientists, or use of an inappropriate model, but because of an insufficient understanding of underlying biological processes. Given their limited understanding of the Hudson River system, however, their evaluation of available methods for mitigating impacts was a reasonable undertaking. Their answers to this more tractable question contributed significantly to the arrangement of the Hudson River Settlement (Barnthouse *et al.*, 1984).

Anthropogenic impacts frequently take the form of disturbances outside the realm of natural fluctuations of a system. Therefore, prediction of impacts is further hindered by the need to extrapolate beyond the normal range of variations into a realm unfamiliar to the scientist. This aspect of uncertainty contributed to the difficulty experienced by Hudson River investigators in characterizing and predicting effects of human disturbance. As much as 10 to 20 years later, as in the case for impacts of cooling water uptake, the long-term effects from the anthropogenic disturbances still cannot be quantified with confidence.

2. *If the data were not adequate for impact estimation, what could have been done differently?*

In every case—power plants, PCBs, and the Westway—results of scientific investigations yielded answers that led to even more questions. An extensive data base on the growth and distribution of striped bass in the estuary did not solve the question of long-term power plant impacts on that population, in large part because the dynamics of striped bass populations could not be understood and verified in the amount of time available. In a similar vein, many measurements and models of PCBs in the Hudson's sediments and water column yielded (and may continue to yield for some time) evidence that may be interpreted by several theories of PCB transport and transformation—each having a different implication for management.

It is clear that much more time and effort could have been expended on all assessments, if available. It is also true that those resources are not likely to become much more available than at present under the current assessment structure. There are several alternatives that could be resorted to. One would be to narrow the scope of the impacts sought, as was done in the power plants and Westway cases. This alternative may yield a quicker, more precise answer in the short-term; but unless the question is chosen well, there is a danger that more important impacts will be overlooked. Another alternative would be to establish a mechanism by which a solid baseline of data could be collected and updated for the entire estuary; specific impact assessments could then make use of that data base, complementing it with studies adapted to the particular situation.

3. *Was the environmental impact assessment (EIA) work subjected to continual peer review, rather than reviewed solely after the fact for publication purposes? Was the work ever reviewed at all?*

As McDowell pointed out in Chapter 3, much of the data collected remained buried in in-house reports and was never analyzed. However, it appears that those data that actually were used in decision-making were fairly well reviewed, often during litigative procedures. In this way, the environmental assessment protocols were a success. In fact, it is because of the extensive reviewing that so many new questions emerged; it is also why studies later in the course of impact assessments contained much greater detail than did earlier investigations.

4. *Did the EIA work carry any regulatory clout? If adverse impacts were predicted, would the regulatory agency of concern be able to alter the design of the proposed project to minimize effects?*

Under certain circumstances, assessment studies did have the ability to affect the outcome of a project proposal. If the results of a study stood up under general extensive review, and if adverse effects were predicted, then changes were made in project designs. To date, however, this has occurred only when both sides in a dispute have felt that they would be better off by entering what inevitably became a compromise agreement. It did not occur when the agency charged with the responsibility to decide on a project also carried out the environmental assessment studies. This was demonstrated in the Westway case, when the U.S. Army Corps of Engineers' own studies predicted adverse environmental impacts; and yet

Synthesis and Evaluation

the Corps tried to issue the final requisite permit for the construction to begin. A special study by the Committee on Government Operations of the U.S. House of Representatives found the state of decision-making in the Westway issue to be highly biased, in part because of the Corps' collaboration with the Federal Highway Administration (FHWA) and New York State's Department of Transportation (DOT)—two groups with vested interests in Westway (Committee on Government Operations, 1984). Their final recommendation to Congress included a proposal to transfer authority to grant dredge-fill permits under section 404 of the CWA from the Corps to the Environmental Protection Agency (EPA).

Congress did not choose to empower NEPA with authority to act on findings of adverse impacts. Therefore, environmental assessment is ultimately part of a political process. Even when scientific investigations are relatively divorced from the political arena (not always true), their results are weighed together with other factors when decisions are made about a given project's merit.

5. *What is to be done now with the collected data, and how can they best be complemented in future monitoring/assessment studies?*

The Hudson River Foundation, created as one of the terms of the Hudson River Settlement, has discussed placing all data collected during the assessment work in a computerized data base system that would be accessible to any interested party (J. Cooper, pers. comm.). Unfortunately, much of the information was archived in obscure places and many of the original samples were discarded. Storage of large amounts of field samples is problematic, but can be an important aspect of impact assessments that extend over a number of years as they have in the Hudson River. If samples are discarded after a short time, they cannot be reanalyzed or verified in the future when refinements in analytical techniques improve the quality of information obtained. This has been a problem for some of the PCB studies (J. Sanders, pers. comm.).

Present-day monitoring of the Hudson River is carried out in several programs that are the responsibility of New York State. These include monitoring young-of-the-year (y-o-y) and juvenile fish entrainment and impingement by power plants, a toxic substances program, and water quality monitoring. These programs have been largely designed to build on earlier studies and to maintain a long-term record of the quality of life in the estuary.

In the remainder of this chapter, we summarize features of environmental assessment and management that can aid impact evaluations in the future. Some of these features arose spontaneously in the case of the Hudson River and other estuaries. Several concepts are drawn from a document (Limburg *et al.*, 1984) describing the major consensus, regarding estuarine impact assessment methods, from a series of workshops held on the subject.

Environmental Assessment of Estuarine Ecosystems: Past, Present, and Future

After more than 12 years of practice, the institutionalized procedure of developing EIAs has come under a great deal of scrutiny, both from the legal (Trubeck, 1977; Anderson, 1973) and scientific perspectives (e.g., Friesema, 1982; Kibby and Glass, 1980; Rosenberg et al., 1981). Rosenberg *et al.* (1981) surveyed over 50 EIA studies in a variety of categories, and judged their success in the following areas: "1) definition of scientific objectives, 2) background preparation, 3) identification of main impacts, 4) prediction of effects, 5) formulation of usable recommendations, 6) monitoring and assessment, 7) sufficient lead time, 8) public participation, 9) adequate funding, and 10) evidence that recommendations were used." Estuarine impact and power plant impact assessments were given average scores in their evaluations; however, in general, the assessments were characterized by poor research design, lack of coordination among studies, questionable ethics, difficulty in accessing literature on similar impacts, etc. (Rosenberg *et al.,* 1981).

In a less rigorous, but nevertheless insightful, critique, Kibby and Glass (1980) examined the specific reasons why so many of the environmental impact statements (EIAs) had so little worth. The major faults of many EISs, according to Kibby and Glass (1980), could be summarized as:

1. Too much collection of irrelevant data;
2. Inclusion of data that were collected but never used in the evaluation process;
3. Presentation of circuitous lines of reasoning that either evaded the issues or even appeared to mislead the reader;
4. Lack of detailed information about certain essential processes; and
5. Lack of time to carry out the assessments.

Interestingly, the collective Hudson River EIAs bore all of these traits. Some of them even persisted well past 1976—the year that Kibby and Glass presented their findings at a symposium. Thus, many of the problems of the EIS procedure appear to be well entrenched and difficult to remove.

Ecosystems Studies for Impact Assessment

The virtue of using ecosystems approaches to impact assessment has been discussed at length in the past decade. Leggett (1981) summarized a workshop debate dealing with population-level vs. community-ecosystem-level approaches to power plant impact assessments. There, the population-level advocates emphasized the "acceptability" of these assessments in court, the greater yield of numerical data per unit time and effort, and the fact that the public relates more readily to a single species issue (usually about fish) than to the ecosystem. On the other side were scientists who advocated community-ecosystem approaches as necessary to understand long-term environmental impacts, because they would be felt by society much longer than immediate economic ones. Therefore, it was argued that the latter approaches could better carry out the spirit of NEPA.

What is meant by a "systems approach" to environmental study? A term borrowed from engineering, a systems approach implies that a certain conceptual framework is provided to organize our understanding of complex situations. It includes: 1) a delineation of boundaries that should be relevant to the problem at hand (i.e., the problem should define the system of interest); 2) questions that are posed to understand the structure of the ecosystem; and 3) the approach that is used to investigate the functions of various parts of the system. For ecosystems, it may be appropriate to evaluate impacts at several different scales (population, community, ecosystem) more or less concurrently.

Current Role of Ecosystem Studies in Estuarine Impact Assessments

If properly executed and couched in an ecosystem perspective, EIS assessments can tell much about what long-term impacts on a system are likely to be. From this, it is possible to estimate effects on communities and populations, sometimes in the shorter term. Limburg *et al.* (1984) give some examples of assessments wherein that approach succeeded fairly well in predicting impacts or in isolating the cause of environmental deterioration, as in the case of the Chesapeake Bay (Chesapeake Bay Program, 1983; Orth and Moore, 1983). There, deterioration occurred over a vast area and a long (30-year) time span; thus, the effects were hardly isolated and could not have been detected by the examination of single populations alone. (Ecosystem monitoring has also proven to be invaluable in detection of the decline of many European and North American forests.) Other estuarine ecosystem assessments that have helped in regional planning include work on: 1) the Narragansett River and Bay (Kremer and Nixon, 1978) (sewage management); 2) the Severn estuary in western England (Longhurst, 1978) (construction of locks for flood control); 3) the Crystal River in Florida (Kemp, 1977; McKellar, 1977) (effects of a nuclear power plant's effluent on estuarine bays); and 4) the James River estuary in Virginia (O'Connor *et al.*, 1983) (fate and transport of Kepone). In the Hudson, the ecosystem studies of the fate of PCBs continue to be crucial to decisions concerning remedial action.

In much of the research done on the Hudson, reference was made to the ecosystem that provides support for organisms and processes. However, with the exception of the PCB case, the systems approach was mostly given perfunctory attention in EIS work before being dismissed in favor of population studies. In Chapter 2, we have assembled much of the existing information on the food web and environmental parameters, from which it is obvious that the Hudson is as diverse and alive as most major east coast estuaries; in fact, it may be better off than others, such as the Chesapeake. Much of the information has come from basic research studies, which reached a peak in the mid-1970s with the momentum generated by such interest groups as the Hudson River Environmental Society. Such studies need to be encouraged, expanded, and updated where necessary. In particular, more ecosystems work is needed in the upper portion of the river (above the Troy Dam).

Research Needs and Useful Approaches

In Chapter 1, we stated that our concept of "ecosystem approach" included the investigation of population-level, community-level, and ecosystem-level properties, where appropriate. In retrospect, most of the scientific investigations carried out for impact assessment on the Hudson could have been incorporated into broader ecosystem studies that would help to address questions of long-term and cumulative impacts. However, there is a noticeable scarcity of published data on *how the Hudson ecosystem works;* most of the assessment studies simply failed (intentionally or unintentionally) to link the facts together into an understandable story.

In this section, we present several methods of evaluating ecosystems for potential impacts. These range from the simple and aggregated to the specific and detailed. As outlined in Limburg *et al.* (1984), the actual assessments may be carried out in a tiered fashion, with certain tests or observations made first, followed by a choice of more involved investigations. Measurements within an ecosystem study should identify effects and/or concentrations and gradients through populations, communities, and ecosystem compartments. Human impacts also must be weighed against the background of natural phenomena. Figure 6.1 is a visual representation of the kinds of groupings relevant to the study of estuarine problems. It is important to note the hierarchical format and exchanges within and between different functional groups. A species population should be understood in the context of its interaction at a higher level of organization; for example, how a dominant polychaete species contributes to nutrient cycling in a benthic community. Assessment should also be made of biotic-abiotic relationships. Temperate estuaries are generally dominated by the physical forces of tides, upriver freshwater flow, and seasonal gradients. To what extent do these abiotic forces produce patterns of adaptation in the biota? To what extent are anthropogenic factors controlling? Where will an anthropogenic change cause a "bottleneck" in the system?

Integrative, ecosystem-level measures received little attention in the Hudson studies. However, such measures can provide a relatively simple way to obtain information about the general status of the ecosystem, particularly when the status is compared over space or time and when the impacts potentially pose large-scale problems. If pathways of energy and/or material transfer were shown to be fundamentally altered, as a result of human activity, such a finding would have major implications for the future of at least a portion of the ecosystem. For instance, impacts resulting in loss of seagrass beds would affect water flow, sediment exchange rates, floral and faunal communities, and human recreational and economic activities.

As examples of useful, albeit aggregated, approaches to ecosystem assessment, community metabolic studies provide a gross measure of energy fixation and its partitioning in the system. This, in turn, indicates the general levels of energy potentially available for processing in the food web, as well as whether the system as a whole is a net yielder or producer of biological capital. Sirois (1973) recommended the diagnostic use of community metabolic studies (production and respiration) for characterizing ecosystem response to pollution stress. He was able to identify stressed communities on the basis of the ratio of production to respiration (P/R) along a gradient from the Tappan Zee (RM 26) to New York Harbor (RM -2). The method can also be successfully used to detect absolute and relative effects of thermal loadings from power plants (e.g., Knight and Coggins, 1982). In a report on near-field effects of once-through cooling at the Roseton Power Plant, LMS (1977) found that measurements of primary productivity (measured as ^{14}C uptake) clearly demonstrated entrainment effects; yet these findings were apparently given little weight in the overall assessment of impacts.

Trophic analyses should be coupled with metabolism studies in order to understand how biological components interact with each other and also with their physical environment. This is very important to fully comprehend transfers of carbon, nutrients, and toxic substances, and also consequences of alterations of these flows. The preliminary trophic analyses that were carried out to estimate PCB transfers in the food web (Hydroscience, 1979) fell short of their goal partly because of poor estimates of biomass in the system. Even the biomasses of major fish stocks in the Hudson have never been estimated, except by the crudest of calculations (Sheppard, 1976).

Many states now require EIAs to include the study of several species that are considered representative of the ecosystem where the impact of a project will be felt (Limburg *et al.*, 1984). We regard this "Representative Important Species" (RIS) approach as a positive move away from single species studies. RIS is by no means a complete assessment, but it can be considered a first step toward an expanded evaluation of the system state. RIS studies should be carried out in such a fashion that broader ecosystems questions, which may involve linkages between organisms and abiotic parts of their environment, can be formulated and addressed. Even representative important components, such as the benthic or submerged

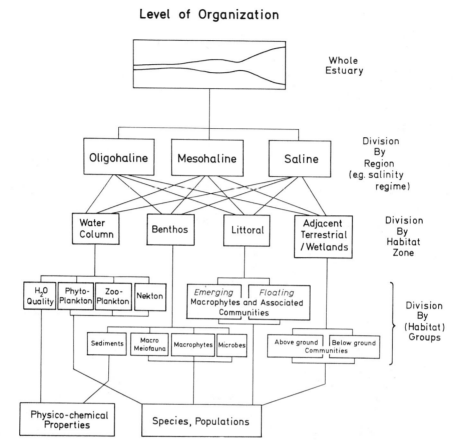

Figure 6.1. Suggested perceptual scales of organization in estuaries and attributes to consider when assessing potential anthropogenic impacts. This is meant as a guide rather than as a strict set of rules; the evaluator should be able to identify those ecosystem components most likely to be affected, and should select for study ecological attributes that will best reflect impact (Limburg et al., 1984).

aquatic vegetation subsystems, could and should be directly assessed for impacts. An example might be the impact of high levels of cadmium (as in Foundry Cove in the lower Hudson) on the ability of benthic fauna to cycle nutrients.

Another way to characterize ecosystems is by means of energy or materials budgets. Budgets go a step beyond trophic analyses in that they involve abiotic components of the ecosystem as well (such as sediments). Knowledge of where energy (as fixed carbon) and major nutrients (nitrogen, phosphorus, and silica) enter and leave the system, and how they are moved about within, is crucial to understanding the ecosystem's func-

Table 6.1. Ecological attributes to consider along with organizational scales as shown in Figure 6.1.

Within the estuarine system
Biological
- Distribution of species, species richness, or some other measure of diversity or community structure;
- Major and minor species constituents (representative important species, endangered and/or rare species, nuisance species);
- Biomass, turnover times and interactions (if any) of dominant species or other species of interest;
- Metabolic processes or indices, e.g., gross and net primary production (P_G and P_N); respiration (R); P/R ratios; bioaccumulation, transformation, or depuration of toxic chemicals; nutrient cycling;
- Behavior capable of altering structure or function of ecosystem component(s).

Chemical
- Availability of nutrients for biological production;
- Nutrient dynamics (cycling through various ionic states and compounds, through system components or parts thereof);
- Mediation of chemical dynamics by physical processes (see below);
- Fate and effect of introduced, toxic substances.

Physical
- Tidal excursion and range;
- Light availability, water transparency and color, compensation depth
- Current velocities;
- Temperature, salinity, pH, alkalinity, etc.

External to the estuary, and/or shared
- Magnitudes and dynamics of fresh and saltwater in- and effluxes;
- Major imports and exports of materials (including species, organic material, chemicals, etc.);
- Anthropogenic influences (examples are: power plants; shoreline development, sewage, dredge-and-fill; agricultural erosion and runoff).

(From Limburg et al., 1984.)

tions. For example, in the Hudson, we know that over 50% of the nitrogen inputs come from sewage sources, and that only 2 to 27% of this is consumed in primary biological production (Garside *et al.*, 1976). The same sort of budgeting is important for tracing the fate of toxic chemicals, as was seen in Chapter 4.

In Chapter 2, we discussed some of the budgets that have been put together for energy (e.g., McFadden *et al.*, 1978; Gladden *et al.*, 1984) and nutrient flows in the estuary. Data are available from Hammond (1975), Simpson *et al.* (1975), and Deck (1981) on nutrient inputs to the estuary; other data describe some of the inputs and transfers to the New York Bight (Mayer, 1982). Yet none completely describe all of the inputs and outputs of the estuary, and little information has been published on the upper river. Furthermore, the role of biota in trapping, mobilizing, or cycling matter in the Hudson ecosystem is far from well understood.

Mathematical models in impact assessment work are widespread and range from the simplest of calculations (e.g., the oxygen sag-curve model to measure BOD impacts) to extremely complex, total ecosystem models (e.g., PEST [Park, *et al.*, 1980]). Entrainment-impingement models, based on paradigms from fisheries science, have been routinely used (and abused, as in the case of the Hudson) to assess power plant impacts on fish populations (for more discussion, see Hall, 1977 and Barnthouse *et al.*, 1984). Other applications have included fate and transport of toxic substances (EXAMS, Burns *et al.*, 1981), hydrodynamic and physical/chemical models for evaluation of thermal plume, wasteload allocation, water diversion, and dredge-and-fill, and models that incorporate trophic aspects of the impacted system with physical and/or chemical phenomena (e.g., Kremer and Nixon, 1978).

Many reviews exist on the usefulness of mathematical models, of which Swartzman *et al.* (1977), Mitsch (1983), Turgeon (1983), Barnthouse *et al.* (1984), and Limburg *et al.* (1984) serve as useful references for estuarine impact assessment models. For all their promise as synthetic tools, models have been plagued by problems of data requirements, uncertainties (what *is* the proper formulation to describe a given impact?), and error due to limitations of the numerical computation procedures used. Thus far, models of fairly well-understood, purely physicochemical processes have progressed more successfully than biological ones, both on the Hudson and elsewhere, although we have seen (Chapter 4) the difficulties that can arise when using physical models to predict effects.

The state-of-the-art of biological modelling is such that much of what is developed for impact assessment is also a testing of theory, rather than straightforward application of reliable algorithms. There are many unresolved questions about the ecology of estuaries, and models must reflect those gaps in scientific understanding. This situation is unlikely to change in the near future; we must learn to live with this fact. For a decision-maker, it may be better to use cautiously the results from a model known to be imperfect, rather than to use nothing at all.

In general, ecosystems studies that have had the greatest success in elucidating environmental problems have used a variety of evaluative techniques, including: 1) field measurements that quantify flows and storages of energy, nutrients, and biomass, as well as physical controlling parameters; 2) experiments, especially meso- and microcosms, that isolate or mimic parts of the real system, but are simple enough to study a particular process; and 3) mathematical models that link together disparate information and can be used to test the consequences of various hypotheses put forth by the investigator. These approaches are more powerful when developed in parallel, so that results from one kind of investigation can help the researcher to clarify, modify, or suggest new hypotheses in concurrent endeavors. Thus, for instance, a project's modelling team can synthesize field-derived and experimentally derived information and suggest what sorts of further measurements would be most useful. Measurements, in turn, can be used to verify or invalidate a model.

One cannot say *a priori* that any of these methods for examining ecosystem structure and function will be the "best" to use in the Hudson or anywhere else. However, it is important to be sure that a general characterization of the ecosystem is on record as a baseline for comparison with subsequent alterations. Otherwise, a fairly complete survey should be included as a first level of an ecological assessment package; such a package could be included in any major impact assessment work. If adequate information already exists about the area under consideration, it may not be necessary to duplicate the work.

Institutional Changes

One way that planning and management authorities deal with the problem of scientific biases is to develop infrastructures that allow scientists to operate more independently than when under contract to parties required to produce the EIS. Then research monies are not contingent upon producing a "desired" result. An independent scientific team may have greater potential for dealing objectively with available scientific data. Such teams, reporting to an autonomous scientific panel, can remove at least those uncertainties that stem from the political arena rather than from scientific constraints, unless the autonomous board itself becomes politicized. For instance, the Hudson's PCB Settlement Advisory Committee, as an independent review body for directing and reviewing research pertinent to remedial action on the problem, can be said to have greatly stimulated the understanding of chemical and sediment movement in the river and estuary.

Other states and regions have taken up this institutional pattern for environmental management and have been fairly successful in bringing together regulators, regulated interests, scientists, decision-makers, and laypeople to work out development plans that everyone can at least live with (Limburg *et al.*, 1984). The State of Maryland's Power Plant Siting Program is a good example. Patterned after programs existing in several European countries, the Maryland program maintains an autonomous board of scientific and technological advisors. Funding comes through the state, but is collected from the utility companies. The program is an apparent success partly because of the general agreement among all parties that the unbiased scientific review process is in the best interest of all parties. Also, since 1972, coastal states have instituted Coastal Zone

Management offices in accordance with the Coastal Zone Management Act; many of those programs have had great influence on the allocation of estuarine resources (Limburg et al., 1984).

As suggested by McDowell in Chapter 3, it might be prudent to restructure environmental management so that the need to place so much weight on scientific research interpretation in a litigative setting is decreased. There are other points in decision-making, particularly in the legal process, where scientific input is needed and welcomed. An opportunity occurs during the creation of laws, when scientists can provide technical information to assist lawmakers in structuring the intent and scope of environmental legislation. Scientists also assist in the formulation of language of proposed legislation and can aid lawmakers in debate of the legislation. Once environmental legislation is passed, scientific input is also necessary for writing rules and criteria for regulatory and enforcement action (Limburg et al., 1984).

Mediation provides another avenue by which science can enter the environmental regulatory process outside of a litigative setting. Settlement of the 17-year Hudson River power plant dispute was brought about through mediation after many years of litigation failed. A critical point in the settlement negotiation was reached through a collaborative modelling effort by expert scientists on opposing sides of the cooling tower issue (Talbot, 1983; Barnthouse et al., 1984). The necessity for shutting down plant operation during critical periods in the life history of Hudson River fish was agreed upon by the scientists as the only feasible alternative to cooling towers that would afford some protection to fish-spawning activities (Talbot, 1983). However, an attempt to mediate the Westway dispute was not as successful. Meetings between the numerous and varied parties interested in Westway failed to result in any compromise plan. Finally, a suggestion was made by the Hudson River Foundation in early 1984 to use an independent mediator to help resolve the PCB disputes.

Another institutional change that has received favorable response is the strategy known as adaptive environmental assessment and management—the precepts of which are developed in Holling (1978). The forcefulness of this approach lies in the underlying philosophy of developing assessment techniques to deal with uncertainty and risk. Adaptive management necessitates the constant collection of information (including baseline studies) to decrease uncertainty, prior to and over the course of the activity; at the same time, it sets up a framework whereby policymakers and/or managers can interact with the scientists who carry out the assessments. The approach integrates data collection, mathematical assessment and optimization techniques, and intense discussion to evaluate and modify options. This is probably the best posture to adopt for most environmental assessment work, given the absence of clear-cut answers to essential questions (Limburg et al., 1984).

Cumulative Impact Assessment: The Way of the Future?

Throughout this book, we have dealt separately with each environmental impact. Yet, if the Hudson is to be managed as an ecosystem, it is important to understand cumulative as well as immediate effects. Based on what is known about the hydrodynamics, sewage-derived and natural nutrient and carbon loadings, PCB and other chemical transformations, and entrainment-impingement effects of power plants, is it possible even in a qualitative manner to say what the additional impact of Westway (filling in of 200 acres on the lower Hudson) would be? Or of an additional power plant in the mid-Hudson region? Or of decommissioning an operating plant?

The answer is, probably not at this point. Although large data bases exist, our review appears to be the only attempt to link together all of the myriad sources of biological, chemical, physical, and social information. However, there is certainly reason to believe that more synthesis might happen in the future, at least with respect to the fate of hazardous chemicals in the system. Coordination of research in a cooperative spirit has occurred in the past on the Hudson, and it recently was called for again in the January 1984 workshop on PCBs (Chapter 4). Also, in this age of rapid information transfer, the establishment of a computerized data storage and retrieval system, which could be generally accessed by remote computers, could prove extremely valuable.

No amount of good will can solve problems without money to pay for the research. Thus, a second, very important factor is the development of funding sources. The latter-day formation of the Hudson River Foundation has done much to refocus interest on the Hudson ecosystem. The Foundation's purpose is to support both basic and applied research on the Hudson River, with emphasis on potential human uses of the estuarine ecosystem (HRF, 1984). Perhaps the Hudson River Foundation, in stewarding the bulk of future research, can successfully orchestrate the necessary research efforts, including among other approaches ecosystem studies *in situ* and synthetic models.

In summary, future impact assessments on the Hudson River should be greatly aided first by learning from past experiences such as those we have tried to document here. Second, although several different regulatory agencies have jurisdiction over the various activities that affect the Hudson, some centralized "book-keeping" mechanism to keep track of cumulative activities is necessary. Third, that information should be organized and made available to researchers, so that constant review and evaluation of the "state-of-the-River" can be accomplished. As long as we continue to utilize the Hudson's resources to the point of scarcity, long-term monitoring programs are needed. These should provide information on the status of the ecosystem, as well as on economically important populations. Only in this way will the successful development of managerial models proceed. Finally, scientific assessments were seen both to suffer and gain from courtroom exposure; and so ways in which time spent in court can be minimized without losing the critical review of data may help scientists and decision-makers alike to get on with the business of assessing and managing the Hudson. More emphasis on scientific input to development of legislation and regulation is recommended. Along with a renewed commitment to integrated studies, integrated planning has been instituted in the form of New York's Coastal Management Program. This program has been set up in accordance with the Federal Coastal Zone Management Act of 1972 (U.S.C. sections 1451 *et seq.*), and it was recently (autumn 1982) approved for New York State. Among other benefits, a 1670-ha sanctuary will be set aside along the Hudson estuary. This will be used for study of the ecosystem and should be extremely useful for baseline work for impact assessments.

This extensive review has led us to conclude that some of the scientific evaluation studies were performed as well as they could have been, given the circumstances, while others fell disappointingly short of that mark. The Hudson River has been the proving ground for much of America's environmental impact assessment, and many of the mistakes made have already served as lessons to decision-makers elsewhere. The mechanisms for managing the estuary have been evolving toward a more holistic perspective; certainly, most environmental investigators dealing with the Hudson today have a broader understanding of potential consequences than they had 15 years ago.

References

Anderson, F.R. 1973. NEPA in the courts: a legal analysis of the National Environmental Policy Act. Resources for the Future, Inc. Johns Hopkins University Press, Baltimore, MD.

Barnthouse, L.W., J. Boreman, S.W. Christensen, C.P. Goodyear, W. Van Winkle, and D.S. Vaughan. 1984. Population biology in the courtroom: the lesson of the Hudson River controversy. Bioscience 34(1):14–19.

Burns, L.A., D.M. Cline, and R.R. Lassiter. 1981. Exposure analysis modeling system (EXAMS) user manual and system documentation report. ERL, U.S. Environmental Protection Agency, Athens, GA.

Chesapeake Bay Program. 1983. Chesapeake Bay: A framework for action. U.S. Environmental Protection Agency, Region 3, Philadelphia, PA. (2 vols.)

Committee on Government Operations. 1984. The Westway project: A study of failure in federal/state relations. 66th Report by Committee on Government Operations, together with dissenting views. 98th Congress, 2d Session. House Report 98-1166. Union Calendar No. 650. U.S. Gov't. Printing Office, Washington, D.C. 57 pp.

Deck, B.L. 1981. Nutrient-element distributions in the Hudson estuary. Ph.D. Dissertation. Columbia University, New York, NY. 396 pp.

Friesema, H.P. 1982. The scientific content of environmental impact statements: workshop conclusions. Northwestern University, Center for Urban Affairs and Policy Research, Evanston, IL. (Working paper.)

Garside, C., T.C. Malone, O.A. Roels, and B.A. Shartstein. 1976. A evaluation of sewage derived nutrients and their influence on Hudson estuary and New York Bight. Est. Coastal Mar. Sci. 4:281–289.

Gladden, J.B., F.C. Cantelmo, J.M. Croom, and R. Shapot. 1984. An evaluation of the Hudson River ecosystem in relation to the dynamics of fish populations. Trans. Am. Fish. Soc. (in press).

Hall, C.A.S. 1977. Models and the decision-making process: the Hudson River power plant case, pp. 345–364 *In* Ecosystem Modeling in Theory and Practice (C.A.S. Hall and J.W. Day, Jr., eds.). Wiley-Interscience, New York, NY.

Hammond, D.E. 1975. Dissolved gases and kinetic processes in the Hudson River. Ph.D. Dissertation. Columbia University, New York, NY.

Holling, C.S. (ed.) 1978. Adaptive environmental assessment and management. IIASA International Series on Applied Systems Analysis. John Wiley & Sons, New York, NY. 377 pp.

HRF. 1984. Annual program plan and solicitation of proposals. Hudson River Foundation, New York. 52 pp.

Hydroscience. 1979. Analysis of the fate of PCB's in the ecosystem of the Hudson estuary. Prepared by Hydroscience, Inc., Westwood, N.J. for N.Y.S. Dept. of Environmental Conservation, Albany, NY.

Kemp, W.M. 1977. Energy analysis and ecological evaluation of a coastal power plant. Ph.D. Dissertation. University of Florida, Gainesville, FL. 560 pp.

Kibby, H. and N. Glass. 1980. Evaluating the evaluations: a review perspective on environmental impact assessment, pp. 40–48 *In* Biological Evaluation of Environmental Impacts. Proceedings of a symposium. Council on Environmental Quality and U.S. Fish and Wildlife Service. FWS/OBS-80/26, Washington, D.C.

Knight, R.L. and W. Coggins. 1982. Record of estuarine and salt marsh metabolism at Crystal River, FL., 1977–1981. Final Report to Florida Power Corporation. Contract QEA-000045. Systems Ecology and Energy Analysis Group, Dept. of Environmental Engineering, University of Florida, Gainesville, FL. 87 pp.

Kremer, J.N. and S.W. Nixon. 1978. A coastal marine ecosystem: simulation and analysis. Ecological Studies 24. Springer-Verlag, New York, NY. 215 pp.

Leggett, W.C. 1981. Moderator's summary—population-level vs. community/ecosystem-level approaches to impact assessment, pp. 75–78 *In* Issues Associated with Impact Assessment (L.D. Jensen, ed.). Proc. 5th National Workshop on Entrainment and Impingement. EA Communications, Sparks, MD.

Limburg, K.E., C.C. Harwell, and S.A. Levin. 1984. Principles for estuarine impact assessment: lessons learned from the Hudson River and other estuarine experiences. Report No. 24. Ecosystems Research Center, Cornell University, Ithaca, NY.

Longhurst, A.R. 1978. Ecological models in estuarine management. Ocean Management 4:287–302.

LMS. 1977. Roseton Generating Station: Near-field effects of once-through cooling system operation on Hudson River biota. Prepared by Lawler, Matusky, and Skelly, Engineers, and by Ecological Analysts, Inc. for Central Hudson Gas and Electric Corp., Poughkeepsie, NY.

McFadden, J.T., Texas Instruments, Inc., and Lawler, Matusky, and Skelly, Engineers. 1978. Influence of the proposed Cornwall pumped storage project and steam electric generating plants on the Hudson River estuary, with emphasis on striped bass and other fish populations. Revised. Prepared for Consolidated Edison Co. of NY, Inc.

McKellar, H.N., Jr. 1977. Metabolism and model of an estuarine bay ecosystem affected by a coastal power plant. Ecol. Modelling 3:85–118.

Mitsch, W.J. 1983. Aquatic ecosystem modeling—its evolution, effectiveness, and opportunities in policy issues. (Mss.)

O'Connor, D.J., J.A. Mueller, and K.J. Farley. 1983. Distribution of Kepone in the James River estuary. J. Environ. Eng. Div., ASCE 109(2):396–413.

Orth, R.J. and K.A. Moore. 1983. Chesapeake Bay: an unprecedented decline in submerged aquatic vegetation. Science 222:51–53.

Park, R.A., C.I. Conolly, J.R. Albanese, L.S. Clesceri, G.W. Hietzman, H.H. Herbrandson, B.H. Indyke, J.R. Loehe, S. Ross, D.D. Sharma, and W.W. Shuster. 1980. Modeling transport and behavior of pesticides and other toxic organic materials in aquatic environments. Report No. 7. Center for Ecological Modeling, Rensselaer Polytechnic Institute, Troy, NY. 165 pp.

Rosenberg, D.M., V.H. Resh, S.S. Balling, M.A. Barnaby, J.N. Collins, D.V. Durbin, T.S. Flynn, D.D. Hart, G.A. Lamberti, E.P. McElravy, J.R. Wood, T.E. Blank, D.M. Schultz, D.L. Marrin, and D.G. Price. 1981. Recent trends in environmental impact assessment. Can. J. Fish. Aquat. Sci. 38:591–624.

Sheppard, J.D. 1976. Valuation of the Hudson River fishery resources: past, present and future. N.Y.S. Dept. of Environmental Conservation, Bureau of Fisheries, Albany, NY. 51 pp. (Mss.)

Simpson, H.J., D.E. Hammond, B.L. Deck, and S.C. Williams. 1975. Nutrient budgets in the Hudson River estuary, pp. 616–635 In Marine Chemistry in the Coastal Environment (T.M. Church, ed.). ACS Symposium Series No. 18.

Sirois, D.L. 1973. Community metabolism and water quality in the lower Hudson River estuary. Hudson River Ecology, 3rd Symp. Paper No. 15. Hudson River Environ. Soc., Bronx, NY.

Swartzman, G., R. Deriso, and C. Cowan. 1977. Comparison of simulation models used in assessing the effects of power-plant-induced mortality on fish populations, pp. 333–361 In Assessing the Effects of Power-Plant-Induced Mortality on Fish Populations (W. Van Winkle, ed.). Pergamon Press, New York, NY.

Talbot, A.R. 1983. Settling things. Six case studies in environmental mediation. The Conservation Foundation and the Ford Foundation. Washington, D.C. 101 pp.

Trubeck, D.M. 1977. Allocating the burden of environmental uncertainty: the NRC interprets NEPA's substantive mandate. Wisc. Law Rev. 747–776.

Turgeon, K.W. (ed.) 1983. Marine ecosystem modelling. Proceedings from a workshop held April 6–8, 1982. U.S. Dept. of Commerce, NOAA, National Environmental Satellite, Data and Information Service. NOAA-S/T 83-38, Washington, D.C. 274 pp.

7
Hudson River Data Base

MARY ANN MORAN, PETER S. WALCZAK,
and KARIN E. LIMBURG

This chapter summarizes the data base literature for research conducted on the Hudson River, and describes the comprehensive bibliography that follows as Bibliographies A and B. The following discussion includes, for the most part, research carried out for the estuarine power plant impact assessments. The literature that was sponsored directly by the utility companies is described first, followed by a summary of the open literature. Research concerning PCBs is summarized in Chapter 4; however, highlights of that extensive literature are included in Table 7.27.

Utility-Sponsored Research

Utility companies operating power plants on the Hudson River have generated an extensive data base on biology and water quality of the river. Since this data base contains information valuable to future environmental assessments and ecosystem analyses of the river, it is worthwhile to summarize the collection and analysis efforts.

Consolidated Edison Co. of NY, Inc., Orange and Rockland Utilities, Inc., and Central Hudson Gas and Electric Corp. have been the three main utilities sponsoring Hudson River research. Most of their efforts, undertaken in response to federal regulatory agencies' requirements that cooling towers be constructed at existing power plants, have been directed toward demonstrating that once-through cooling does not significantly alter the Hudson ecosystem. The utilities' research programs have been involved in the establishment of baseline studies at sites of existing and proposed power generating facilities.

The majority of data collection was handled by environmental and engineering consulting firms hired by the utility companies. The major consultants involved in amassing the Hudson River data base were: 1) New York University Medical Center; 2) Quirk, Lawler, and Matusky, Engineers (QLM), which later became Lawler, Matusky, and Skelly, Engineers (LMS); 3) Texas Instruments, Inc. (TI); and 4) Ecological Analysts, Inc. (EA). A bibliography of reports generated from utility research appears after this chapter (Bibliography A). Complete citations for references mentioned in the following sections in this chapter can be found in the bibliography.

Three categories of utility-sponsored research will be discussed: fisheries, non-fisheries (biological), and water quality. Fisheries data sets are the most extensive, reflecting the fact that once-through cooling effects on Hudson River fish was the primary environmental issue facing the utility companies.

Fisheries Data Sets

An extensive review of the Hudson River fisheries data sets was recently compiled by Ronald J. Klauda, Paul H. Muessig, and John A. Matousek (Klauda *et al.*, in press) based on the authors' employment experience with consulting firms during the data collection period. What follows in this section is, in large part, a summary of Klauda, Muessig, and Matousek's work. For a complete discussion of the fisheries data sets, including information on computer files, data formatting, and study design strengths and weaknesses, the reader is referred to their paper.

Fisheries data collection began in the mid-1960s. By 1974, the program had expanded from near-field (power plant vicinity) sampling at fixed stations to an intensive river-long collection effort. Some of the data sets and analyses were published in a collection of technical reports submitted by the consulting firms to the utility companies, although according to Klauda *et al.* (in press), much of the fisheries data have never been analyzed. Major data sets are maintained by the utilities in SAS-formatted computer files. Fisheries technical reports are included in Bibliography A, and are organized by the consulting firms authoring the reports.

A major thrust of the fisheries data collection effort was the investigation of distribution, abundance, and species composition of Hudson River fish. Near-field sampling was conducted at existing and proposed power plant locations (Table 7.1). Far-field studies on fish distribution (Table 7.2) were conducted throughout the estuary, with major emphasis on the region of the Hudson between the George Washington Bridge and Troy. Data collection included species identification, location, catch per effort, and length and weight measurements of the fish caught. The distribution,

Table 7.1. Near-field data sets on distribution, abundance, and species composition of Hudson River fish.

Location (consulting firm)	Yr	Primary species	Data
Indian Point Plant (Texas Instruments)	1972–1980	Striped bass, white perch, Atlantic tomcod, plus other common species	1. Catch per effort 2. Lengths and weights 3. Water chemistry 4. Beach seine, bottom trawl, surface trawl 5. Seven fixed seine stations 6. Seven fixed trawl stations 7. Weekly survey, day, Apr.–Dec. (surface trawl July–Dec.)
Indian Point Plant (Ecological Analysts, Inc.)	1977–1979	Ichthyoplankton	1. Seven standard stations 2. Day/night distribution 3. Depth distribution 4. Seasonal succession and abundance 5. Concurrent sampling at three stations on transect at plant, and in plant intake and discharge, striped bass only 6. Length data 7. Water chemistry
Bowline Point Plant (Lawler, Matusky, and Skelly)	1971–1980	White perch, striped bass, Atlantic tomcod, hogchoker, alewife, blueback herring, American shad, bay anchovy, plus other common species	1. Bottom trawl, surface trawl, beach seine, gill net, and trap net 2. Three fixed bottom trawl stations 3. Eight fixed beach seine stations 4. Three fixed surface trawl stations 5. Water chemistry

Table 7.1. Continued

Location (consulting firm)	Yr	Primary species	Data
Lovett Plant (Lawler, Matusky, and Skelly)	1972–1976	White perch, striped bass, bay anchovy, hogchoker, blueback herring, plus other common species	6. Species identification, length, and weight analysis 7. Seasonally stratifed sampling program 8. Diel sampling during several years 1. Bottom trawl, surface trawl, and beach seine 2. Two fixed trawl stations and two fixed beach seine stations 3. Water chemistry 4. Species identification, length and weight analysis
Roseton, Danskammer Point Plants (Ecological Analysts, Inc.)	1976–1977	Striped bass, white perch, clupeids, white catfish, spottail shiner	1. Distribution and abundance in and around the thermal discharges vs. control sites 2. Electroshocker, seines, traps, body temperature, taggings 3. Diversity 4. 1976, May-Nov., 15 sample dates 5. 1977, Mar.-Nov.; 24 sample dates
Roseton, Danskammer Point Plants (Lawler, Matusky, and Skelly)	1971–1979	White perch, blueback herring, Atlantic tomcod, hogchoker, alewife, spottail shiner, plus other common species	1. Bottom trawl, surface trawl, beach seine, trap net, gill net 2. Two long-term trawl stations and four long-term seine stations 3. Large study area (15 km) with nine trawl stations, and eight seine stations 4. Species identification, length, and weight analysis

Area	Years	Species	Methods
Kingston area (Lawler, Matusky, and Skelly)	1971–1973	Blueback herring, white perch, spottail shiner, plus other common species	5. Sexual differentiation of major species during 1973–1975 6. Diel sampling program 1. Bottom trawl, surface trawl, beach seine, gill net, trap net 2. Species identification, length and weight analysis 3. Three fixed trawl stations and two fixed seine stations 4. Diel sampling program 5. Water chemistry
Ossining area (Texas Instruments)	1972–1973	Striped bass, white perch, Atlantic tomcod, plus other common species	1. Catch per effort, no./m^3 2. Lengths and weights 3. Water chemistry 4. Plankton nets, beach seines, bottom trawl, box traps, gill nets 5. Several stations 6. May 1972 to Apr. 1973
Albany Plant (Lawler, Matusky, and Skelly)	1975	White perch, blueback herring, American eel, spottail shiner, plus other common species	1. Bottom trawl, surface trawl, beach seine, gill net 2. Two beach seine stations, six trawl and gill net stations 3. River samples collected during four monthly periods; Apr., June, Aug., Oct. 4. Species identifications, length and weight analysis 5. Diel sampling program 6. Water chemistry

From Klauda et al. (in press).

Table 7.2. Far-field data sets on distribution, abundance, and species composition of Hudson River fish. All data sets were collected by Texas Instruments.

Survey	Yr	Primary species	Data
Ichthyoplankton	1973–1980	Striped bass, white perch, Atlantic tomcod	1. No./m^3 by life stage 2. Lengths for striped bass larvae in 1976–1980 3. Water chemistry 4. Epibenthic sled, Tucker trawl 5. Stratified random design 6. Weekly runs, day or night, Apr.-July (Feb.-Apr. for Atlantic tomcod)
Beach Seine	1973–1980	Striped bass, white perch	1. Catch per effort by age/length group 2. Lengths and weights for juvenile and yearling striped bass and white perch 3. Lengths for juveniles of selected species 4. Stratified random design 5. Weekly or biweekly runs, day, Apr.-Dec. (some night samples in 1973–1974) 6. Water chemistry 7. 30.5-m seine
Fall Shoals	1973–1980	Striped bass, white perch, Atlantic tomcod	1. No./m^3 by age/length group (except 1973) 2. Lengths and weights for juvenile striped bass and white perch 3. Lengths for juveniles of selected species 4. Biweekly runs, night, Aug.-Dec. 5. Water chemistry

Interregional Trawl	1973–1980	Striped bass, white perch, Atlantic tomcod	1. Catch per effort by age/length group 2. Lengths and weights for juvenile striped bass and white perch 3. Lengths for juveniles of selected species 4. 35–40 fixed station 5. Biweekly runs, day, Apr.-Dec. 6. Water chemistry
Try Trawl	1978–1980	Striped bass, white perch, Atlantic tomcod	1. Catch per effort by age/length group 2. Lengths and weights for juvenile striped bass and white perch 3. Lengths for juveniles of selected species 4. Stratified random design in depths of 1.5–6 m 5. Biweekly runs, day, Apr.-Dec. 6. Water temperature only
Southern (Lower) Estuary	1974–1975	Striped bass, white perch, Atlantic tomcod	1. Catch per effort by age/length group 2. Non-random design to maximize catch 3. Approximate biweekly runs, day 4. Several gear used 5. June 1974 to July 1975
Mark-Recapture	1972–1980	Striped bass, white perch, Atlantic tomcod	1. Marked fish releases and recoveries 2. Juvenile and adult striped bass 3. Juvenile and yearling white perch 4. Adult Atlantic tomcod
Relative Contribution	1974–1975	Striped bass	1. Stock discrimination 2. Age, length, weight, sex, maturity of adults 3. Biochemical, meristic and morphometric characters

From Klauda et al. (in press).

abundance, and species composition data set is quite extensive, with at least 65,000 samples collected between 1972 and 1980 (Klauda et al., in press).

In addition to the distribution and abundance data set, the 1971–1980 utility data can be grouped into five more categories (Klauda et al., in press):

1. Biological characteristics (length, weight, age, fecundity, stomach contents) of striped bass, white perch, Atlantic tomcod, bluefish, blueback herring, and alewife (collected by TI and LMS).
2. Physiology and behavior (thermal tolerance and preference, respiration, growth, chemical discharge effects) of striped bass, white perch, Atlantic tomcod, alewife, white catfish, and spottail shiner (collected by TI and EA).
3. Entrainment (passage of organisms through the power plant cooling system) and impingement (trapping of organisms against intake screens) of striped bass, white perch, clupeids, bay anchovy, Atlantic tomcod, alewife, blueback herring, American shad, and other species. (This is quite an extensive data set, and includes abundance of entrained and impinged fish as well as survival studies; collected by TI, LMS, and EA).
4. Gear performance and special studies (collected by TI, LMS, and EA).
5. Culturing and stocking of Hudson River striped bass (collected by TI and EA).

These data sets, including the years of the study, species surveyed, and strengths and weaknesses of study design, are described in detail in Klauda et al. (in press).

Non-Fisheries Data Sets

Utilities' biological sampling programs included collection of non-fisheries data sets. Phytoplankton, microzooplankton, macrozooplankton, and benthos constituted the major non-fish categories; bacteria and periphyton were also occasionally sampled.

Collection of non-fisheries data was much less extensive than fisheries sampling, since the planktonic and benthic communities did not feature as prominently in the impact assessment issues. River-long non-fisheries programs were rarely undertaken; most sampling efforts were confined to three main areas centered on power plant locations: Indian Point vicinity, Bowline-Lovett vicinity, and Roseton-Danskammer vicinity.

The major impetus for non-fisheries data collection was the determination of power plant effects on entrainable organisms. Screens covering the water intake pipes kept debris and larger organisms from being swept into the power plants with cooling water. Smaller planktonic organisms, however, were vulnerable to being drawn into the plant and passing through the cooling system.

Research on entrainment effects was generally designed with either of two objectives:

1. To determine the tolerance of entrained organisms and their survival rate following entrainment; and
2. To describe the communities of organisms (phytoplankton, microzooplankton, and macrozooplankton) subject to entrainment.

The body of literature stemming from the first objective is of less general interest for ecosystem modelling and future impact assessments and will not be considered in this summary. However, references to entrainment survival reports are included in Bibliography A. The latter data collection effort—characterization of plankton communities at power plant locations—will be treated in more detail. An additional research program of importance in collection of non-fisheries data was the investigation of effects of the thermal plume created by power plant discharge of cooling water. This research, also discussed in the following sections, involved the characterization of the benthic invertebrate community in power plant vicinities. Table 7.3 summarizes all of the major collection programs for non-fisheries data.

A. Indian Point Near-Field Studies. Non-fisheries data collection at Indian Point began in 1971, when researchers at the New York University Medical Center were hired by Consolidated Edison to study power plant effects on non-screenable organisms. An attempt was made to quantitatively describe the communities of organisms that were in the vicinity of Indian Point, and were therefore vulnerable to entrainment. These near-field studies examined species composition, abundance, and distribution of phytoplankton, microzooplankton, and macrozooplankton. Data were collected by New York University until 1977, and then by EA beginning in 1978 (New York University, 1971, 1973, 1974a, 1974b, 1976a, 1976b, 1977, 1978b; EA, 1980b, 1980c, 1981a).

Near-field sampling at Indian Point occurred within a few miles of the power plant, usually river mile (RM) 39 to 43. Seven stations were sampled beginning in April or May of each year and extending to November or December. Plankton nets served as the major collecting gear, and both daytime and nighttime samples were taken. The collection schedule varied, but it was usually weekly or monthly.

Phytoplankton sampling occurred from 1971–1975, and included calculation of abundance and identification to genus. During 1971 through 1979, microzooplankton and macrozooplankton were sampled and species inventory and abundance information were compiled. Exceptions to the zooplankton sampling scheme occurred in 1973 when no micro- or macro-

Table 7.3. Summary of Hudson River non-fisheries data sets collected as part of utility-sponsored research programs.

Organism	Indian Pt. near-field studies (NYU, EA)[a]	Indian Pt. ecological studies (TI)	Bowline/ Lovett near-field studies (LMS)	Bowline/ Lovett entrainment abundance studies (EA)	Roseton/ Danskammer near-field studies (LMS)	Roseton/ Danskammer habitat selection studies (LMS)	Kingston aquatic ecology studies (LMS)
Bacteria	1971						
Phytoplankton	1971–1975		1971–1977	1975–1976	1971–1977		1970–1973
Microzooplankton	1971, 1972, 1974–1979		1971–1977		1971–1978		1970–1973
Macrozooplankton	1971, 1972, 1974–1979		1973–1977		1971–1978		1970–1973
Benthos		1972–1974	1969–1976	1976	1971–1978	1969, 1975–1977	1971–1973
Periphyton						1973–1977	
Macrophytes						1973–1977	

[a] Consulting firms collecting non-fisheries data: NYU, New York University Medical Center; EA, Ecological Analysts, Inc.; TI, Texas Instruments, Inc.; LMS, Lawler, Matusky, and Skelly, Engineers. Publications reporting non-fisheries data are listed in Bibliography A according to the consulting firm conducting the research.

zooplankton samples were taken, and in 1977 when abundance was computed only for major taxonomic groups and most common species.

Microbial sampling was also a part of the Indian Point near-field studies, but only during January, February, and September 1971. Water samples were analyzed for ATP (adenosine triphosphate) content, and bacterial plate counts were made.

B. Indian Point Ecological Studies. A concurrent sampling program, referred to as the Indian Point Ecological Studies, was conducted in the Indian Point vicinity to determine the effects of plant operation on screenable organisms. TI began a fisheries study of the ecological consequences of impingement (see the "Fisheries" section in this chapter), but also included in this study was an investigation of the biological impact of thermal and chemical effluents on benthos (TI, 1972b, 1973b, 1974f, 1975e, 1975k).

In 1972, grab samples and dredge samples of the benthos were taken at monthly intervals (April to December) at seven stations. Data analysis involved descriptions of community composition, diversity, biomass, relative abundance, and seasonal distribution of benthos in the Indian Point area. In 1973 and 1974, benthic community structure was compared between control areas and areas within the influence of the thermal plume. Detailed studies of the population biology of the isopod *Cyathura polita* were undertaken in 1973 and 1974. *Cyathura polita* was chosen as a research focus because of its low mobility and its high abundance in the Indian Point area.

C. Bowline-Lovett Near-Field Studies. Hudson River aquatic ecology studies in the Bowline and Lovett power plant vicinity (RM 36 to 42) were undertaken by LMS to describe the communities subjected to power plant influence (LMS, 1974a, 1975f, 1976g, 1977a, 1978g, 1980b; LMS and EA, 1977a). Non-fisheries data sets were collected from 1971 until 1977. The most complete summary of sampling methodology is given in LMS and EA (1977a). LMS (1980b) provides a good summary of data collected during all years of the non-fisheries sampling effort at Bowline and Lovett.

Methods of phytoplankton data collection varied throughout the seven years of sampling (Table 7.4). Net samples were taken in 1971 and 1972, and whole water samples were collected from 1973–1977. Sampling occurred once or twice per month at anywhere from two to seven stations and one to three water depths. For most years, species inventories were compiled and data were summarized by total abundance, chlorophyll *a* concentration, and abundance of the five most important species (Table 7.5).

Community analyses were performed on the 1976 phytoplankton data (LMS and EA, 1977a); these included diversity measures, cluster analysis, and niche breadth analysis. Statistical analyses of phytoplankton

Table 7.4. Phytoplankton sampling methods at Bowline Point and Lovett Generating Stations, 1971–1977.

Yr	No. of stations sampled	Sampling frequency	Collection gear	Sampling method	Collection depth(s)	Time
1971	1 1 1	Aug. 9 June 10 Aug. 9	Juday Plankton trap 76-μm mesh	Net	Surface[a]	Day
1972	5	Once/mo June, July Twice/mo Sept., Oct. Thrice/mo Aug.	Wisconsin plankton net Net 18-μm mesh	Net	Surface[a]	Day
1973	5	Once/mo May, Nov., Dec. Twice/mo June-Oct.	Plastic bucket 18-μm mesh net for filtration	Whole water (May-Dec.) Filtrate (nanoplankton) (May-Oct.) Net (May-Oct.)	Surface[a]	Day
1974	6	Once/mo Apr. Twice/mo May-Dec.	PVC Niskin water bottle	Whole water	50% and 1% light transmittance 1 m above bottom	Day
1975	5 1 1	Once/mo Jan.-Feb. (two stations only) Twice/mo Mar.-Dec. Once/mo Jan.-Dec. Twice/mo Mar.-Aug.	PVC Niskin water bottle	Whole water (composite)	Surface[a] 1% light 1 m above bottom	Day
1976	4 1 1	Once/mo Feb., July Twice/mo Mar.-June Sept.-Dec. Thrice/mo Aug. Once/mo Feb. Twice/mo Mar.-June,	Teel centrifugal pump	Whole water	Oblique tow (bottom-surface)	Night
1977	3	Once/mo Mar. Twice/mo Apr.-Dec.	Teel centrifugal pump	Whole water	Oblique tow (bottom-surface)	Night

From Lawler, Matusky, and Skelly, and Ecological Analysts, Inc. (1977a).
[a] 0.3 to 0.6 m below surface.

Table 7.5. Laboratory analysis of Bowline Point and Lovett Generating Stations phytoplankton samples, 1971–1977.

Yr	Samples analyzed	Biomass	Chlorophyll	Phaeophytin
1971	All collections analyzed no replicates	—	—	—
1972	All collections analyzed no replicates	—	—	—
1973	Once/mo (May, July-Oct.) Twice/mo (June) net, filtrate, whole water samples analyzed, no replicates analyzed Once/mo (Nov., Dec.) whole water replicates analyzed	—	—	—
1974	Once/mo duplicates analyzed	Once/mo ash-free dry weight Three size fractions (<20 µm, 20–76 µm, >76 µm)	Twice/mo Three size fractions (<20 µm, 20–76 µm, >76 µm)	—
1975	Once/mo duplicates analyzed	Twice/mo ash-free dry weight two replicates analyzed	Twice/mo two replicates	Twice/mo two replicates
1976	Twice/mo no replicates (Lovett samples not analyzed)	Twice/mo dry weight and ash-free dry weight two replicates	Twice/mo two replicates	Twice/mo two replicates
1977	All collections analyzed no replicates	Twice/mo dry weight and ash-free dry weight two replicates	Twice/mo two replicates	Twice/mo two replicates

From Lawler, Matusky, and Skelly, and Ecological Analysts, Inc. (1977a).

abundance by sampling data and station for 1971 through 1977 are summarized in LMS (1980b).

Microzooplankton were sampled from 1971–1977, once or twice per month from approximately April through December. Sampling methods and stations varied (Table 7.6). Results were presented as species inventories and as abundance and percent composition summaries by major taxonomic group. The 1976 data were analyzed with a diversity index, cluster analysis, and niche breadth analysis (LMS and EA, 1977a). An ordination technique was used for 1976 and 1977 data in an attempt to assess the "health" of the microzooplankton communities at sampling stations by comparing standing crop and niche partitioning (LMS, 1980b).

Macrozooplankton sampling in the Bowline-Lovett area was less extensive than phytoplankton and microzooplankton sampling. Data were collected in 1973 from March to August, and in 1974–1977 during non-winter months (Table 7.7). Species inventories were reported for each year, along with abundance and percent composition summaries by major taxonomic group. Cluster analysis and diversity measurements were performed on 1976 data (LMS and EA, 1977a) and community ordination was performed on 1976 and 1977 data (LMS, 1980b).

Benthic sampling occurred with Ekman grab samplers (1969–1971) or Ponar grab samplers (1972–1977). Station number varied from four to nine, and sampling was done on an irregular monthly schedule (Table 7.8). Data were reported as species inventories; abundance, biomass, and percent composition were summarized by major taxonomic group. Community analysis included diversity measures and "classification" analysis through use of the percent similarity index.

Supplementary field studies during 1976 in the Bowline-Lovett vicinity involved periphyton and benthos colonization of artificial substrates (LMS, 1977a).

D. Bowline-Lovett Entrainment Abundance Studies. Phytoplankton samples were collected by EA at the intake and discharge canals at Bowline Point and Lovett power plants (EA, 1975, 1976a, 1976b, 1976g, 1976h, 1977c, 1977f, 1977g). Collections at Bowline were made in 1975 (June, August, and November) and in 1976 (April to November). Lovett collections were made seasonally in 1975. The samples were analyzed for primary productivity (^{14}C), ATP (Bowline only), and chlorophyll *a*. Microzooplankton and macrozooplankton were also sampled at the Bowline intake and discharge, but only data on percent survival are reported (EA, 1981b).

E. Roseton-Danskammer Near-Field Studies. LMS conducted the near-field studies at Roseton and Danskammer power plants (RM 62 to 68) (LMS, 1975g, 1975h, 1978a, 1978b, 1979b, 1979c, 1980f; LMS and EA, 1977b). Non-fisheries data collection began in 1971 and was carried out

Table 7.6. Microzooplankton sampling methods at Bowline Point and Lovett Generating Stations, 1971–1977.

Yr	No. of stations sampled	Sampling frequency	Collection gear	Collection depth(s)	Time
1971	2	June, Aug.	Juday trap, vertical tows 12.4-cm mouth diameter	Bottom-surface	Day
1972	5	Once/mo May (two stations only), June-Oct. (five stations)	Wisconsin net 76-μm mesh vertical tows	Bottom-surface	Day
1973	5	Twice/mo June-Dec.	12.4-cm mouth diameter Wisconsin net 76-μm mesh vertical tows June-Aug.: sequential Sept.-Dec.: simultaneous	3.-m surface 6.-m surface-bottom-surface	Day
1974	6	Once/mo Apr., Oct.-Dec. Twice/mo May-Sept.	12.7-cm mouth diameter Clarke-Bumpus net 150-μm mesh Apr.-Dec.: horizontal tows, 5 min at 70 cm/s May 24 to Dec.: oblique tows, 150-μm mesh Aug. 30 to Dec.: oblique tows, 80-μm mesh	1%, 50% light transmittance level, 1 m above bottom Surface-bottom-surface	Day
1975	6	Once/mo Jan.-Feb. (two stations only) Twice/mo Mar.-Dec.	12.7-cm mouth diameter Clarke-Bumpus net 80-μm mesh oblique tows	Surface-bottom-surface	Day
1976	1	Once/mo Jan.-Dec.			
	4	Twice/mo Mar.-June, Sept.-Dec. Once/mo Thrice/mo Aug.	Centrifugal pump 0.5-m net, 80-μm mesh filtering 500 gal oblique tows, 7–10 min at 50 cm/s boat speed	Bottom-surface	Night
	2	Same except deleted after June 3			
1977	3	Twice/mo Mar.-Dec.	Centrifugal pump 0–5-m net, 80-μm mesh filtering 500 gal oblique tows, 7-10 min at 50 cm/s boat speed	Bottom-surface	Night

From Lawler, Matusky, and Skelly and Ecological Analysts, Inc. (1977a).

Table 7.7. Macrozooplankton sampling methods at Bowline Point and Lovett Generating Stations, 1973–1977.

Yr	No. of stations sampled	Sampling frequency	Collection gear	Time
1973	1–11	Weekly, Mar.-May Monthly, June-Aug.	571-μm plankton net	Day (Mar.-May) Day/night (May-Aug.)
1974	1–6	Twice/mo Apr.-Aug.	571-μm plankton net, epibenthic sled	Day/night
1975	6–7	Once/mo Mar.-Dec.	571-μm plankton net	Day/night (Mar.-Aug.) Day (Sept.-Dec.)
1976	4	Twice/mo Feb.-Dec.	571-μm plankton net	Night
1977	3	Twice/mo Mar.-Dec.	571-μm plankton net	Night

From Lawler, Matusky, and Skelly and Ecological Analysts, Inc. (1977a).

until 1978. LMS and EA (1977b) gives a detailed description of methodology.

The objective of the phytoplankton sampling scheme was to provide a description of the community and to detail temporal variation. The number of stations varied from two in 1971 to eight in 1975–1977, and the sampling interval varied from one week to one month. Net phytoplankton samples were collected from 1971–1973, and whole water samples were collected beginning in 1974 (Table 7.9). Data reporting included a species inventory by year, abundance information on the dominant species, and seasonal succession analysis by major algal group (by abundance and biovolume; Table 7.10). LMS and EA (1977b) contains diversity, cluster, and niche breadth analyses for 1976 data and primary productivity estimates for 1975 and 1976.

Microzooplankton (Table 7.11) and macrozooplankton (Table 7.12) samples were collected once or twice monthly from 1971–1978, although the 1978 data were not analyzed (Baslow and Logan, 1982). The number of stations sampled varied from two to eight, and microzooplankton collections were made at the surface only (1972) or at surface, middepth, and bottom (1973–1978). Data were analyzed and presented in the same manner as for phytoplankton.

Benthic sampling occurred in the Roseton-Danskammer area from 1971–1978, with a variable sampling scheme (Table 7.13). Collection generally occurred from spring through fall, sampling interval varied from one week to one month, and station number varied from three to eight.

Species lists, abundance, and biomass were compiled for each year, although identification prior to 1975 was taken only to major taxonomic group. Some community analysis techniques were applied to the 1975 and 1976 data.

A supplementary analysis of only the 1973 near-field data was conducted by Lane and Wright as consultants to LMS (LMS, 1979g). Niche analysis theory was used to calculate measures of community structure and community stability of phytoplankton and rotifers.

F. Roseton-Danskammer Habitat Selection Studies. Thermal plume areas at the Roseton power plant were sampled by LMS from 1973–1977 (LMS, 1978a, 1978b, 1978j, 1979b, 1979c, 1979d, 1979e). Non-fisheries data sets collected included plume and non-plume benthic, macrophyte, and periphyton communities.

Benthic invertebrates were sampled with artificial substrates left in the river for colonization. Sampling occurred from April through December at six stations in 1975, 1976, and 1977. Results were reported by species lists, and abundance and biomass by major taxon. In one of the few utility studies of Hudson River macrophytes, plants were sampled once each year in August from 1973 through 1977 at nine stations in the Roseton area. A species list was prepared and biomass data were collected for each species. Periphyton were quantified on artificial substrates and on sampled macrophytes. Artificial substrates (glass slides) were suspended in the river at a variable number of stations at two week intervals (non-winter months from 1974 through 1978). Species lists were compiled and standing crop, chlorophyll *a* concentration, and ash-free dry weight were determined.

An earlier study of the Roseton and Danskammer benthos was conducted by QLM (1969f) as part of a preconstruction survey of the Roseton site. Dredge samples were taken weekly or biweekly from May to October 1969.

G. Kingston-Ossining Aquatic Ecology Studies. LMS (1975b, 1975e) conducted baseline ecological studies at the site of a proposed (but never constructed) Central Hudson Gas and Electric Corp. power plant at Kingston (RM 95). Preliminary plankton studies were carried out from 1970–1972. In 1973, whole water and net samples of phytoplankton and microzooplankton were collected monthly from June to December. An inventory of species present and abundance by major taxa were compiled. Macrozooplankton samples were collected from March through August at 12 stations.

Benthic sampling was conducted at Kingston during 1971, 1972, and 1973. Ponar grab samples were taken monthly during non-winter months. Abundance of major taxonomic groups and affinity between samples were computed, and seasonal variation in abundance was examined.

Table 7.8. Benthos sampling methods at Bowline Point and Lovett Generating Stations, 1969–1977.

Yr	Transects/stations sampled	Sampling dates	Collection gear	Collection depth(s)
1969	4 stations in projected discharge area	Aug. (22, 29) Sept. (5, 12, 19, 26) Oct. (3, 10)	6 × 6 in Ekman grab	20–28 ft
1970	4 stations in projected discharge area	June (2, 9, 12, 16, 23, 30) July (7, 14, 21) Aug. (7, 14, 25) Sept. (7, 14, 21) Oct. (1, 13, 29)	6 × 6 in Ekman grab	20–28 ft
Preliminary 1971	3 river stations 5 pond stations	Apr. (27) May (10, 17, 25) June (10, 17) July (8)	6 × 6 in Ekman grab	Variable
1971–1972	7 river stations 4 pond stations	1971 Dec. (3) 1972 May (12) Oct. (16) Aug. (12) Dec. (13)	6 × 6 in Ponar grab	Variable

Year	Stations	Dates	Sampler	Depth
1973	8 river transects 4 pond stations	June (8) Sept. (9) Dec. (7)	6 × 6 in Ponar grab	River 5–20 ft Pond 10–45 ft
1974	8 river transects 4 pond stations	Monthly Apr.–Dec.	6 × 6 in Ponar grab	Same as 1973
1975	8 river transects 5 pond stations	Monthly Mar.–Dec.	6 × 6 in Ponar grab	Same as 1973
1976	14 river transects 1 pond station	Apr. (26) July (19) Sept. (21) Dec. (16)	6 × 6 in Ponar grab	Same as 1973
1977	4 river transects	Apr. (18) July (25) Oct. (26) Dec. (20)	6 × 6 in Ponar grab	20 ft

From Lawler, Matusky, and Skelly and Ecological Analysts, Inc. (1977a).

Table 7.9. Phytoplankton sampling methods at Roseton and Danskammer Generating Stations, 1971–1978.

Yr	No. of stations sampled	Sampling frequency	Collection gear	Sampling method	Collection depth
1971	2	Once/mo Aug., Nov.-Dec. Twice/mo Oct.	Wisconsin plankton net 18-μm mesh	Net	Surface
1972	4	Once/mo May-June, Sept.-Dec. Twice/mo July-Aug.	Wisconsin plankton net 18-μm mesh	Net	Surface
1973	6	Once/mo May-Dec.	Van Dorn bottle Wisconsin plankton net	Whole water Nanoplankton (filtrate); net	Surface
1974	5 1	Twice/mo Jan., Apr.-Dec., Mar. Weekly, Jan., Mar.-Dec.	Van Dorn bottle	Whole water	Surface
1975	8	Twice/mo Jan., Apr.-Dec. Once/mo Feb.-Mar.	300-ml sampling bottle	Whole water	Surface
1976	8	Twice/mo Jan. Once/mo Feb. Every 2 wk Mar.-Nov.	300-ml sampling bottle	Whole water	Surface
1977	8	Every 2 wk Mar.-Dec.	300-ml sampling bottle	Whole water	Surface
1978	3	Twice/mo Apr.-Dec.	300-ml sampling bottle	Whole water	Surface

From Lawler, Matusky, and Skelly and Ecological Analysts, Inc. (1977b); Lawler, Matusky, and Skelly (1978b, 1979b, 1979c, 1980f).

Table 7.10. Laboratory analysis of Roseton and Danskammer Generating Stations phytoplankton samples, 1971–1977.

Yr	Samples analyzed	Biovolume	Chlorophyll	Phaeophytin
1971	All collections analyzed	—	—	—
1972	All collections analyzed	—	—	—
1973	All collections analyzed, duplicate analyses on one sample	—	—	—
1974	Once/mo or once/wk	Once/mo	—	—
1975	All collections analyzed, no replicates		Twice/mo two replicates	Twice/mo two replicates
1976	All collections analyzed, no replicates	Twice/mo[a]	Twice/mo two replicates	Twice/mo two replicates
1977	All collections analyzed, no replicates	Twice/mo[a]	Twice/mo two replicates	Twice/mo two replicates
1978	All collections analyzed, no replicates	Twice/mo[a]	Twice/mo two replicates	Twice/mo two replicates

From Lawler, Matusky, and Skelly and Ecological Analysts, Inc. (1977b).
[a] Literature values used if not enough cells available for measurement.

Table 7.11. Microzooplankton sampling methods at Roseton and Danskammer Generating Stations, 1971–1977.

Yr	No. of stations sampled	Sampling frequency	Collection gear	Collection depths	Time
1971	2	Once/mo Aug., Nov.-Dec. Twice/mo Oct.	12.4-cm mouth diameter Wisconsin net 76-μm mesh vertical tows	Bottom-surface	Day
1972	4	Once/mo May (two stations only), June, Sept., Oct., Dec. (all four stations) Twice/mo July, Aug. Once/mo Apr.-June	12.4-cm mouth diameter Wisconsin net 76-μm mesh vertical tows	9.1-m surface	Day
1973	7	July 17, 18	12.4-cm mouth diameter Wisconsin net 76-μm mesh vertical tows 12.7-cm mouth diameter Clarke-Bumpus 150-μm mesh horizontal tows	Surface, mid-depth, bottom	Day/night
1974		Once/mo Aug.-Dec.	12.7-cm mouth diameter Clarke-Bumpus 76-μm mesh horizontal tows	Surface, mid-depth, bottom	Day/night
1975	8	Once/mo Feb., Mar., Apr. Twice/mo May-Dec. Twice/mo Jan. (two stations only)	Apr.-Dec. Centrifugal pump 76-μm mesh Mar. (only) 12.4-cm mouth diameter Wisconsin net 76-μm mesh vertical tow (day only)	Surface, mid-depth, bottom	Day/night
1976	8	Twice/mo Mar.-May, July-Oct. Thrice/mo June, Nov.	Centrifugal pump 76-μm mesh	Surface, mid-depth, bottom	Day/night
1977	8	Twice/mo Mar.-Dec.	Centrifugal pump 76-μm mesh	Surface, mid-depth, bottom	Day/night
1978	3	Twice/mo Apr.-Dec.	Centrifugal pump 76-μm mesh	Surface, mid-depth, bottom	Day/night

From Lawler, Matusky, and Skelly and Ecological Analysts, Inc. (1977b); Lawler, Matusky, and Skelly (1980f).

Table 7.12. Macrozooplankton sampling methods at Roseton and Danskammer Generating Stations, 1973–1977.

Yr	No. of stations sampled	Sampling frequency	Collection gear	Time
1973	1–6	Twice/mo Mar.–Aug.	571-μm plankton net	Day/night
1974	4	Once/mo Apr.–Dec.	571-μm plankton net, epibenthic sled	Day/night (Apr.–Aug.) Day (Sept.–Dec.)
1975	1–4	Twice/mo Feb.–Aug. Once/mo Jan., Sept.–Dec.	571-μm plankton net, epibenthic sled	Day/night
1976	1–4	Once/mo Jan.–Dec.	571-μm plankton net, epibenthic sled	Day/night
1977	3	Twice/mo Apr., July, Aug. Once/wk May, June	571-μm plankton net, epibenthic sled	Day/night
1978	3	Twice/mo Mar., Apr., July, Aug. Once/wk May, June Once/mo Sept.–Dec.	571-μm plankton net, epibenthic sled	Day/night

From Lawler, Matusky, and Skelly and Ecological Analysts, Inc. (1977b); Lawler, Matusky, and Skelly (1979c, 1980f).

Table 7.13. Benthic sampling methods at Roseton and Danskammer Generating Stations, 1971–1977.

Yr	No. of transects sampled	Sampling frequency	Collection gear	Collection depths
Preliminary 1971	3	May (7, 12, 19, 26) June (2, 8, 22) July (13)	6 × 6 in Ekman grab	Three depths per transect (variable)
1971–1972	4	Once/mo Oct., Dec. 1971 June, Aug., Oct., Dec. 1972	6 × 6 in Ponar grab	10, 20, 30 ft at each transect
1973	4	Once/mo June-Dec.	6 × 6 in Ponar grab	10, 20, 30 ft at each transect
1974	8	Once/mo Mar.-Dec.	6 × 6 in Ponar grab	10, 20, 30 ft at four transects 20 ft at four transects
1975	8	Alternate monthly Apr.-Dec. Once/mo Mar.	6 × 6 in Ponar grab	10, 20, 30 ft at four transects 20 ft at four transects
1976	8	Alternate monthly Apr.-Dec. Once/mo Mar.	6 × 6 in Ponar grab	10, 20, 30 ft at four transects 20 ft at four transects
1977	8	Alternate monthly Apr.-Dec.	6 × 6 in Ponar grab	10, 20, 30 ft at four transects 20 ft at four transects
1978	3	Alternate monthly Apr.-Dec.	6 × 6 in Ponar grab	3, 6, 9 ft

From Lawler, Matusky, and Skelly and Ecological Analysts, Inc. (1977b); Lawler, Matusky, and Skelly (1978b, 1979b, 1980f).

H. River-Long Surveys. Seasonal longitudinal plankton surveys were undertaken by researchers at New York University Medical Center in 1973 (New York University Medical Center, 1976a). Microzooplankton and phytoplankton were surveyed at 16 river stations located at approximately 10-mile intervals from the Battery to just above the Troy Dam. By 1975, macrozooplankton sampling had been added to the research program. (Information on sampling after 1975 and reporting and analysis of data were not found in any New York University Medical Center reports.)

Water Quality Data Sets

Water quality data were collected concurrently with the biological sampling program. Water temperature, dissolved oxygen content, salinity, and pH were measured each time a plankton or fish sample was collected. Generally, water quality information was published with the corresponding biological data in the numerous technical reports prepared by TI, EA, LMS, and New York University Medical Center. Water sampling during the far-field biological sampling (conducted mainly by TI) may be of special interest, since the entire estuary was repeatedly and sequentially sampled for dissolved oxygen (DO), salinity (conductivity), temperature, and pH. These data were published in "water quality data displays" (TI, 1977v, 1977w, 1977x, 1977y, 1977z, 1978k, 1979l, 1980g).

Parameters other than temperature, DO, pH, and salinity were occasionally sampled, although collection was generally limited to power plant vicinities. At Indian Point, nutrient (nitrite, nitrate, ortho- and total phosphate), chlorides, hardness, and trace metal data were collected monthly in 1971 and 1972 as part of the near-field studies conducted by New York University Medical Center (New York University Medical Center, 1973). An atlas of chemical and physical parameters for the Indian Point area, with monthly means, ranges, and diel variations in temperature, DO, pH, salinity, and organic carbon, was published by TI in 1973 (TI, 1973a).

In the Bowline Point and Lovett vicinity, extensive water chemistry sampling occurred as part of the near-field studies conducted by LMS from 1971 through 1980 (LMS, 1974a, 1975f, 1976g, 1977a, 1978g, 1979d, 1981d, 1981e; LMS and EA, 1977a). Parameters sampled varied through the years; 1973 and 1974 had the most complete program (Table 7.14). Water temperature, DO, pH, total solids, volatile solids, suspended solids, nitrate, orthophosphate, and silica were collected throughout the 10-year sampling period. Sampling occurred at only one station; generally, samples were taken once each month, although data collection during 1971–1973 was fairly irregular and concentrated in the non-winter months.

At Roseton and Danskammer, water chemistry sampling also occurred as part of the near-field studies program (LMS, 1975g, 1975h, 1978a,

Table 7.14. Water quality sampling at Bowline Point and Lovett Generating Stations, 1971–1980.

Parameter	Yr(s) measured	Method[a]
Temperature	1971–1980	ASTM
Dissolved oxygen (DO)	1971–1980	EPA
Chlorides	1971–1976	SM
pH	1971–1980	SM
Alkalinity	1971–1976	SM
Sulfate (SO_4)	1971–1976	SM
Biochemical oxygen demand (BOD)	1971–1976	SM
Chemical oxygen demand (COD)	1971–1976	SM
Total solids (TS)	1971–1980	SM-EPA
Total dissolved solids (TDS)	1971–1980	EPA
Total volatile solids (TVS)	1971–1980	SM-EPA
Total suspended solids (TSS)	1971–1976	EPA
Organic nitrogen	1971–1976	EPA
Ammonia nitrogen (NH_3-N)	1971–1977	EPA
Nitrite nitrogen (NO_2-N)	1976	EPA
Nitrate nitrogen (NO_3-N)	1971–1980	EPA
Total Kjeldahl nitrogen (TKN)	1971–1976	SM
Total phosphorus (TP)	1971–1977	SM
Orthophosphate phosphorus (OP)	1971–1980	EPA
Phenol	1971, 1973–1976	SM
Total coliform	1973–1976	SM
Fecal coliform	1973–1976	SM
Total organic carbon (TOC)	1973	EPA
Silica (converted to silicon)	1975–1980	SM
Specific conductance (at 25°C)	1971, 1973–1980	SM
Cadmium (Cd)	1972–1974	AAS
Calcium (Ca)	1972	AAS
Chromium (Cr)	1971–1975	AAS
Copper (Cu)	1972, 1976	AAS
Iron (Fe)	1971, 1973–1976	AAS
Lead (Pb)	1971, 1972, 1974–1976	AAS
Magnesium (Mg)	1972	AAS
Manganese (Mn)	1972–1974	AAS
Molybdenum (Mo)	1973–1974	AAS
Nickel (Ni)	1972–1974	AAS
Potassium (K)	1972	AAS
Silver (Ag)	1973–1974	AAS
Sodium (Na)	1972, 1974	AAS
Zinc (Zn)	1971–1976	AAS
Turbidity	1971, 1972, 1974–1977	EPA
Total hardness	1972	ASTM
Color	1972	SM

From Lawler, Matusky, and Skelly and Ecological Analysts, Inc. (1977a).
[a] AAS, Atomic absorption spectrophotometer; SM, Standard methods (American Public Health Association); EPA, U.S. Environmental Protection Agency; ASTM, American Society for Testing and Materials.

1978b, 1979b, 1979c, 1980f, 1980g, 1981a, 1982b; LMS and EA, 1977b). Data were collected from 1971–1981, with the most complete list of parameters being sampled during 1972 and 1974 (Table 7.15). Stations and station number varied through the years. Sampling was usually conducted on a monthly schedule, although irregular monthly and seasonal sampling occurred in the earlier years.

Water quality sampling at the site of the proposed Kingston plant (RM 95) occurred in 1973 (May to December). Monthly readings of alkalinity, biological oxygen demand (BOD), chemical oxygen demand (COD), solids (total, volatile, and suspended), NH_3, NO_3, organic nitrogen, total phosphorus, phenol, chloride, sulfate, and cadmium were taken (LMS, 1975b, 1975e).

Open Literature

Bibliography B

In-house reports, open literature publications, and government documents constitute a substantial resource for information on the Hudson River ecosystem. Bibliography B, following this chapter, lists publications from these sources. Physical, chemical, and biological aspects of the Hudson are covered in approximately 1000 entries.

The bibliography is arranged alphabetically within subject categories. Hudson River biological research (section I) is divided into phytoplankton, macrophytes, invertebrates, fish (by major species), other vertebrates, and parasites and diseases. Physical and chemical publications (section II) are organized into categories of geology, hydrology, temperature, salinity, water chemistry, and water quality. Hudson River pollution literature (sewage, PCBs, radionuclides, and other toxic materials) is covered in section III of the bibliography. Also catalogued are regulatory information, impact assessment statements, and studies and bibliographies that may provide further sources of Hudson-related topics (sections IV through VIII). Multiple listing of a single publication occurs when literature falls within more than one subject category.

A cross-section of the bibliographic entries has been reviewed in order to characterize the data base represented by the non-utilities literature. Tables 7.16 through 7.27 list the reports reviewed (organized by subject matter) and summarize the data presented within. The majority of the studies reviewed were found to be limited in scope to single sites on the river. This contrasts with the availability of river-long surveys from the utility-sponsored research efforts. Open literature publications also generally involved data collection for one year or less.

Table 7.15. Water quality sampling at Roseton and Danskammer Generating Stations, 1971–1981.

Parameter	Yr(s) measured	Method[a]
Temperature	1971–1981	ASTM
Dissolved oxygen (DO)	1972, 1974–1981	EPA
Chlorides	1971–1978	SM
pH	1971–1981	SM
Alkalinity	1971–1978	SM
Sulfate (SO_4)	1971–1978	SM
Biochemical oxygen demand (BOD)	1971–1979	SM
Chemical oxygen demand (COD)	1971–1979	SM
Total solids (TS)	1971–1978	SM-EPA
Total dissolved solids (TDS)	1971–1978	EPA
Total volatile solids (TVS)	1971–1978	SM-EPA
Total suspended solids (TSS)	1971–1978	EPA
Organic nitrogen	1972, 1974–1976	EPA
Ammonia nitrogen (NH_3-N)	1971–1979	EPA
Nitrate nitrogen (NO_3-N)	1971–1979	EPA
Total Kjeldahl nitrogen (TKN)	1971–1979	SM
Total phosphorus (TP)	1971–1979	SM
Orthophosphate phosphorus (OP)	1971, 1972, 1975–1978	EPA
Phenol	1971–1979	SM
Total coliform	1974	SM
Fecal coliform	1974	SM
Total organic carbon (TOC)	1971, 1974	EPA
Specific conductance (at 25°C)	1972, 1974–1981	SM
Cadmium (Cd)	1972	AAS
Calcium (Ca)	1972	AAS
Chromium (Cr)	1971–1978	AAS
Copper (Cu)	1972	AAS
Iron (Fe)	1972	AAS
Lead (Pb)	1972	AAS
Magnesium (Mg)	1972	AAS
Manganese (Mn)	1972	AAS
Nickel (Ni)	1972	AAS
Potassium (K)	1972	AAS
Sodium (Na)	1972	AAS
Zinc (Zn)	1971–1978	AAS
Turbidity	1972, 1974–1981	EPA
Total hardness	1972	ASTM
Color	1972	SM

From Lawler, Matusky, and Skelly and Ecological Analysts, Inc. (1977b); Lawler, Matusky, and Skelly (1979b, 1979c, 1980f, 1980g, 1981a, 1982b).

[a] AAS, Atomic absorption spectrophotometer; SM, Standard methods (American Public Health Association); EPA, U.S. Environmental Protection Agency; ASTM, American Society for Testing and Materials.

Open Literature

Highlights of Hudson River Literature

The publications detailed in this review of Hudson River open literature were selected from the open literature bibliography (Bibliography B). They are of particular interest for natural history information, ecosystem-level approaches, or historic comparisons.

Phytoplankton (Table 7.16) at various locations in the Hudson River were characterized by Heffner (1973), Howells and Weaver (1969), and Storm and Heffner (1976). A species list with distribution information covering the entire estuary was compiled by Fredrick et al. (1976). The most complete description of natural history and distribution of Hudson River phytoplankton has been compiled in a biological atlas by Boyce Thompson Institute (Weinstein, 1977). (The atlas also covers macrophyte, invertebrate, and fish species, with information on longitudinal and seasonal distribution.) Malone and others (Malone, 1975; Malone, 1977; Malone and Chervin, 1979; Malone et al., 1979; Neale et al., 1981) have investigated phytoplankton biology from a system-level perspective (productivity, abiotic influences). Most of this work, however, has been confined to the lower estuary and New York Bight.

Macrophytes (Table 7.16) along the shores and shallows of the Hudson were described several decades ago by Muenscher (1937) and Foley (1948). These older surveys list plant species and provide some information on community structure. Buckley and Ristich (1976) compared the early studies with their more recent macrophyte survey. Macrophyte research for the Hudson River, however, is far from extensive; there has not been sufficient quantification of present-day marshland and tidal flat areas. Historic information on the extent of Hudson wetlands, current descriptions of macrophyte communities, and the status of nuisance plants also needs research attention.

Invertebrate communities (Table 7.17) and their distribution along physical gradients in the Hudson have been described by Ristich et al. (1977). Rathjen and Miller (1957) studied ichthyoplankton life history and Dovel (1981) reported on composition, distribution, and abundance of ichthyoplankton. Ichthyoplankton (striped bass only) and young-of-the-year of Hudson River fish were studied to evaluate the fisheries effects of the proposed pumped-storage project at Cornwall (Carlson and McCann, 1969). The data base in this report is good, but errors are evident in calculations of Hudson flow rate. A computer simulation study of the distribution of Hudson River ichthyoplankton was undertaken by Englert et al. (1974).

Hudson River fisheries literature is reviewed in Tables 7.18 through 7.22. Historical data on species composition of the Hudson River fish

---

Tables 7.16–7.27 appear following references.

fauna is found in Mearns (1898), Breeder (1948), Hudson River Valley Commission (1966), Moore (1936), Curran and Ries (1936), Greeley and Bishop (1932), and Greeley (1935, 1936). Smith (1976) describes the present fish fauna. Reports on the occurrence of exotics and subtropical fish species in the Hudson include Alevras (1973), Dew (1974), Meyers (1930), Tabery *et al.* (1978), Reider (1979), and Young *et al.* (1982).

The first comprehensive fish survey of the Hudson River watershed was done by the New York State Department of Environmental Conservation (DEC). The upper Hudson was surveyed in 1932, the Mohawk-Hudson watershed in 1934, and the lower watershed in 1936. Information on the scope of this study with particular reference to the lower Hudson watershed is presented in Moore (1936).

Perlmutter *et al.* (1966, 1967, 1969) studied the distribution and abundance of Hudson shore zone fish during summer months. Although the data are limited in extent, they include useful species lists, rank order lists, and length-frequency distributions. Heller and Hermo (1969) studied diurnal abundance of shore zone fish. Diurnal distribution patterns of impinged fish at Indian Point were studied by Muessig (1976). Esser (1982) reviewed the long-term changes in fish abundance and distribution in the Hudson-Raritan Estuary covering the period 1900–1977.

Documentation of the historical commercial fisheries on the Hudson River may be found in Goode (1887) and McDonald (1887). Additional comment on the historical fisheries is given in Moore (1936), Curran and Ries (1936), Hudson River Valley Commission (1966), and Squires (1981). McHugh and Ginter (1978) described the different phases of the New York Bight commercial fisheries, including general but valuable comment on Hudson-related species.

Greeley (1936) was the first to study the biology and growth of some important Hudson River fish. Townes (1936) collected data on the feeding behavior of Hudson species, and The Oceanic Society (1980) conducted a detailed Hudson River fishery management study.

Many of the papers referred to in the above sections were published as proceedings from a series of symposia on Hudson River ecology. The Hudson River Environmental Society organized four symposia that took place between 1966 and 1976. The proceedings are valuable sources of information not just on biological aspects of the river, but on physical, and chemical aspects as well (Hudson River Environmental Society, 1966, 1969, 1974, 1976). Additional sources of information on physical and chemical parameters are summarized in Tables 7.23 (geology and hydrology) and 7.24 (temperature and salinity).

Although water quality studies (Table 7.25) have largely concentrated on single Hudson River sites, there have been some river-long water sampling efforts. The Hudson River Research Council (1980) conducted two coordinated sampling surveys in April 1977 and August 1978. Salinity, water temperature, and DO were measured simultaneously from the

Battery (RM 0) to Troy (RM 152) at several hundred points along the estuary. The DEC (1947–1978) has conducted unpublished water quality monitoring at 21 sites, beginning as early as 1947 at some locations. The U.S. Geological Survey is also maintaining long-term sampling stations along the Hudson and on several of its tributaries (USGS, 1946–present).

Nutrient budgets for the Hudson have been estimated by Simpson *et al.* (1975) (phosphorus, silica) and Deck (1981) (phosphorus, nitrogen, silica). The influence of sewage inputs to nutrient levels in the lower Hudson is covered in Garside *et al.* (1976).

Recent concern over pollutants in the Hudson River is responsible for a growing literature covering the sources, physical transport, and biological fate of recognized pollutants. Table 7.26 summarizes information sources on toxics, radionuclides, and metals; Table 7.27 is a selective description of PCB literature.

A coordinated ecosystem approach to New York Bight research has been undertaken by the National Oceanographic and Atmospheric Administration (NOAA) under the Marine Ecosystem Analysis (MESA) project at the State University of New York at Stony Brook. Research includes programs in physical oceanography, geology, water chemistry, biology, and toxicology. Although mostly centered south of the Hudson estuary, some of this work provides insight into the functioning of the lower Hudson (Mayer, 1982). The Marine Sciences Research Center, also located at the State University of New York at Stony Brook, is in the process of compiling a bibliography of the New York Bight and surrounding areas (New York Harbor, Raritan Bay, and Long Island Sound). The lower Hudson River will also be included in this annotated, computerized bibliography.

Tables 7.16–7.27 follow the references.

References[1]

Klauda, R.J., P.H. Muessig, and J.A. Matousek. 1985. Fisheries data sets compiled by utility-sponsored research in the Hudson River Estuary. Proceedings of a Fisheries Workshop. Hudson River Environ. Soc., New Paltz, NY (in press).

Baslow, M.H. and D.T. Logan. 1982. The Hudson River ecosystem. A case study. Report to the Ecosystems Research Center. Cornell University, Ithaca, NY. 460 pp.

---

[1]See Bibliographies A and B for other scientific reports and publications referenced in this chapter.

Table 7.16. Hudson River open literature highlights: Algae and macrophytes.

Reference	Description	Yr(s) of study	Sampling intensity	Sample location[a]
				0 25 50 75 100 150 BT YK TZ CH IP WP CW PK HP KG SG CS AL
Neale et al. (1981)	Effects of freshwater flow on phytoplankton and salinity	1977–1978	Irreg. weekly, Feb.–Sept.	Lower New York Bay
Howells and Weaver (1969)	Phytoplankton communities at Indian Point and further north	1968–1969	Monthly, Apr. '68–Aug. '69	XX (IP) XX XX (HP, KG)
Storm and Heffner (1976)	Phytoplankton abundance, chlorophyll a, water quality	1973–1974	Seasonal	XXXXXXXXXXXXXXXXXXXXXXXXXXXXXX (all)
Heffner et al. (1976)	Phytoplankton sampling methodologies encouraged for H.R.	General background	No data	
O'Reilly et al. (1976)	Annual primary productivity	1973–1974	Monthly	Lower New York Bay
Fredrick et al. (1976)	Phytoplankton distribution and species list	1972–1976	Summary of 5 studies	XXXXXXXXXXXXXXXXXXXXXXXXXXXXXX (all)
Williams and Scott (1962)	Diatoms of major U.S. waterways	1960–1961	Biweekly	XX (PK)
Malone (1977)	Environmental controls of phytoplankton productivity	1973–1974	Monthly	XXX (BT) and south
Weiss (1978)	Distribution of diatoms in H.R. sediments	1973–1975	Monthly	XXXXXXXXXXXXXXXXXXXXXXXXXXXX and New York Bight
USGS (1976–1981)	Phytoplankton species lists at Green Island	1975–1980		X (AL)
Kiviat (1976)	Distribution of threatened plant, Golden-club			
Buckley and Ristich (1976)	Distribution of rooted vegetation, comparison to earlier macrophyte surveys	1968–1974		XXXXXXXXXXXX
McVaugh (1936)	Partial list of aquatic plants	1936		XX

[a] Numbers indicate river miles from the Battery in Manhattan. BT = Battery, YK = Yonkers, TZ = Tappan Zee, CH = Croton-Harmon, IP = Indian Pt., WP = West Point, CW = Cornwall, PK = Poughkeepsie, HP = Hyde Park, KG = Kingston, SG = Saugerties, CS = Coxsackie, AL = Albany.

Upper Hudson	Concurrent phys./chem.	Methods given	Species	Abundance	Diurnal	Seasonal	Spatial	Growth/ prim. prod.	Feeding
	Flow, salinity, depth	X	Phytopl.					Prod.	
	N, PO$_4$, temp., elements	X	Phytopl.	X		X	X		
X	Detailed	X	Phytopl.	X		X	X	X	
			Phytopl.						
	Detailed	X	Nannopl. netpl.	X		X	X	X	
			Algae				X		
		X	Diatoms	X			X		
		X	Phytopl.					X	
	Temp., D.O., salinity		Diatoms				X		
			Phytopl.	X					
		X	*Orontium aquaticum*	X			X		
	Salinity	X	Macrophytes				X		
			Aquatic and upland plants	X			X		

Table 7.16. *Continued*

Reference	Reproduction				Exploitation		Survival[b]			
	Maturity	Fecundity	Behavior	Time	Sport	Comm.	Pred./par.	Disease	Ent./imp.	Toxicity
Neale et al. (1981)										
Howells and Weaver (1969)									X	X
Storm and Heffner (1976)										X
Heffner et al. (1976)										
O'Reilly et al. (1976)										
Fredrick et al. (1976)										
Williams and Scott (1962)										
Malone (1977)										
Weiss (1978)										
USGS (1976–1981)										
Kiviat (1976)										
Buckley and Ristich (1976)										
McVaugh (1936)										

[b]Pred./par. = Survival in presence of predator(s) or parasite(s), Ent./imp. = Entrainment and impingement.

Table 7.17. Hudson River open literature highlights: Invertebrates.

Reference	Description	Yr(s) of study	Sampling intensity	Sample location[a]												
				0		25		50		75		100				150
				BT	YK	TZ	CH	IP	WP	CW	PK	HP	KG	SG	CS	AL
Hirshfield et al. (1966)	Plankton tows, comparison to 1936 survey	1964	Weekly, July, Aug.	X	X	X	X	X	X	X						
Howells et al. (1969)	Inventory and distribution of invertebrates	1965–1969	'65–'68, Summer, '69 all year	X	X	X	X	X	X	X	X	X	X			
Pearce (1974)	Pollution effects on invertebrates		Review report	X	and New York Bay, Raritan Bay											
Occhiogrosso et al. (1976)	Metal (cadmium) effects on invertebrate densities	1973–1974	Biweekly (Apr.-Oct.), mo (Nov.-Dec., Mar.)					X	X							
Menzie et al. (1976)	Chironomid identification and ecology	1975	Seasonal					X	X							
Simpson (1976)	Water quality inferences from invertebrate communities	1972–1973	Summer									X	X	X	X	X
Ristich et al. (1977)	Spatial diversity of invertebrates	1972	May-Oct., 4 samples	X	X	X	X	X	X	X						
Burbanck (1962)	Range, habitat, preference of H.R. *Cyathura* biotype	1960	One sample, Oct.									X (Beacon)				
Weiss et al. (1978)	Bivalves and Formanifera in sediments	1973–1975	Monthly	X	X	X	X	X	X	X	X	X	X			
				and New York Bight												
Williams et al. (1975)	Benthos of Ossining area, comparison to earlier studies	1972–1973	Monthly, non-winter					X	X							
Crandall (1977)	Epibenthic invertebrates of Croton Bay	1974	Monthly, Apr.-Dec.					X	X							
Zubarik (1976)	Entrainment of zooplankton	1972–1975	Monthly					X (IP)								
Ginn et al. (1976)	Thermal effects on *Gammarus* survival and reproduction							X (IP)								

[a] See Footnote a in Table 7.16 for explanation.

Table 7.17. *Continued*

Reference	Upper Hudson	Concurrent phys./chem.	Methods given	Species	Abundance	Distribution[c] Diurnal	Seasonal	Spatial	Growth/ prim. prod.
Hirschfield et al. (1966)		Salinity	X	Inverts fish	X			X	
Howels et al. (1969)		X	X	Benthic, pelagic Inverts	X		X	X	
Pearce (1974)				Benthic inverts				X	
Occhiogrosso		X	X	Inverts	X			X	
Menzie et al. (1976)			X	Chironomids	X		X	X	
Simpson (1976)	X		X	Macroinverts	X			X	
Ristich et al. (1977)			X	Macroinverts	X			X	
Burbanck (1962)				*Cyathura polita*	X			X	
Weiss et al. (1978)		Temp., D.O. salinity		Formanifera bivalves				X	
Williams et al. (1975)		Sediment, redox	X	Benthos	X		X		
Crandall (1977)		Salinity, temp.	X	Epibenthos	X				
Zubarik (1976)		X	X	Microzooplankton	X				
Ginn et al. (1976)				*Gammarus*					

[c] Abbreviations for life-stages: E = egg, L = larvae, J = juvenile, A = adult, S = spawner.

| | | Reproduction | | | Exploitation | | Survival[b] | | | |
| | | | | | | | Pred./ | | Ent./ | |
Feeding	Maturity	Fecundity	Behavior	Time	Sport	Comm.	par.	Disease	imp.	Toxicity
Fish stomachs										
						X			X	
										Cd
X	X		X							
				X						
									X	
		X	X						X	

[b]See Footnote b in Table 7.16 for explanation.

Table 7.18. Hudson River open literature highlights: Striped bass.

Reference	Description	Yr(s) of study	Sampling intensity	Sample location[a] (0: BT YK TZ CH IP WP CW PK HP KG SG CS AL :150)
Klauda et al. (1980)	Juvenile abundance speculation	1972–1980	Review report	H.R. compared to other rivers
Raney and DeSylva (1953)	Striped bass races, H.R. vs. Chesapeake	1950–1952		XXXXXXXXXXXXXXXXXX (TZ–AL)
Rathjen and Miller (1957)	Life history of H.R. striped bass	1955	Irreg. daily, May–June	XXXXXXXXXXXXXXXXX (TZ–AL)
McCann and Carlson (1969)	Effects of Cornwall Plant on striped bass	1965–1968	Semi-weekly, day/night, Apr.–Nov.	XXXXXXXXXXXXXXXXXXXXXXXX (TZ–AL)
Clark (1968)	Seasonal movements of striped bass	1959–1963	Seasonal	
Hickey et al. (1977)	Skeletal abnormalities in striped bass	1972–1973	Nov. '72	XXXXXXXXXXXXXXXXXXXXX (YK–AL)
Rachlin et al. (1978)	Karyotypic analysis			XX (CW)
Hogan and Williams (1976)	Gill parasite on H.R. striped bass	1972–1973		XXXXXXXX (YK–IP)
Wallace (1975)	Analysis of striped bass models	1970–1974		
Koo (1970)	Background of landing statistics	1930–1966	Yearly	Atlantic coast
Whipple (1982)	Striped bass physiological ecology, comparison of H.R. and San Francisco	1982	One sample date	XX (WP)
VanWinkle et al. (1979)	Periodicities in striped bass fisheries data	1930–1974	Yearly	Atlantic region
Dahlberg (1979)	Survival of striped bass eggs and larvae	Review report		Hudson discussed
Merriman (1941)	Atlantic coast striped bass studies	1936–1941		XXXXXXXXXXXXXXXXXXXXXXXXXXX (YK–AL)
Shaefer (1968)	Feeding habits of striped bass	1964	Biweekly, Apr.–Nov.	New York Bight
Austin and Custer (1977)	Striped bass migration in Long Island Sound	1966–1972		

[a] See Footnote *a* in Table 7.16 for explanation.

Upper Hudson	Concurrent phys./chem.	Methods given	Species	Abundance	Distribution			Growth/ prim. prod.	Feeding
					Diurnal	Seasonal	Spatial		
		X	Striped bass	J,A				X	
		X	Striped bass			A	A		
	Temp. salinity	X	Striped bass	E,L,J	X	E,L,J, A,S	E,L,J, A,S	L,J	
	Salinity	X	Striped bass, others	E,L,J	E,L,J	E,L,J	E,L,J	X	
		X	Striped bass			X	X		
			Striped bass						
			Striped bass, white perch, tomcod					X	
			Striped bass				X		
		X	Striped bass	A		A	A		
			Striped bass						
		X	Striped bass	A			A		
		X	Striped bass, others		E				
	Salinity	X	Striped bass	J,A			J,A,S	J	J
		X	Striped bass						J,A
			Striped bass			X	X		

Table 7.18 *Continued*

Reference	Reproduction				Exploitation		Survival[b]			
	Maturity	Fecundity	Behavior	Time	Sport	Comm.	Pred./par.	Disease	Ent./imp.	Toxicity
Klauda et al. (1980)	X	X			X	X	X		X	X
Raney and DeSylva (1953)										
Rathjen and Miller (1957)				X						
McCann and Carlson (1969)				X					X	
Clark (1968)										
Hickey et al. (1977)								X		
Rachlin et al. (1978)										
Hogan and Williams (1976)							X			
Wallace (1975)									X	
Koo (1970)						X				
Whipple (1982)							X			X
VanWinkle et al. (1979)						X				
Dahlberg (1979)							X			
Merriman (1941)				X						
Shaefer (1968)										
Austin and Custer (1977)										

[b]See Footnote *b* in Table 7.16 for explanation.

Table 7.19. Hudson River open literature highlights: Tomcod.

Reference	Description	Yr(s) of study	Sampling intensity	Sample location 0 25 50 75 100 150 BT YK TZ CH IP WP CW PK HP KG SG CS AL	Upper Hudson	Concurrent phys./chem.	Methods given	Species	Abundance
Nittel (1976)	Feeding habits in the H.R.	1973–1974	Dec., May–Oct. (monthly)	XXXXXXXXXXXXX			X	Tomcod	
Dew and Hecht (1976)	Production and population dynamics of tomcod	1973–1975	Spring months	XXXXXXXXXXXXX		Cond.	X	Tomcod	L
Smith et al. (1979)	Tumors in H.R. tomcod	1977–1978	Monthly, Dec.–Feb.	XXXXXXXXXXXXXXXXXX			X	Tomcod	
Grabe (1978)	Feeding habits of tomcod	1973–1975	June–Dec., 1/mo (A), 2/mo (J)	XX		Temp., D.O., salinity	X	Tomcod	

Table 7.19. *Continued*

Reference	Distribution			Growth/ prim. prod.	Feeding	Reproduction				Exploitation		Survival			
	Diurnal	Seasonal	Spatial			Maturity	Fecundity	Behavior	Time	Sport	Comm.	Pred./ par.	Disease	Ent./ imp.	Toxicity
Nittel (1976)					J,A										
Dew and Hecht (1976)			L	X					X						
Smith et al. (1979)													X		X
Grabe (1978)					X										

Table 7.20. Hudson River open literature highlights: Shad.

Reference	Description	Yr(s) of study	Sampling intensity	Sample location[a]
				0 — 25 — 50 — 75 — 100 — 150 BT YK TZ CH IP WP CW PK HP KG SG CS AL
Leggett and Whitney (1972)	General review report on shad migration, H.R., Conn. River, Columbia River			
Talbot (1954)	Shad population estimate, effects of pollution and fishing	1950–1951		XXXXXXXXXXXXXXXXXXXXXXXXXXXXX
Nichols (1958)	Effects of N.Y. and N.J. pound net catches on shad	1956	Weekly	XXXXXXXXXXXXXXXXXXXXXXXXX
Talbot and Sykes (1958)	Atlantic coast shad migration, tagging study	1938–1956		Atlantic coast, including the Hudson
Walburg and Nichols (1967)	Biology and management of shad	1960		XXXXXXXXXXXXXXXXXXXXXXXXXXXXX
Walburg (1957)	Feeding and growth of juvenile shad	1954	Aug.	
Nichols (1966)	Comparison of meristic characteristics of H.R. shad to other rivers	1950, '51, '54, '57, '58		XX XX
Talbot (1956)	Discussion of shad decline on east coast	1915–1955		XXXXXXXXXXXXXXXXXXXXXXXXXXXXX
Fischler (1959)	Contributions of H.R. shad to Atlantic coast, meristic analyses	1956	Monthly, May, June	XXXX
Lehman (1953)	Fecundity of H.R. shad	1951	Apr.	
Davis (1957)	Fecundity of H.R. shad as compared to other rivers	1955	Apr.	
Medeiros (1975)	H.R. shad fishery management			
Burdick (1954)	Factors influencing shad abundance	1924–1951	Yearly	XXXXXXXXXXXXXXXXXXXXXXXXXXXXX
Sykes and Talbot (1958)	Shad life history and migrations	1938–1956		Atlantic coast, including the Hudson
Baskous and Lanahan (1973)	Shad prices			XXXXXXXXXXXXXXXXXXXXXXXXXXXXX

[a] See Footnote *a* in Table 7.16 for explanation.

Table 7.20. *Continued*

Reference	Upper Hudson	Concurrent phys./chem.	Methods given	Species	Abundance	Distribution[c]			Growth/ prim. prod.
						Diurnal	Seasonal	Spatial	
Leggett and Whitney (1972)		Temp.		Shad			X		
Talbot (1954)		D.O., B.O.D., salinity, colif.	X	Shad	X		S	S	
Nichols (1958)			X	Shad	X			X	
Talbot and Sykes (1958)			X	Shad			X	X	
Walburg and Nichols (1967)		X		Shad	X		X	E,L,A,S	X
Walburg (1957)			X	Shad					X
Nichols (1966)			X	Shad					X
Talbot (1956)			X	Shad	X				
Fischler (1959)				Shad				X	X
Lehman (1953)			X	Shad					X
Davis (1957)			X	Shad					X
Medeiros (1975)			X	Shad	X		X	X	
Burdick (1954)		Flow, O$_2$	X	Shad	X				X
Sykes and Talbot (1958)			X	Shad			X	X	
Baskous and Lanahan (1973)				Shad, striped bass					

[c]See Footnote *c* in Table 7.17 for explanation.

| Feeding | Reproduction | | | | Exploitation | | Survival[b] | | | |
	Maturity	Fecundity	Behavior	Time	Sport	Comm.	Pred./par.	Disease	Ent./imp.	Toxicity
	X			X						
	X	X		X		X				
						X				
X	X	X	X	X		X	X			
X										
		X								
		X								
	X	X	X	X		X				
	X					X				
		X		X		X				
					X	X				

[b]See Footnote *b* in Table 7.16 for explanation.

Table 7.21. Hudson River open literature highlights: Sturgeon.

Reference	Description	Yr(s) of study	Sampling intensity	Sample location[a]													Upper Hudson	Concurrent phys./chem.	Methods given	Species
				0	25		50		75		100			150						
				BT	YK	TZ	CH	IP	WP	CW	PK	HP	KG	SG	CS	AL				
Vladykov and Greeley (1963)	Taxonomy and natural history of sturgeon																			*Acipenser brevirostris, A. oxyrhynchus*
Dovel (1978)	Literature review for sturgeon	1975–1978																		Sturgeon
Hoff and Klauda (1979)	Sturgeon data collected by Texas Instruments, including impingement information	1972–1979	Daily (impingement), seasonal				X	X	X	X	X	X	X	X	X	X		Temp., D.O., cond., turbid.	X	*Acipenser brevirostris*
Kioski et al. (1971)	Record-size shortnose sturgeon caught	1970	1/29/70	X	X													Temp., D.O., salinity		*Acipenser brevirostris*
Crandall (1978)	General report on sturgeon studies																			Sturgeon

Table 7.21. *Continued*

Reference	Abundance	Distribution			Growth/ prim. prod.	Feeding	Reproduction				Exploitation		Survival[b]		
		Diurnal	Seasonal	Spatial			Maturity	Fecundity	Behavior	Time	Sport	Comm.	Pred./ par.	Disease	Ent./ imp. Toxicity
Vladykov and Greeley (1963)	X		X	X	X	X	X	X	X	X		X			
Dovel (1978)	X			X	X										
Hoff and Klauda (1979)				X			X	X							
Kioski et al. (1971)			S	S					X			X			
Crandall (1978)															

[a,b]See footnote in Table 7.16 for explanation.

Table 7.22. Hudson River open literature highlights: Fish (general).

Reference	Description	Yr(s) of study	Sampling intensity	Sample location[a] (BT YK TZ CH IP WP CW PK HP KG SG CS AL)
Bath et al. (1976)	H.R. fish species list			
Smith (1976)	Species list by habitat type			XXXXXXXXXXXXXXXXXXXXXXXXXXXXXXXXXXXX
Greeley and Bishop (1932)	Frequency, life history of fish in upper H.R.	1931		
Greeley	Fish species and distribution	1936		XXXXXXXXXXXXXXXXXXXXXXXXXXXXXXXXXXXX
Perlmutter et al. (1966)	Length-frequency distributions	1966	Monthly, June-Aug.	XXXXXXXXXXXXXXXXXXXXXXXXXXX
Clark and Smith (1969)	Natural history of migratory fish of the H.R.	Review report		XXXXXXXXXXXXXXXXXXXXXXXXXXXXXXXXXXXX
Perlmutter et al. (1969)	Fish species lists	1965–1969	Weekly, or biweekly, summer mo	XXXXXXXXXXXXXXXXXXXXX
Esser (1982)	Long-term changes in H.R. finfishes	1800–1977	Review report	XXXXXXXXXXXXXXXXXXXXXXXXXXXXXXXXXXXX
Dovel (1981)	Ichthyoplankton of lower H.R.	1972	1/wk, Apr.-Aug., 2/wk, Aug.-Nov.	XXXXXXXXXXXXXXXXXXXXXX
Jensen (1977)	New York marine fish	Review report		XXXX
Moore (1936)	Lower H.R. stocking, '26–'35, historical comments on sturgeon, shad	1936		XXXXXXXXXXXXXXXXXXXXXXXXXXXXXXXXXXXX
Heller and Hermo (1969)	Diurnal population fluctuations	1969	Irreg. monthly, June-Oct.	XX (around HP/KG) XX (around SG)
Englert et al. (1976)	Computer simulation of distribution of early life stages	1974–1975	Texas Inst. 1974 data	XXXXXXXXXXXXXXXXXXXXXXXXXXXXXXXX
Grim (1966)	Study design of N.E. Biologists, Inc., H.R. work	1965–1966	Seasonal	XXXXXXXXXXXXXXXXXXXXXXXXXXXXXXXX
Muessig and Mayercek (1976)	Diel distribution patterns of impinged fish	1974	Daily, Apr.-Oct., biweekly seines	XX (IP)
Casne and Cannon (1976)	Fish in thermal discharges		7 samples	XX (HP/KG) XX (SG/CS)
Jensen (1970)	Thermal loading	Review report		XXXXXXXXXXXXXXXXXXX
Wallace (1978)	Anomalies of larval transport affecting entrainment	1966, 1973, 1974	Irreg. weekly Apr.-July '66, weekly Apr.-Aug. '73, '74	XXXXXXXXXXXXXXXXXXXXXXXXXXXXXXXXXXXX
Dorfman (1973)	Serum protein research on white perch	1971	1 day	XX
Woolcott (1962)	Infraspecific variation in white perch			XXXXXXXXXXXXXXXXXXXX
Holsapple and Foster (1975)	Reproduction in white perch	1972		XXXXXXXX
Alevras (1973)	Lookdown found at Indian Point	1971	One date	XX (IP)
Tabery et al. (1978)	Lizardfish larvae found in H.R.	1974	July	XXX
Koski (1978)	Hogchoker life history	1969–1970	Irreg. monthly	XXXXXX
Koski (1974)	Sinstrality and albinism in three hogchokers	1969–1971		

[a] See Footnote a in Table 7.16 for explanation.

Open Literature   221

Upper Hudson	Concurrent phys./chem.	Methods given	Species	Abundance	Distribution[c]			Growth/ prim. prod.	Feeding
					Diurnal	Seasonal	Spatial		
X							X		
X							X		
		X					X		
		X	18 most common species	X			X		
	Temp., salinity		SB, ST, eel, river herr., others	X		X	X	X	X
		X	Most common species	X			X		
			SH, SB, ST, OS, WK, BL, SC, FL	X		X	X		
	Temp., salinity	X	All species	X	E,L,J, A,S	E,L,J, A,S	X		
			SB, AO, AB	X					
			ST, SH, salmon	X					
	Tide	X	All species caught	X	X		X		
			Juvenile striped bass	E,L			E,L		
	X	X	Striped bass, others						
	Temp.	X	WP, TC, BH, AN, SH, SB, BL, HG	X	X	X			
	X	X	17 species	X			X		
	Water temp.	X	SB, WP, SH, AO, AB, oysters				X	X	
			SB, SH, WP			L	L		
	O₂, salinity	X	WP						
		X	WP				X	X	
	Temp. salinity	X	WP			S	S		
	Temp. salinity		*Selene vomer*				X		
			Synodus foetens				L		
			Hogchoker						X

[c]See Footnote c in Table 7.17 for explanation.

Table 7.22. *Continued*

Reference	Reproduction				Exploitation		Survival[b]			
	Maturity	Fecundity	Behavior	Time	Sport	Comm.	Pred./par.	Disease	Ent./imp.	Toxicity
Bath et al. (1976)										
Smith (1976)										
Greeley and Bishop (1932)										
Greeley (1936)										
Perlmutter et al. (1966)										
Clark and Smith (1969)	X	X	X	X		X				
Perlmutter et al. (1969)										
Esser (1982)				X		X				
Dovel (1981)			X	X						
Jensen (1977)					X	X				PCBs
Moore (1936)										
Heller and Hermo (1969)										
Englert et al. (1976)										
Grim (1966)										
Muessig and Mayercek (1976)									X	
Casne and Cannon (1976)										
Jensen (1970)										
Wallace (1978)										
Dorfman (1973)										
Woolcott (1962)										
Holsapple and Foster (1975)	X	X		X						
Alevras (1973)									X	
Tabery et al. (1978)										
Koski (1978)	X			X						
Koski (1974)										

[b]See Footnote *b* in Table 7.16 for explanation.

Table 7.23. Hudson River open literature highlights: Geology and hydrology.

Reference	Description	Yr(s) of study	Sampling intensity	Sample location (0–150: BT YK TZ CH IP WP CW PK HP KG SG CS AL)
Olsen et al. (1978)	Estuary sediment patterns			XXXXXXXXXXXXXX
USGS (1955–1982)	Flow and water quality data	1955–present	1/mo	Scattered stations
Busby and Darmer (1970)	Flow and water quality data, H.R. and tributaries	1967	Once in Feb.	XXXXXXXXXXXXXXXXXXXXX
Stewart (1958)	Flow volumes, upstream currents			XXXXXXXXXXXXX
Schureman (1934)	Detailed charts of tides and currents in H.R.			XXXXXXXXXXXXXXXXXXXXXXXXXXXXXXXX
Johnson (1966)	Geology, mineral resources			XXXXXXXXXXXXXXXXXXXXXXXXXXXXXXXX
McCrone and Koch (1966)	Cation exchange of H.R. sediments	1966	July, Aug.	XXXXXXXXXXXXXXXXXXXXXXXX
Hohman and Parke (1966)	Dye tracer study	1965	10 days	XXXXXXXXXXXXXXXXXXXXXXXXXXXXX
Panuzio (1966)	Flow characteristics, tides, sedimentation			XXXXXXXX And New York Bay
Hooper (1966)	Cornwall Plant model, effects on flow patterns			XX
Busby (1966)	Description of salt front, tidal flow			XXXXXXXXXXXXXXXXXXXXXXXXXX
Darmer (1969)	Flow, salinity, salt front movement	1966–1969		XXXXXXXXXXXXXXXXXXXXXXXXXX
Neal (1969)	Predictive model of flow impacts			Replica constructed
Sanders (1974)	Geomorphology of H.R.			XXXXXXXXXXXXXXXXXXX
Abood (1974)	Mathematical treatment of H.R. circulation			XXXXXXXXXXXXXXXXXXXXXXXX

Table 7.23. *Continued*

Reference	Upper Hudson	Concurrent biological	Methods given	D.O.	Temp.	pH	Prod./CHL-a	Salinity	Alkalinity	Turbidity	Carbon
Olsen et al. (1978)											
USGS (1955–1982)	X			X	X	X	X	X	X	X	X
Busby and Darmer (1970)								X			
Stewart (1958)											
Schureman (1934)											
Johnson (1966)											
McCrone and Koch (1966)			X								
Hohman and Parke (1966)			X		X						
Panuzio (1966)					X	X		X		X	
Hooper (1966)			X								
Busby (1966)	X										
Darmer (1969)								X			
Neal (1969)											
Sanders (1974)											
Abood (1974)			X					X			

Nitrogen	Phos-phorus	Silica	Cadmium	Mercury	Other metals	Coliform	B.O.D.	Radio-nuclides	PCBs	Other organics
					X			X		
X	X	X	X	X	X	X				

Table 7.24. Hudson River open literature highlights: Temperature and salinity.

Reference	Description	Yr(s) of study	Sampling intensity	Sample location (0 BT, 25 YK, TZ, CH, IP, 50 WP, CW, PK, 75 HP, KG, 100 SG, CS, 150 AL)
Wrobel (1974)	Thermal balance, wastewater plumes	1968–1969	Seasonal	XXXXXXXXXXXXXXXXXXXXXXXXXXXXXX
Abood et al. (1976)	H.R. ambient temperature distribution			
Sorge (1969)	Thermal discharges	1969		Power plant locations
Jensen (1970)	Thermal loading in marine region			XXXXXXXXXX
Lawler and Abood (1969)	Thermal studies, theoretical and field			XXXXXXXXXXXXXXXXXXXXXXXXXXXXX
Hunkins (1981)	Salt transport in the estuary			XXXXXX
Simpson et al. (1973)	Description and simple modelling of salt movement	1972	1 day	XXXXXXXX

Table 7.24. *Continued*

Reference	Phosphorus	Silica	Cadmium	Mercury	Other metals	Coliform	B.O.D.	Radio-nuclides	PCBs	Other organics
Wrobel (1974)										
Abood et al. (1976)										
Sorge (1969)										
Jensen (1970)										
Lawler and Abood (1969)										
Hunkins (1981)										
Simpson et al. (1973)										

Open Literature

Upper Hudson	Concurrent biological	Methods given	D.O.	Temp.	pH	Prod./ CHL-*a*	Salinity	Alka- linity	Turbidity	Carbon	Nitrogen
X				X							
				X							
				X							
				X							
				X							
							X				
				X			X				

Table 7.25. Hudson River open literature highlights: Water quality.

Reference	Description	Yr(s) of study	Sampling intensity
Hetling (1976)	Trends in municip. and industrial wastewater loading, 1952–1974		
Mytelka (1976)	Water quality model, no data presented		Outline only
Flemming et al. (1976)	Impact of achieving Clean Water Act goals	1977	
Storm and Heffner (1976)	Phytoplankton abund., chlorophyll a, water quality	1973, 1974	Seasonal
Howells et al. (1970)	General water quality review	1963–1970	
Garside et al. (1976)	Effects of nutrients from sewage	1972–1974	Monthly or semiyearly
Mt Pleasant and Bagley (1975)	Index of 44 water quality parameters, very general		
Swanson and Devine (1982)	Ocean dumping policy, review article	1977–1981	
Tofflemire and Hetling (1969)	List of pollution sources		
Maylath (1969)	Summary of water quality surveillance program, no data given		Biweekly or monthly
Molof (1969)	Nutrient loads, model test, general		
Simpson et al. (1975)	Nutrient distributions, ecosystem approach	1973–1974	Monthly (irreg.)
Ketchum (1974)	Nutrient and pollution levels in lower Hudson	1964	Once
Intorre and DelRienzo (1974)	Power plant industrial wastes, no data given		
Lang (1974)	Present and future sewage treatment facilities, no data		
Rehe and Kohn (1976)	Water quality index	1970–1974	
Giese and Barr (1967)	Flow and water quality of H.R. and tributaries, tidal flow measurements	Data collected previously	
N.Y.S. Dept. of Health (1952–1963)	Classification of water qual. standards in H.R. watersheds	1951–1960	Variable
U.S. Dept. of Interior (1969)	Evaluation of combined sewer overflows		
U.S. Dept. of Interior (1969)	Water quality data, upper New York Bay and Harbor	1969	2 days in Aug.
N.Y.S. Dept. of En. Con. (unpublished data)	Routine water quality sampling by NYSDEC	1947–1983	
Mt. Pleasant (1966)	General treatment of poll. categories and sources		
Gross (1974)	Analysis of waste deposits and sediment in New York metropolitan area		
O'Connor and St. John (1967)	Model to reproduce observ. water quality	1969	Aug.
Hudson River Res. Council (1980)	Two coordinated river-long water qual. sample periods	1977, 1978	Apr. '77, Aug. '78

Open Literature

Sample location[a]						Upper Hudson	Concurrent biological	Methods given	D.O.
0	25	50	75	100	150				
BT YK	TZ CH	IP WP	CW PK	HP KG	SG CS AL				
XXXXXXXXXXXXXXXXXXXXXXXXXXXXXX						X		X	
XXXXXXXXXXXXXXXXXXXX									X
And lower New York Bay and Bight									
XXXXXXXXXXXXXXXXXXXXXXXXXXXXXXX							X		X
And New York Harbor									
XXXXXXXXXXXXXXXXXXXXXXXXXXXXXXXX						X	Phytoplankton	X	X
XXXXXXXXXXXXXXXXXXXXXXXXXXXXXXXX						X	X		
XXX							X	X	
Mostly in New York Bight									
XXXXXXXXXXXXXXXXXXXXXXXXXXXXXXX						X			
New York Bight									
XXXXXXXXXXXXXXXXXXXXXXXXXXXXXXX									X
XXXXXXXXXXXXXXXXXXXXXXXXXXXX									X
XXX									X
Mainly New York Harbor									
XXXXXXXXXXXXXXXXXXX									
XXXX									X
And New York Bay and Bight									
				XXXXX				X	
XXXXXXXXXXXXXXXXXXXXXXXXXXXXXX						X			
XXXXXXXXXXXXXXXXXXXXXXXXXXXXXXX						X			X
XXXXXXXXXXXXXXXXXXXXXXXXXXXXXXX									X
XX									
XXXXXXXXXXXXXXXXXXXXXXXXXX									X
XXXXXXXXXXXXXXXXXXXXXXXXXXXX									X
XX									
And New York Harbor									
			XXXXXXXXXXX						X
XXXXXXXXXXXXXXXXXXXXXXXXXXXXXXXX						X			X

[a] See Footnote *a* in Table 7.16 for explanation.

Table 7.25 *Continued*

Reference	Temp.	pH	Prod./CHL-*a*	Salinity	Alkalinity	Turbidity	Carbon	Nitrogen
Hetling (1976)						X		X
Mytelka (1976)	X		X	X				X
Flemming *et al.* (1976)								
Storm and Heffner (1976)	X	X	X	X	X	X	X	X
Howells *et al.* (1970)								X
Garside *et al.* (1976)	X		X	X				X
Mt Pleasant and Bagley (1975)								
Swanson and Devine (1982)								
Tofflemire and Hetling (1969)								X
Maylath (1969)	X	X		X	X	X		X
Molof (1969)		X						X
Simpson *et al.* (1975)								
Ketchum (1974)	X		X	X				
Intorre and DelRienzo (1974)								
Lang (1974)						X		
Rehe and Kohn (1976)								
Giese and Barr (1967)				X	X			
N.Y.S. Dept. of Health (1952–1963)	X	X		X	X	X	X	
U.S. Dept. of Interior (1969)								
U.S. Dept. of Interior (1969)	X	X		X				
N.Y.S. Dept. of En. Con. (unpublished data)	X	X		X	X	X		X
Mt. Pleasant (1966)								
Gross (1974)								
O'Connor and St. John (1967)	X							
Hudson River Res. Council (1980)	X			X				

Open Literature

Phos-phorus	Silica	Cadmium	Mercury	Other metals	Coliform	B.O.D.	Radio-nuclides	PCBs	Other organics
X						X			
X					X	X			
						X			
X	X	X		X					
X				Mn			X		X
X	X								
X					X	X			
X					X	X			
X					X	X			
X	X								
X									
				X			X		X
				X Fe					
					X	X			
						X			
						X			
X					X	X			
					X				
							X		

Table 7.26. Hudson River open literature highlights: Toxics, radionuclides, and metals.

Reference	Description	Yr(s) of study	Sampling intensity
Pakkala et al. (1972)	Selenium content of fish	1969	
Kniep et al. (1969)	Concentrations of trace elements, radionuclides, and pesticides	1962–1969	Monthly (irreg.)
Stevens (1969)	General review of oil poll. sources and effects		
Klinkhammer and Bender (1981)	Metal mass balances		
Pressman and Kirschner (1976)	Heavy metals from sewage outfalls		
Hazen and Kniep (1976)	Cadmium in sediments of Foundry Cove	1974–1976	
Occhiogrosso et al. (1976)	Heavy metal effects on benthos of Foundry Cove	1973–1974	Biweekly or monthly
Wrenn et al. (1976)	Model of radiocesium transport in H.R.	1971–1973	
Simpson et al. (1976)	Radionuclides in sediments	1971–1975	Monthly (irreg.)
Mehrle et al. (1982)	Metals, PCB, DDT effects on striped bass bone develop.	1978	July, Aug., Sept.
Rehwoldt et al. (1971)	Toxicity of Cu, Ni, Zn to H.R. fish		
Pakkala et al. (1972)	Arsenic in New York State fish	1969	
Tong et al. (1972)	Trace metals in New York State fish	1972	One specimen
Litchfield et al. (1982)	Effects of cadmium on microbes in New York Bight	1977–1978	Seasonal
Baker et al. (1976)	PCB and trace metals in waterfowl		
McCrone (1966)	Chemical content of H.R. sediments	1964–1965	July, Aug., Sept.
Paschoa et al. (1976)	Natural radiation dose to *Gammarus*	1969–1970 1972–1973	
Wills (1966)	Description of pesticides, toxicity, routes, effects		
Davies et al. (1966)	Sources of radiation in the Hudson	1964–1965	3-wk composite

Open Literature   233

Sample location[a]					Upper Hudson	Concurrent biological	Methods given	D.O.
0	25	50	75	100	150			
BT YK	TZ CH IP	WP CW PK	HP KG	SG CS	AL			
				XXXX			X	X
XXXXXXXXXXXXXXXXXXXXXXXX								
XXXXXXXXXX								
XXXX								
		X						
		X				Invertebrates		X
XXXXXXXXXXXXXXXXX								
XXXXXXXXXXXXXXXX								
		X				X	X	
			X			X	X	
						X		
						X		
New York Bight						X		
XXXX					XX	X		
	XXXXXXXXXXXXXXXXXXXXXXXX					X		X
		XX				X		
	XXXXXXXXXXXXXXXXXXXXXXXXX							

[a]See Footnote a in Table 7.16 for explanation.

Table 7.26. *Continued*

Reference	Temp.	pH	Prod./CHL-*a*	Salinity	Alkalinity	Turbidity	Carbon	Nitrogen
Pakkala *et al.* (1972)								
Kniep *et al.* (1969)								
Stevens (1969)								
Klinkhammer and Bender (1981)								
Pressman and Kirschner (1976)								
Hazen and Kniep (1976)								
Occhiogrosso *et al.* (1976)	X	X			X		X	
Wrenn *et al.* (1976)								
Simpson *et al.* (1976)								
Mehrle *et al.* (1982)								
Rehwoldt *et al.* (1971)	X	X			X			
Pakkala *et al.* (1972)								
Tong *et al.* (1972)								
Litchfield *et al.* (1982)								
Baker *et al.* (1976)								
McCrone (1966)	X	X		X	X			X
Paschoa *et al.* (1976)								
Wills (1966)								
Davies *et al.* (1966)								

Phos-phorus	Silica	Cadmium	Mercury	Other metals	Coliform	B.O.D.	Radio-nuclides	PCBs	Other organics
		X		X			X		X
									X
X		X		X					
		X	X	X					
		X							
		X		X					
							X		
							X		
		X		Ar,Pb Se,Zn Cu,Ni Zn Ar				X	X
		X	X	X Ni					
		X		Ni					
		X		Pb,Be As Hb					X
	X			X			X		
							X		
		X		X			X		X
									X

Table 7.27. Hudson River open literature highlights: PCBs.

Reference	Description
Ahmed (1976)	Review of PCBs in environment (includes Hudson R.)
Apicella and Zimmie (1978)	Sediment and PCB transport model for upper Hudson
Armstrong and Sloan (1980)	Trends in levels of PCBs and other contaminants in fish
Bopp et al. (1981)	PCBs in sediments of tidal Hudson
Bopp et al. (1982)	PCB and radionuclide chronologies in Hudson estuarine sediments
Boyle and Highland (1979)	Review article—PCB discovery and hearings, among other things
Brown et al. (1982)	PCB bioaccumulation model for plankton; calibrated to experimental data; trophic model
Brown et al. (1984)	Summary of trends of PCB concentrations in water, fish, zooplankton, and macroinvertebrates
Buckley (1982)	PCB accumulation in foliage near dumpsites vs. background; differential concentration rates
Buckley (1983)	Decline in background levels of PCBs for diff. spp. of plants; suggest broad trend
Buckley and Tofflemire (1983)	Plants near tailwater of dam have relatively high PCB conc.; greater volatilization of PCBs from river due to turbulent diffusion
EPA (1981a)	Draft environmental impact statement, PCB remedial action project
EPA (1981b)	Supplemental DEIS; includes comments from public review period
EPA (1982)	Final EIS for PCB reclamation project
Hetling et al. (1978)	Summary, initial PCB study results (sediments, physicochem. properties, distrib. in biota, etc.
Hetling et al. (1979)	Management alternatives for PCBs in upper Hudson
Horn et al. (1978)	Summary of PCB studies up to 1979
Horn et al. (1983)	PCB levels in striped bass in estuary; decline to <5 ppm for first time
Hullar et al. (1976)	PCB data in fish, sediments, and water and wastewater
Hydroscience (1979)	Model of PCB transport thru food web in estuary
Klauda et al. (1981)	PCB accumulation in tomcod collected near power plants in estuary
LMS (1979)	Upper Hudson sediment and PCB transport modelling study
Merhle et al. (1982)	Effects of toxics on bone development in striped bass from 4 estuarine locations
MPI (1980)	Draft EIS for PCB hot-spot dredging program—alternatives considered
Nadeau and Davis (1974)	Report of 1st EPA investigation of PCB concentrations in vicinity of GE
NRC (1979)	Report on status of PCBs in U.S.; estimated cost/benefit of HR dredging
NUS (1983)	Draft feasibility study for funding PCB cleanup under CERCLA (Superfund)
NYSDEC (1979)	Detailed work plan for implementing PCB Settlement Agreement
NYSDEC (1981a,b; 1982a)	Summaries of toxic substances in fish and wildlife of New York State
NYSDEC (1982b)	Detailed outline for environmental monitoring for PCB reclamation
NYU Medical Center (1982)	Biology of PCBs in Hudson estuarine zooplankton
Peters and O'Connor (1982)	PCB uptake by zooplankton and fish compared w.r.t. direct absorption vs. ingestion in food
Powers et al. (1982)	Toxic effects of PCB desorbed from clays and susp. particulates to phytoplankton
Sanders (1982)	Summary of PCB concentrations in sediments
Shen and Tofflemire (1979)	Volatilization of PCBs and other chlorinated hydroC's from dumpsites; experimental
Simpson (1982)	PCB concentrations in aquatic invertebrates and caddisfly larvae
Sloan and Armstrong (1981)	Patterns of PCB concentrations in migratory fish
Sloan et al. (1983)	Trends in PCB concentrations thru time for water, fish, and invertebrates in Hudson
Smith et al. (1979)	PCB concentrations and the presence of hepatomas in H.R. tomcod
Tofflemire and Quinn (1979)	Detailed mapping of PCBs in upper Hudson sediments
Tofflemire et al. (1982)	Experimental and field data on PCB volatilization from sediment and water
Turk (1980)	Discussion of PCB transport modes in upper Hudson; risk of drinking water contam., etc.
Turk and Troutman (1981)	PCB resuspension and transport in upper Hudson; loading rates to estuary
Schroeder and Barnes (1983)	Same as above; 5-yr trend analysis

Open Literature   237

		Sample Location[a]
Yr(s) of study	Sampling intensity	0 25 50 75 100 150 BT YK TZ CH IP WP CW PK HP KG SG CS AL
20-yr. projection (based on data from 1970s)		
1970–1979	not stated (but at least once/yr)	XXXX X X X
1977–1980?	not stated	XXXXXXXXXXXXXXXXXXXXXXXXXXXXX
1977–1980?	not stated	XX X X X X
n.a.	n.a.	
n.a.	n.a.	
1976–1981	Varied	XXXXXXXXXXXXXXXXXXXXXXXXXXXX
1979	not stated	
1978–1980	Collections in Sept.	
1980–1981	Aug./Sept.	
1980	Based on DEC-coordinated research	
1981		
1981		
1974–1977	Varied	X X X X
1974–1978	Varied	
1977–1978	Varied	
1983	Spring	XX X X X X
1975	Once/site	XXXXXXXXXXXXXXXXXXXXXXXXX
Data before 1978		General estuarine model
1978	Jan.-Feb.	Estuary
1978	7/20/78; 8/22/78; 9/26/78	X
1977–1978		
1974	8/12–8/13	
1978		
1983	Aug. '83 (plus data base review) (Outlines sampling, analyses) (List of funded research)	
1979–1982		XXXXXXXXXXXXXXXXXXXXXXXXXX
	Varied	XXXXXXXXXXXXXXXXXXXXXXXXX
1978–1981		XXXXXXXXXXXXXXXXXXXXXXXXXX
	In vitro	X
?	Once?	X (also in New York Bight)
1977–1981	Review	XXXXXXXXXXXXXXXXXXXXXXXXXX
		XXXXXXXXXXXXXXXXXXXXXX
1978–1981	Spring	XXXXXXXXXXXXXXXXXXXXXX
1977–1981	Varied	X X X X
12/77–2/78		XXXXXXXXXXX
1977–1978		
1979?		
1977	Several X per mo (Mar.-Sept.)	
1977	Same as above (all year)	
1977–1981	Same as above	

[a]See Footnote *a* in Table 7.16 for explanation.

Table 7.27. *Continued*

Reference	Upper Hudson	Biological (B) and/or Physical (P)/Chemical (C)	Methods given
Ahmed (1976)		General review	n.a.
Apicella and Zimmie (1978)	X	Only phys./chem.	Procedure for computations included
Armstrong and Sloan (1980)	X	Biol.	General
Bopp et al. (1982)		Only phys./chem.	X
Bopp et al. (1982)		Only phys./chem.	X
Boyle and Highland (1979)		B,P,C	n.a.
Brown et al. (1982)		Equilibrium chemistry (partition coefficients)	
Brown et al. (1984)	X	B,P,C,	X
Buckley (1982)	X	B	X
Buckley (1983)	X	B	X
Buckley and Tofflemire (1983)	X	B,P	X
EPA (1981a)	X	Discussed generally	
EPA (1981b)	X		
EPA (1982)	X		
Hetling et al. (1978)	X	B,P,C	No—only a data summary
Hetling et al. (1979)	X	B,P,C	
Horn et al. (1978)	X	B,P,C	
Horn et al. (1983)		B	
Hullar et al. (1976)	X	B,P,C	
Hydroscience (1979)		Physicochemical relations modelled	
Klauda et al. (1981)		B	X
LMS (1979)	X	P,C,	X
Merhle et al. (1982)	X	B	X
MPI (1980)	X	Engineering studies	
Nadeau and Davis (1974)	X	B,P,C,	X
NRC (1979)	X		
NUS (1983)	X	Review of B,P,C,	Scanty
NYSDEC (1979)			
NYSDEC (1981a,b; 1982a)	X	B	
NYSDEC (1982b)	X	B,P,C,	X
NYU Medical Center (1982)		B	X
Peters and O'Connor (1982)	X	B,P,C (partition coefficients)	X
Powers et al. (1982)		B,P,C,	X
Sanders (1982)	X	P (distributions, loading rates, transport)	
Shen and Tofflemire (1979)	X	Phys./chem. measures and calculations	
Simpson (1982)	X	B	X
Sloan and Armstrong (1981)	X	B	X
Sloan et al. (1983)	X	B,P,C	X
Smith et al. (1979)		B	Scanty
Tofflemire and Quinn (1979)	X	P,C,	X
Tofflemire et al. (1982)	X	P,C,B	X
Turk (1980)	X	P,C	
Turk and Troutman (1981)	X	P,C	X
Schroeder and Barnes (1983)	X	P,C	X

[b]n.a. = not applicable.

Major spp. or taxa reported (if biol.)	PCB concentrations reported	Effects or impacts reported	Time (T) or spatial (S) trends
Humans, fish	X	X	
	PCB loadings to estuary were projected over a 20-year time span		
Striped bass, largemouth bass, goldfish, white perch, brown bullhead	X		X
n.a.	Concentrations in sediments		S (includes budget)
n.a.	Same as above (also radionuclides)	Effect of removal of Ft. Edward dam	T,S
Humans, fish, birds	Few	X	
6 spp. of zooplankton	X (and also calculated)		
Striped bass, largemouth bass, goldfish, brown bullhead, pumpkinseed, zooplankton, macroinvertebrates			T,S
Red clover, goldenrod, field corn, sumac, trembling aspen, white pine	X		T
Above spp. plus other grasses, maple, pitch pine, red oak, white ash, alfalfa, rye, and more			T,S
Aspen, sumac, goldenrod	X		T,S
	X	Projected impacts discussed	T,S
Many species	X	X	T,S
	X		T, S
Fish, snapping turtles, invertebrates, terr. plants	X	Mass balance and movement discussed	S
Striped bass	X	Discussion of reopening of fishery	T,S
Many spp. of fish	X		S
Striped bass, small fish, plankton, benthos	Calculated	Estimated no. of yrs to reopen fishery	T
Atlantic tomcod	X	Histopathological effects on livers	
	Calculated		T,S
Striped bass	X (also As, Pb, Se, Zn, and Cd)	On bone development	
		X	T,S
Rock bass, yellow perch, shiner, snails	X		S (limited)
	X	Costs of remedial action	
Many spp.	X	X (Useful analyses)	T,S
			T,S
Fish and wildlife	X		S
Fish, invertebrates	X		T,S
Macrozoopl.: Gammarus, Neomysis, Leptodera, Crangon; Microzoopl.	X		T,S
Striped bass, zooplankton, algae, macrophyte	X		
Phytoplankton (Thalassiosira)	X	Inhibition of growth and ^{14}C fixation	
	In sediments		T,S
	X		
Invertebrates, caddisfly larvae	X		T,S
Migratory/marine fish	X		T,S
Fish (9 spp.) and invertebrates	X		T
Tomcod	X	Effects on livers noted	
	X		S
Terrestrial plants	X		
	X		S
	X		T,S
	X		T,S

Hudson River Bibliography

PETER S. WALCZAK, MARY ANN MORAN,
JANET M. BUCKLEY, EDWARD H. BUCKLEY,
and KARIN E. LIMBURG

Bibliography A: Utility-Sponsored Literature

Abood, K.A., E.A. Maikish, and R.R. Kimmel. 1976. Investigations of ambient temperature distribution in the Hudson River. Prepared by Lawler, Matusky, and Skelly, Engineers.
Alden Research Laboratories. 1966. Hudson River model study in the area of the Cornwall pumped storage project. Prepared for Consolidated Edison Co. of NY, Inc.
Alden Research Laboratories. 1969. Indian Point cooling water studies: Model No. 2. Prepared for Consolidated Edison Co. of NY, Inc.
Alden Research Laboratories. 1972. Hydrothermal model studies for Roseton Generating Station. Prepared for Central Hudson Gas and Electric Corp.
Alden Research Laboratories. 1973a. Indian Point intake streamline studies. Prepared for Consolidated Edison Co. of NY, Inc.
Alden Research Laboratories. 1973b. Hydrothermal model studies Roseton phase II, Danskammer power plant. Prepared for Central Hudson Gas and Electric Corp.
Barnett, R.W. and R. Tillman. 1973. Entrainment investigation at a fossil-fueled, steam-electric plant. Prepared by Dutchess Community College, Poughkeepsie, NY for Central Hudson Gas and Electric Corp.
Bechtel Associates Professional Corp. 1977. Cooling tower study. Bowline Point Stations, Units 1 and 2, West Haverstraw, NY. Prepared for Orange and Rockland Utilities, Inc.
Central Hudson Gas and Electric Corp. 1972. Environmental impact evaluation of proposed steam-electric generating station, town of Ulster, NY. Prepared in consultation with Quirk, Lawler, and Matusky, Engineers.
Central Hudson Gas and Electric Corp. 1974a. Demonstration by CHG&E with

regard to proposed NPDES permit for the Danskammer Point Generating Station. Application No. 2. SD OXW 2000485.

Central Hudson Gas and Electric Corp. 1974b. Demonstration by CHG&E with regard to proposed NPDES permit for Roseton Generating Station. Application No. 2. SD OXW 2000943.

Central Hudson Gas and Electric Corp. 1975. Plan of study and demonstration for the Roseton and Danskammer Generating Stations pursuant to 40 CFR, 122.5(b) (2).

Central Hudson Gas and Electric Corp. 1976. Danskammer Point Generating Station thermal discharge and river surface analyses.

Central Hudson Gas and Electric Corp. 1977. Roseton Generating Station, engineering, environmental (nonbiological), and economic aspects of a closed cycle cooling system. Ebasco, Consolidated Edison Co. of NY, Inc., Ecological Analysts, Inc., and Lawler, Matusky, and Skelly, Engineers (Contributors).

Consolidated Edison Co. of NY, Inc. 1971a. Applicants environmental report, Indian Point Nuclear Generating Plant, Unit 2. NRC Docket No. 50-247.

Consolidated Edison Co. of NY, Inc. 1971b. Applicants environmental report, Indian Point Nuclear Generating Plant, Unit 3. NRC Docket No. 50-283.

Consolidated Edison Co. of NY, Inc. 1972a. Applicants environmental report, Indian Point Nuclear Generating Plant, Unit 3. NRC Docket No. 50-286.

Consolidated Edison Co. of NY, Inc. 1972b. Final environmental statement related to operation of Indian Point Nuclear Generating Plant, Unit 2. Sept. 1972.

Consolidated Edison Co. of NY, Inc. 1973a. Applicants environmental report, Indian Point Nuclear Generating Plant, Unit 1. NRC Docket No. 50-3.

Consolidated Edison Co. of NY, Inc. 1973b. The biological effects of chemical discharges at Indian Point.

Consolidated Edison Co. of NY, Inc. 1974. The biological effects of fish impingement on the intake screens at Indian Point. App. BB. Indian Point Unit 3 environmental statement. Jan. 15, 1973.

Consolidated Edison Co. of NY, Inc. 1977a. Environmental report—eliminate requirement for termination of operation with once-through cooling system. NRC Docket No. 50-247.

Consolidated Edison Co. of NY, Inc. 1977b. Indian Point Nuclear Generating Station: near-field effects of once-through cooling system operation on Hudson River biota.

Consolidated Edison Co. of NY, Inc. 1977c. Indian Point Nuclear Generating Station thermal survey program. Routine monthly thermal monitoring. October 1976 survey report No. 1.

Consolidated Edison Co. of NY, Inc. 1978. Indian Point Generating Station 316 (a) demonstration.

Consolidated Edison Co. of NY, Inc. 1979. Indian Point Nuclear Generating Station. Report on the evaluation of hydro-thermal studies.

Consolidated Edison Co. of NY, Inc. and the Power Authority of the State of NY. 1977a. Indian Point Units 2 and 3. Near field effects of once-through cooling system on Hudson River biota.

Consolidated Edison Co. of NY, Inc. and the Power Authority of the State of NY. 1977b. Indian Point Units 2 and 3. Engineering, environmental (nonbiological) and economic aspects of a closed-cycle cooling system.

Bibliography A: Utility-Sponsored Literature

Consolidated Edison Co. of NY, Inc and the Power Authority of the State of NY. 1984. Indian Point Generating Station entrainment abundance and outage evaluation. 1983 annual report. Prepared by EA Engineering, Science, and Technology, Inc., Middletown, NY.

Ecological Analysts, Inc. 1975. 1975 Bowline/Lovett entrainment abundance and survival raw data summaries. Prepared for Orange and Rockland Utilities, Inc.

Ecological Analysts, Inc. 1976a. 1976 Bowline entrainment abundance and survival raw data summaries. Prepared for Orange and Rockland Utilities, Inc.

Ecological Analysts, Inc. 1976b. Bowline Point Generating Station entrainment survival and abundance studies. 1975 annual interpretive report. Prepared for Orange and Rockland Utilities, Inc.

Ecological Analysts, Inc. 1976c. 1975 Bowline Point Generating Station impingement and entrainment survival studies. Annual report. Prepared for Orange and Rockland Utilities, Inc.

Ecological Analysts, Inc. 1976d. Bowline Point Generating Station impingement recirculation and survival studies, Nov. 1975 to Apr. 1976. Prepared for Orange and Rockland Utilities, Inc.

Ecological Analysts, Inc. 1976e. Danskammer Point Generating Station entrainment survival and abundance studies; Vol. I. 1975 annual interpretive report. Prepared for Central Hudson Gas and Electric Corp.

Ecological Analysts, Inc. 1976f. Danskammer Point Generating Station impingement and entrainment survival studies. 1975 annual report. Prepared for Central Hudson Gas and Electric Corp.

Ecological Analysts, Inc. 1976g. 1976 Lovett entrainment abundance and survival raw data summaries. Prepared for Orange and Rockland Utilities, Inc.

Ecological Analysts, Inc. 1976h. Lovett Generating Station entrainment survival and abundance studies. 1975 annual interpretive report. Prepared for Orange and Rockland Utilities, Inc.

Ecological Analysts, Inc. 1976i. Preliminary data summary report for the 1975 studies at the Roseton and Danskammer Point Generating Stations. Prepared for Central Hudson Gas and Electric Corp.

Ecological Analysts, Inc. 1976j. Roseton Generating Station entrainment survival and abundance studies. 1975 annual interpretive report. Prepared for Central Hudson Gas and Electric Corp.

Ecological Analysts, Inc. 1976k. Roseton Generating Station impingement and entrainment survival studies. 1975 annual report. Prepared for Central Hudson Gas and Electric Corp.

Ecological Analysts, Inc. 1977a. Behavior of migrating adult striped bass in relationship to thermal discharge. Prepared for Orange and Rockland Utilities, Inc.

Ecological Analysts, Inc. 1977b. Bowline Point Generating Station dye recirculation study. Prepared for Orange and Rockland Utilities, Inc.

Ecological Analysts, Inc. 1977c. Bowline Point Generating Station entrainment and abundance studies. 1976 annual interpretive report. Prepared for Orange and Rockland Utilities, Inc.

Ecological Analysts, Inc. 1977d. Bowline Point Generating Station entrainment and impingement studies. 1976 annual report. Prepared for Orange and Rockland Utilities, Inc.

Ecological Analysts, Inc. 1977e. Danskammer Point Generating Station impingement survival studies. 1976 annual report. Prepared for Central Hudson Gas and Electric Corp.

Ecological Analysts, Inc. 1977f. Hudson River plants entrainment abundance and survival raw data summaries, 1975. Prepared for Central Hudson Gas and Electric Corp.

Ecological Analysts, Inc. 1977g. Hudson River power plants entrainment abundance and survival raw data summaries, 1976. Prepared for Central Hudson Gas and Electric Corp.

Ecological Analysts, Inc. 1977h. Hudson River thermal effects studies for representative species. Second progress report to Central Hudson Gas and Electric Corp.

Ecological Analysts, Inc. 1977i. Impingement survival studies at Roseton and Danskammer Point Generating Stations. Progress report prepared for Central Hudson Gas and Electric Corp.

Ecological Analysts, Inc. 1977j. Lovett Generating Station dye recirculation study. Prepared for Orange and Rockland Utilities, Inc.

Ecological Analysts, Inc. 1977k. Lovett Generating Station entrainment and impingement studies. 1976 Annual report. Prepared for Orange and Rockland Utilities, Inc.

Ecological Analysts, Inc. 1977l. Lovett Generating Station entrainment survival and abundance studies. Data collected for Orange and Rockland Utilities, Inc.

Ecological Analysts, Inc. 1977m. Preliminary investigations into the use of a continuously operating fine mesh traveling screen to reduce ichthyoplankton entrainment at Indian Point Generating Station. Prepared for Consolidated Edison Co. of NY, Inc.

Ecological Analysts, Inc. 1977n. Roseton Generating Station ichthyoplankton entrainment abundance progress report. Prepared for Central Hudson Gas and Electric Corp.

Ecological Analysts, Inc. 1977o. Second progress report on the Hudson River thermal effect studies for representative species. Prepared for Central Hudson Gas and Electric Corp.

Ecological Analysts, Inc. 1977p. Survival of entrained ichthyoplankton and macroinvertebrates at Hudson River power plants. Prepared for Central Hudson Gas and Electric Corp., Consolidated Edison Co. of NY, Inc., and Orange and Rockland Utilities, Inc.

Ecological Analysts, Inc. 1977q. The effects of intakes and associated cooling water systems on phytoplankton and aquatic invertebrates of the Hudson River. Prepared for Central Hudson Gas and Electric Corp., Consolidated Edison Co. of NY, Inc., and Orange and Rockland Utilities, Inc.

Ecological Analysts, Inc. 1977r. Thermal tolerance laboratory data. Collected for Central Hudson Gas and Electric Corp.

Ecological Analysts, Inc. 1978a. Bowline Point Generating Station 316(a) demonstration. Prepared for Orange and Rockland Utilities, Inc.

Ecological Analysts, Inc. 1978b. Bowline Point Generating Station entrainment abundance studies. 1977 annual report. Prepared for Orange and Rockland Utilities, Inc.

Ecological Analysts, Inc. 1978c. Bowline Point Generating Station entrainment survival studies. 1977 annual report. Prepared for Orange and Rockland Utilities, Inc.

Ecological Analysts, Inc. 1978d. Bowline Point Generating Station reimpingement studies. 1977 annual report. Prepared for Orange and Rockland Utilities, Inc.

Ecological Analysts, Inc. 1978e. Entrainment abundance studies at the Roseton and Danskammer Point Generating Stations. Prepared for Central Hudson Gas and Electric Corp.

Ecological Analysts, Inc. 1978f. Field studies of the effects of thermal discharge from Roseton and Danskammer Point Generating Stations on Hudson River fishes. Prepared for Central Hudson Gas and Electric Corp.

Ecological Analysts, Inc. 1978g. Hudson River thermal effects studies for representative species. Final report. Prepared for Central Hudson Gas and Electric Corp., Consolidated Edison Co. of NY, Inc., and Orange and Rockland Utilities, Inc.

Ecological Analysts, Inc. 1978h. Impingement survival studies at the Roseton and Danskammer Point Generating Stations. Progress report prepared for Central Hudson Gas and Electric Corp.

Ecological Analysts, Inc. 1978i. Indian Point Generating Station entrainment survival and related studies. 1977 annual report. Prepared for Consolidated Edison Co. of NY, Inc.

Ecological Analysts, Inc. 1978j. Lovett Generating Station entrainment abundance studies. 1977 annual report. Prepared for Orange and Rockland Utilities, Inc.

Ecological Analysts, Inc. 1978k. Lovett Generating Station entrainment survival studies. 1977 annual report. Prepared for Orange and Rockland Utilities, Inc.

Ecological Analysts, Inc. 1978l. Roseton and Danskammer Point Generating Stations 316(a) demonstration. Prepared for Central Hudson Gas and Electric Corp.

Ecological Analysts, Inc. 1978m. Roseton Generating Station entrainment abundance studies. 1976 annual report. Prepared for Central Hudson Gas and Electric Corp.

Ecological Analysts, Inc. 1978n. Roseton Generating Station entrainment survival studies. 1976 annual report. Prepared for Central Hudson Gas and Electric Corp.

Ecological Analysts, Inc. 1978o. Roseton Generating Station entrainment survival studies. 1977 annual report. Prepared for Central Hudson Gas and Electric Corp.

Ecological Analysts, Inc. 1978p. Thermal effects literature review for Hudson River representative important species. Prepared for Central Hudson Gas and Electric Corp., Consolidated Edison Co. of NY, Inc., and Orange and Rockland Utilities, Inc.

Ecological Analysts, Inc. 1979a. Bowline Point Generating Station entrainment abundance and survival studies. 1978 annual report. Prepared for Orange and Rockland Utilities, Inc.

Ecological Analysts, Inc. 1979b. Bowline Point impingement survival studies, 1975–1978. Overview report. Prepared for Orange and Rockland Utilities, Inc.

Ecological Analysts, Inc. 1979c. Effects of heat shock on predation of striped bass larvae by yearling white perch. Prepared for Central Hudson Gas and Electric Corp.

Ecological Analysts, Inc. 1979d. Evaluation of the effectiveness of a continuously operating fine mesh traveling screen for reducing ichthyoplankton entrainment at the Indian Point Generating Station. Prepared for Consolidated Edison Co. of NY, Inc.

Ecological Analysts, Inc. 1979e. Indian Point Generating Station entrainment survival and related studies. 1978 annual report. Prepared for Consolidated Edison Co. of NY, Inc.

Ecological Analysts, Inc. 1980a. Danskammer Point Generating Station entrainment survival studies. 1978 annual report. Prepared for Central Hudson Gas and Electric Corp.

Ecological Analysts, Inc. 1980b. Indian Point Generating Station entrainment and near field river studies. 1977 annual report. Prepared for Consolidated Edison Co. of NY, Inc.

Ecological Analysts, Inc. 1980c. Indian Point Generating Station entrainment and near field river studies. 1978 annual report. Prepared for Consolidated Edison Co. of NY, Inc.

Ecological Analysts, Inc. 1980d. Roseton Generating Station entrainment abundance studies. 1978 annual report. Prepared for Central Hudson Gas and Electric Corp.

Ecological Analysts, Inc. 1980e. Roseton Generating Station entrainment survival studies. 1978 annual report. Prepared for Central Hudson Gas and Electric Corp.

Ecological Analysts, Inc. 1981a. Indian Point Generating Station entrainment and near field river studies. 1979 annual report. Prepared for Consolidated Edison Co. of NY, Inc.

Ecological Analysts, Inc. 1981b. Bowline Point Generating Station entrainment abundance and survival studies. 1979 annual report with overview of 1975–1979 studies. Prepared for Orange and Rockland Utilities, Inc.

Ecological Analysts, Inc. 1981c. Bowline Point Generating Station entrainment abundance studies. 1980 annual report. Prepared for Orange and Rockland Utilities, Inc.

Ecological Analysts, Inc. 1981d. Indian Point Generating Station entrainment survival and related studies. 1979 annual report. Prepared for Consolidated Edison Co. of NY, Inc.

Ecological Analysts, Inc. 1982. Bowline Point Generating Station entrainment abundance and impingement survival studies. 1981 annual report. Prepared for Orange and Rockland Utilities, Inc.

EG&G Environmental Consultants. 1977a. Thermal surveys at the Bowline Point Generating Station.

EG&G Environmental Consultants. 1977b. Thermal surveys at the Lovett Generating Station.

Equitable Environmental Health, Inc. 1975. Hart Island aquatic ecology survey: Final report. Prepared for the Power Authority of the State of New York.

Feldman, A. 1971. Hudson River Roseton study. A microfloral population investigation of the Hudson River in the vicinity of Roseton. Final report. Prepared by Dutchess Community College, Poughkeepsie, NY for Central Hudson Gas and Electric Corp.

Hudson River Policy Committee. 1969. Hudson River fisheries investigations, 1965–1968. Evaluation of a proposed pumped storage project at Cornwall, New

York in relation to fish in the Hudson River. N.Y.S. Dept. of Environmental Conservation. Presented to Consolidated Edison Co. of NY, Inc.

Kretser, W.A. 1968. Aspects of striped bass larvae *Morone saxatilus* captured in the Hudson River near Cornwall, NY, Apr. to Aug. 1968. Prepared for Consolidated Edison Co. of NY, Inc.

Lanza, G.R., G.J. Lauer, T.C. Ginn, P.C. Storm, and L. Zubarik. 1974. Biological effects of simulated discharge plume entrainment at Indian Point Nuclear Power Station, Hudson River estuary. New York University Medical Center.

LaSalle Hydraulic Laboratory. 1976a. Cornwall pumped storage scheme hydraulic model study of Hudson River flows around intake screen structure. Prepared for Consolidated Edison Co. of NY, Inc.

LaSalle Hydraulic Laboratory. 1976b. Indian Point generating plants. Hydraulic model study of Hudson River flows around cooling water intakes. Prepared for Consolidated Edison Co. of NY, Inc.

Lauer, G.J. 1973. Biological effects of chemical discharges at Indian Point. Report to Consolidated Edison Co. of NY, Inc.

Lawler, Matusky, and Skelly, Engineers. 1974a. 1973 Hudson River aquatic ecology studies. Bowline Point and Lovett Generating Stations. Report to Orange and Rockland Utilities, Inc.

Lawler, Matusky, and Skelly, Engineers. 1974b. Cornwall gear evaluation study. Prepared for Consolidated Edison Co. of NY, Inc.

Lawler, Matusky, and Skelly, Engineers. 1974c. Danskammer Point fish impingement study. Prepared for Central Hudson Gas and Electric Corp.

Lawler, Matusky, and Skelly, Engineers. 1975a. Albany Steam Electric Generating Station impingement survey. Prepared for Niagara Mohawk Power Corp.

Lawler, Matusky, and Skelly, Engineers. 1975b. Aquatic ecology studies, Kingston, NY 1973. Prepared for Central Hudson Gas and Electric Corp.

Lawler, Matusky, and Skelly, Engineers. 1975c. Chemical discharge and toxicity report for the Roseton and Danskammer Point Generating Stations. Prepared for Central Hudson Gas and Electric Corp.

Lawler, Matusky, and Skelly, Engineers. 1975d. Ecological studies of the Hudson River in the vicinity of Lloyd, NY, 1973–1974. Report to N.Y.S. Energy Research and Development Authority.

Lawler, Matusky, and Skelly, Engineers. 1975e. 1974 Hudson River aquatic ecology studies at Kingston, NY. Prepared for Central Hudson Gas and Electric Corp.

Lawler, Matusky, and Skelly, Engineers. 1975f. 1974 Hudson River aquatic ecology studies—Bowline Point and Lovett Generating Stations. Prepared for Orange and Rockland Utilities, Inc.

Lawler, Matusky, and Skelly, Engineers. 1975g. 1973 Hudson River aquatic ecology studies at Roseton and Danskammer Point. Prepared for Central Hudson Gas and Electric Corp.

Lawler, Matusky, and Skelly, Engineers. 1975h. 1974 Hudson River aquatic ecology studies at Roseton and Danskammer Point. Prepared for Central Hudson Gas and Electric Corp.

Lawler, Matusky, and Skelly, Engineers. 1975i. Hudson River study: Aquatic factors governing the siting of power plants. Prepared for Empire State Electric and Energy Research Corp.

Lawler, Matusky, and Skelly, Engineers. 1975j. Report on development of a real-time two dimensional model of the Hudson River striped bass population. Prepared for Consolidated Edison Co. of NY, Inc.

Lawler, Matusky, and Skelly, Engineers. 1975k. Thermal survey at Roseton/Danskammer Point Generating Stations—Sept. 1974. Prepared for Central Hudson Gas and Electric Corp.

Lawler, Matusky, and Skelly, Engineers. 1975l. Thermal survey at Roseton/Danskammer Point Generating Stations—Apr. 1975. Prepared for Central Hudson Gas and Electric Corp.

Lawler, Matusky, and Skelly, Engineers. 1975m. Thermal survey at Roseton/Danskammer Point Generating Stations—July 1975. Prepared for Central Hudson Gas and Electric Corp.

Lawler, Matusky, and Skelly, Engineers. 1975n. Velocity studies at intake of Central Hudson Gas and Electric and Orange and Rockland Power Plants on the Hudson (draft). Project 327-02/03. Prepared for Central Hudson Gas and Electric Corp. and Orange and Rockland Utilities, Inc.

Lawler, Matusky, and Skelly, Engineers. 1976a. Albany Steam Electric Generating Station 316(a) demonstration submission. EPA NPDES Permit.

Lawler, Matusky, and Skelly, Engineers. 1976b. Analysis of the Lovett Generating Station thermal discharges (July 1975). Prepared for Orange and Rockland Utilities, Inc.

Lawler, Matusky, and Skelly, Engineers. 1976c. Danskammer Point thermal analysis report. Evaluation of influence on fish populations. Prepared for Central Hudson Gas and Electric Corp.

Lawler, Matusky, and Skelly, Engineers. 1976d. Danskammer Point reduced flow study evaluation of the influence of reduced flow on fish impingement. Prepared for Central Hudson Gas and Electric Corp.

Lawler, Matusky, and Skelly, Engineers. 1976e. Environmental impact assessment. Hudson River water quality analysis. Prepared for National Commission on Water Quality. PB 251099. NTIS, Springfield, VA.

Lawler, Matusky, and Skelly, Engineers. 1976f. Hudson River ambient temperature survey: Aug. to Sept. 1975. Prepared for the Central Hudson Gas and Electric Corp., Consolidated Edison Co. of NY, Inc., Niagara Mohawk Power Corp., and Orange and Rockland Utilities, Inc.

Lawler, Matusky, and Skelly, Engineers. 1976g. 1975 Hudson River aquatic ecology studies Bowline Point and Lovett Generating Stations. Prepared for Orange and Rockland Utilities, Inc.

Lawler, Matusky, and Skelly, Engineers. 1976h. Preliminary evaluations of the effectiveness of a chain barrier and of reduction in cooling water flow in reducing impingement at the Danskammer Point Generating Station. Prepared for Central Hudson Gas and Electric Corp.

Lawler, Matusky, and Skelly, Engineers. 1976i. The Aug. 1976 thermal survey at the Lovett Generating Station. Prepared for Orange and Rockland Utilities, Inc.

Lawler, Matusky, and Skelly, Engineers. 1976j. Thermal survey at Roseton/Danskammer Point Generation Stations—Aug. 1976. Prepared for Central Hudson Gas and Electric Corp.

Lawler, Matusky, and Skelly, Engineers. 1976k. Thermal surveys at the Bowline

Point and Lovett Generating Stations (Apr., June, Aug., Oct., 1975). Prepared for Orange and Rockland Utilities, Inc.

Lawler, Matusky, and Skelly, Engineers. 1976l. 1976 Roseton intake avoidance study. Preliminary report on an evaluation of intake avoidance as a function of size for white perch. Prepared for Central Hudson Gas and Electric Corp.

Lawler, Matusky, and Skelly, Engineers. 1977a. 1976 Hudson River aquatic ecology studies at Bowline Point Generating Station. Prepared for Orange and Rockland Utilities, Inc.

Lawler, Matusky, and Skelly, Engineers. 1977b. 1976 Hudson River aquatic ecology studies at Lovett Generating Station. Prepared for Orange and Rockland Utilities, Inc.

Lawler, Matusky, and Skelly, Engineers. 1977c. Thermal survey at Roseton/Danskammer Point Generating Stations—Dec. 1976. Prepared for Central Hudson Gas and Electric Corp.

Lawler, Matusky, and Skelly, Engineers. 1977d. 1976–1977 littoral survey—rock substrates periphyton. Prepared for Central Hudson Gas and Electric Corp.

Lawler, Matusky, and Skelly, Engineers. 1977e. Viability studies of impinged Atlantic tomcod. Roseton and Danskammer Point Generating Stations, 1974–1975. Data collected for Central Hudson Gas and Electric Corp.

Lawler, Matusky, and Skelly, Engineers. 1977f. Influence of Indian Point Unit 2 and other steam electric generating plants on the Hudson River estuary, with emphasis on striped bass and other fish populations. Supplement 1. Prepared for Consolidated Edison Co. of NY, Inc.

Lawler, Matusky, and Skelly, Engineers. 1977g. Dissolved oxygen loss due to once-through cooling at the Roseton Generating Station. Prepared for Central Hudson Gas and Electric Corp.

Lawler, Matusky, and Skelly, Engineers. 1977h. Roseton fish migration route study—Evaluation of river cross-sectional distribution patterns of fish larvae and fish eggs. Prepared for Central Hudson Gas and Electric Corp.

Lawler, Matusky, and Skelly, Engineers. 1977i. Evaluation of the influence of a chain barrier and of reduced flow on size distribution of impinged fish. Prepared for Central Hudson Gas and Electric Corp.

Lawler, Matusky, and Skelly, Engineers. 1978a. Annual progress report, 1974. Prepared for Central Hudson Gas and Electric Corp.

Lawler, Matusky, and Skelly, Engineers. 1978b. Annual progress report, 1975. Prepared for Central Hudson Gas and Electric Corp.

Lawler, Matusky, and Skelly, Engineers. 1978c. Mitigation of impingement impact at Hudson River power plants. Prepared for Consolidated Edison Co. of NY, Inc.

Lawler, Matusky, and Skelly, Engineers. 1978d. 1977 impingement studies at the Lovett Generating Station. Prepared for Orange and Rockland Utilities, Inc.

Lawler, Matusky, and Skelly, Engineers. 1978e. Ichthyoplankton gear comparison study. Prepared for Orange and Rockland Utilities, Inc.

Lawler, Matusky, and Skelly, Engineers. 1978f. Bowline Point Generating Station hydrothermal analysis. Prepared for Orange and Rockland Utilities, Inc.

Lawler, Matusky, and Skelly, Engineers. 1978g. 1977 Hudson River aquatic ecology studies at the Bowline Point Generating Station. Prepared for Orange and Rockland Utilities, Inc.

Lawler, Matusky, and Skelly, Engineers. 1978h. 1977 Hudson River aquatic ecology studies at the Lovett Generating Station. Prepared for Orange and Rockland Utilities, Inc.

Lawler, Matusky, and Skelly, Engineers. 1978i. Relationship between population density and growth for young-of-the-year striped bass and white perch. Prepared for Central Hudson Gas and Electric Corp., Consolidated Edison Co. of NY, Inc., and Orange and Rockland Utilities, Inc. Submitted in the EPA-utilities Adjudicatory Hearing No. C/II-WP-77-01 EXb. UT-50.

Lawler, Matusky, and Skelly, Engineers. 1978j. Roseton and Danskammer Point Generating Stations habitat selection studies. Prepared for Central Hudson Gas and Electric Corp.

Lawler, Matusky, and Skelly, Engineers. 1978k. Roseton and Danskammer Point Generating Stations hydrothermal analysis. Prepared for Central Hudson Gas and Electric Corp.

Lawler, Matusky, and Skelly, Engineers. 1978l. The relationship of temperature to growth and abundance of *Morone* and other fish species and invertebrates in the Hudson River estuary (draft). Prepared for Central Hudson Gas and Electric Corp., Orange and Rockland Utilities, Inc., and Consolidated Edison Co. of NY, Inc.

Lawler, Matusky, and Skelly, Engineers. 1979a. Empirical evidence of control mechanisms in the Cape Fear, Chesapeake, and Hudson River estuaries (draft). Chapter Two of a compensation report. Prepared for Carolina Power and Light Co.

Lawler, Matusky, and Skelly, Engineers. 1979b. Annual progress report, 1976. Prepared for Central Hudson Gas and Electric Corp.

Lawler, Matusky, and Skelly, Engineers. 1979c. Annual progress report, 1977. Prepared for Central Hudson Gas and Electric Corp.

Lawler, Matusky, and Skelly, Engineers. 1979d. 1978 Hudson River aquatic ecology studies at the Bowline Point Generating Station. Prepared for Orange and Rockland Utilities, Inc.

Lawler, Matusky, and Skelly, Engineers. 1979e. 1978 Hudson River aquatic studies at the Lovett Generating Station. Prepared for Orange and Rockland Utilities, Inc.

Lawler, Matusky, and Skelly, Engineers. 1979f. Mitigation of entrainment and impingement impact at Hudson River power plants. Prepared for Consolidated Edison Co. of NY, Inc.

Lawler, Matusky, and Skelly, Engineers. 1979g. Application of the niche analysis to four biotic communities at the Roseton-Danskammer Point Generating Stations, Hudson River, in 1973. Research Grant No. 12-400-010. Vol. I and II. Prepared by P.A. Lane and J.A. Wright for LMS and Central Hudson Gas and Electric Corp.

Lawler, Matusky, and Skelly, Engineers. 1979h. 1978 impingement studies at the Lovett Generating Station. Prepared for Orange and Rockland Utilities, Inc.

Lawler, Matusky, and Skelly, Engineers. 1979i. Fish impingement monitoring screen carryover study. Roseton Electric Generating Station, June 1978 to Feb. 1979. Prepared for Central Hudson Gas and Electric Corp.

Lawler, Matusky, and Skelly, Engineers. 1980a. Texas Instruments and Lawler, Matusky, and Skelly, Engineers ichthyoplankton (1975–1979) and river fish

(1977–1979) comparison in the Croton/Haverstraw Bay region of the Hudson River. Prepared for Orange and Rockland Utilities, Inc.

Lawler, Matusky, and Skelly, Engineers. 1980b. Evaluation of lower trophic level aquatic communities in the vicinity of the Bowline Point Generating Station, 1971–1977. Prepared for Orange and Rockland Utilities, Inc.

Lawler, Matusky, and Skelly, Engineers. 1980c. 1979 impingement studies at the Lovett Generating Station. Prepared for Orange and Rockland Utilities, Inc.

Lawler, Matusky, and Skelly, Engineers. 1980d. Trawl comparison in the northern Newburgh Bay area of the Hudson River, Yankee trawl vs. Otter trawl. Prepared for Central Hudson Gas and Electric Corp.

Lawler, Matusky, and Skelly, Engineers. 1980e. Ichthyoplankton entrainment survival study at Roseton Electric Generating Station. Prepared for Orange and Rockland Utilities, Inc.

Lawler, Matusky, and Skelly, Engineers. 1980f. Annual progress report, 1978. Prepared for Central Hudson Gas and Electric Corp.

Lawler, Matusky, and Skelly, Engineers. 1980g. Annual progress report, 1979. Prepared for Central Hudson Gas and Electric Corp.

Lawler, Matusky, and Skelly, Engineers. 1980h. 1979 Hudson River aquatic ecology studies at the Lovett Generating Station. Prepared for Orange and Rockland Utilities, Inc.

Lawler, Matusky, and Skelly, Engineers. 1981a. Annual progress report, 1980. Prepared for Central Hudson Gas and Electric Corp.

Lawler, Matusky, and Skelly, Engineers. 1981b. Preliminary reports on 1981 impingement. Prepared for Central Hudson Gas and Electric Corp.

Lawler, Matusky, and Skelly, Engineers. 1981c. Trawl methodology study: constant distance vs. constant time. Prepared for Central Hudson Gas and Electric Corp.

Lawler, Matusky, and Skelly, Engineers. 1981d. 1979 Hudson River aquatic ecology studies at Bowline Point Generating Station. Prepared for Orange and Rockland Utilities, Inc.

Lawler, Matusky, and Skelly, Engineers. 1981e. 1980 Bowline Point aquatic ecology studies. Prepared for Orange and Rockland Utilities, Inc.

Lawler, Matusky, and Skelly, Engineers. 1981f. Impingement studies at the Bowline Point Generating Station. Prepared for Orange and Rockland Utilities, Inc.

Lawler, Matusky, and Skelly, Engineers. 1981g. 1980 impingement studies at the Lovett Generating Station. Prepared for Orange and Rockland Utilities, Inc.

Lawler, Matusky, and Skelly, Engineers. 1981h. Impingement studies at the Lovett Generating Station. Prepared for Orange and Rockland Utilities, Inc.

Lawler, Matusky, and Skelly, Engineers. 1982a. Biological evaluation of a 0.85 cm mesh barrier net deployed to mitigate fish impingement at the Bowline Point Generating Station cooling water intakes. Prepared for Orange and Rockland Utilities, Inc.

Lawler, Matusky, and Skelly, Engineers. 1982b. Annual progress report, 1981. Impingement abundance sampling at Roseton and Danskammer Point Generating Stations. Prepared for Central Hudson Gas and Electric Corp.

Lawler, Matusky, and Skelly, Engineers and Ecological Analysts, Inc. 1977a. Bowline Point Generating Station: Near-field effects of once-through cooling system operation on Hudson River biota. Prepared for Orange and Rockland Utilities, Inc.

Lawler, Matusky, and Skelly, Engineers and Ecological Analysts, Inc. 1977b. Roseton Generating Station: Near-field effects of once-through cooling system operation on Hudson River biota. Prepared for Central Hudson Gas and Electric Corp.

McFadden, J.T. (ed.) 1977a. Influence of the proposed Cornwall pumped storage project and steam electric generating plants on the Hudson River estuary, with emphasis on striped bass and other fish populations. Prepared for Consolidated Edison Co. of NY, Inc.

McFadden, J.T. (ed.) 1977b. Influence of Indian Point Unit 2 and other steam electric generating plants on the Hudson River estuary, with emphasis on striped bass and other fish populations. Prepared for Consolidated Edison Co. of NY, Inc.

McFadden, J.T. and J.P. Lawler (eds.). 1977c. Supplement I to influence of Indian Point Unit 2 and other steam electric generating plants on the Hudson River estuary, with emphasis on striped bass and other fish populations. Prepared for Consolidated Edison Co. of NY, Inc.

McFadden, J.T. (ed.) 1978. Influence of the proposed Cornwall pumped storage project and steam electric plants on the Hudson River estuary with emphasis on striped bass and other fish populations (revised). Prepared for Consolidated Edison Co. of NY, Inc.

Meldrim, J.W., J.J. Gift, and B.R. Petrosky. 1971. A supplementary report on temperature shock studies with the white perch (*Morone americana*). Prepared by Ichthyological Associates, Inc., Ithaca, NY for Consolidated Edison Co. of NY, Inc.

New York University Medical Center. 1971. Ecological studies of the Hudson River near Indian Point, Apr. 1969 to Apr. 1970. Annual report. Prepared for Consolidated Edison Co. of NY, Inc.

New York University Medical Center. 1973. Hudson River ecosystem studies. Effects of entrainment by the Indian Point Power Plant on Hudson River estuary biota. Progress report for 1971 and 1972. Prepared for Consolidated Edison Co. of NY, Inc.

New York University Medical Center. 1974a. Hudson River ecosystem studies. Effects of entrainment by the Indian Point Power Plant on biota in the Hudson River estuary. Progress report for 1973. Prepared for Consolidated Edison Co. of NY, Inc.

New York University Medical Center. 1974b. Hudson River ecosystem studies. Semiannual progress report, Jan. 1 to June 30, 1974. Prepared for Consolidated Edison Co. of NY, Inc.

New York University Medical Center. 1976a. Hudson River ecosystem studies. Effects of entrainment by the Indian Point Power Plant on biota in the Hudson River estuary. Progress report for 1974. Prepared for Consolidated Edison Co. of NY, Inc.

New York University Medical Center. 1976b. Hudson River ecosystem studies. Effects of entrainment by the Indian Point Power Plant on biota in the Hudson River estuary. Addenda to the 1973 report. Prepared for Consolidated Edison Co. of NY, Inc.

New York University Medical Center. 1976c. Mortality of striped bass eggs and larvae in nets. Prepared for Consolidated Edison Co. of NY, Inc.

New York University Medical Center. 1976d. The effects of changes in hydrostatic pressure on some Hudson River biota. Progress report for 1975. Prepared for Consolidated Edison Co. of NY, Inc.

New York University Medical Center. 1976e. Hudson River ecosystem studies. Effects of temperature and chlorine on entrained Hudson River organisms. Progress report for 1975. Prepared for Consolidated Edison Co. of NY, Inc.

New York University Medical Center. 1977. Hudson River ecosystem studies. Effects of entrainment by the Indian Point Power Plant on biota in the Hudson River estuary. Progress report for 1975. Prepared for Consolidated Edison Co. of NY, Inc.

New York University Medical Center. 1978a. Distribution and abundance of striped bass ichthyoplankton in the Hudson River at Indian Point, and at the Indian Point Power Plant intakes. Prepared for Consolidated Edison Co. of NY, Inc.

New York University Medical Center. 1978b. Hudson River ecosystem studies. Effects of entrainment by the Indian Point Power Plant on biota in the Hudson River estuary. Progress report for 1976. Prepared for Consolidated Edison Co. of NY, Inc.

Niagara Mohawk Power Corp. 1976. 316(a) demonstration submission, NPDES permit NY 0005959. Albany Steam Electric Generating Station. Prepared by Lawler, Matusky, and Skelly, Engineers.

Normandeau Associates, Inc. 1984a. Hudson River ecological study in the area of Indian Point: 1983 annual report. Prepared for Consolidated Edison Co. of NY, Inc., and the Power Authority of the State of New York.

Normandeau Associates, Inc. 1984b. Abundance and stock characteristics of the Atlantic tomcod (*Mircrogadus tomcod*) spawning population in the Hudson River, winter 1983-1984. Prepared under contract with Consolidated Edison Co. of NY, Inc.

Normandeau Associates. 1984c. Precision and accuracy of stratified sampling to estimate fish impingement at Indian Point Units 2 and 3. Prepared for Consolidated Edison Co. of NY, Inc. and the Power Authority of the State of New York.

Orange and Rockland Utilities, Inc. 1971. Project description—Bowline Point Station Units 1 and 2, Haverstraw, NY. Prepared by Bechtel Associates.

Orange and Rockland Utilities, Inc. 1973a. Bowline Generating Station— Haverstraw, NY. Draft of final environmental statement (revised).

Orange and Rockland Utilities, Inc. 1973b. Impingement data for the period Dec. 1, 1972 through Dec. 31, 1973 for the Bowline Point Generating Station.

Orange and Rockland Utilities, Inc. 1973c. Impingement data for the period Dec. 1, 1972 through Dec. 31, 1973 for the Lovett Generating Station.

Orange and Rockland Utilities, Inc. 1974a. Bowline 316(a) demonstration.

Orange and Rockland Utilities, Inc. 1974b. Impingement data for the period Jan. 1, 1974 through Dec. 31, 1974 for the Bowline Generating Station.

Orange and Rockland Utilities, Inc. 1974c. Impingement data for the period Jan. 1, 1974 through Dec. 31, 1974 for the Lovett Generating Station.

Orange and Rockland Utilities, Inc. 1975a. Impingement data for the period Jan. 1, 1975 through Mar. 31, 1975 for the Bowline Generating Station.

Orange and Rockland Utilities, Inc. 1975b. Impingement data for the period Jan. 2, 1975 through Mar. 31, 1975 for the Lovett Generating Station.

Orange and Rockland Utilities, Inc. 1975c. Work scope for the 1975 aquatic studies to be conducted for Orange and Rockland Utilities, Inc. at the Bowline Point and Lovett Generating Stations. Submitted to N.Y.S. Dept. of Environmental Conservation.

Orange and Rockland Utilities, Inc. 1976a. Bowline Point plume surveys. Prepared by the Calspan Corp. Report NA-5800-M-2.

Orange and Rockland Utilities, Inc. 1976b. Alternative intake design with fish handling capabilities, Bowline Point Generating Station Units 1 and 2. Prepared by Stone and Webster Engineering Corp., Boston, MA.

Orange and Rockland Utilities, Inc. 1977. Bowline Point Generating Station. Engineering, environmental (non-biological), and economic aspects of a closed-cycle cooling system.

Parkinson, F.E. 1976. Cornwall pumped storage scheme hydraulic model study of Hudson River flows around screen structures. LaSalle Hydraulic Laboratories. Prepared for Consolidated Edison Co. of NY, Inc.

Perlmutter, A., E.E. Schmidt, R. Heller, H.C. Ford, and S. Sininsky. 1968. Distribution and abundance of fish along the shores of the lower Hudson during the summer of 1967. Ecological survey of the Hudson Progress Report 3. 42 pp. Prepared for Consolidated Edison Co. of NY, Inc.

Prescott, G.W. 1966. Algal sampling of the Hudson River. Summary report to Consolidated Edison Co. of New York, Inc. for Quirk, Lawler, and Mutusky, Engineers. 17 pp.

Quirk, Lawler, and Matusky, Engineers. 1966a. Algal sampling of the Hudson River. Summary report to Consolidated Edison Co. of NY, Inc. 26 pp. Supplement by G.W. Prescott. 6 pp.

Quirk, Lawler, and Matusky, Engineers. 1966b. Effect of contaminant discharge at Indian Point on Hudson River water intake at Chelsea, NY. Prepared for Consolidated Edison Co. of NY, Inc.

Quirk, Lawler, and Matusky, Engineers. 1968a. Effect of Indian Point cooling water discharge on Hudson River temperature distribution. Prepared for Consolidated Edison Co. of NY, Inc.

Quirk, Lawler, and Matusky, Engineers. 1968b. Effect of Lovett Unit 5 cooling water discharge on Hudson River temperature distribution. Prepared for Orange and Rockland Utilities, Inc.

Quirk, Lawler, and Matusky, Engineers. 1969a. Effect of Indian Point cooling water discharge on Hudson River temperature distribution. Prepared for Consolidated Edison Co. of NY, Inc.

Quirk, Lawler, and Matusky, Engineers. 1969b. Effect of submerged discharge of Indian Point cooling water on Hudson River temperature distribution. Prepared for Consolidated Edison Co. of NY, Inc.

Quirk, Lawler, and Matusky, Engineers. 1969c. Influence of Hudson River net non-tidal flow on temperature distribution. Prepared for Consolidated Edison Co. of NY, Inc.

Quirk, Lawler, and Matusky, Engineers and Oceanographic Analysts, Inc. 1969d. Effect of Bowline cooling water discharge on Hudson River temperature distribution and ecology. Prepared for Orange and Rockland Utilities, Inc.

Quirk, Lawler, and Matusky, Engineers and Oceanographic Analysts, Inc. 1969e. Effect of Lovett Plant cooling water discharge on Hudson River temperature distribution and ecology. Prepared for Orange and Rockland Utilities, Inc.

Quirk, Lawler, and Matusky, Engineers and Oceanographic Analysts, Inc. 1969f. Effect of Roseton Plant cooling water discharge on Hudson River temperature distribution and ecology. Report to Central Hudson Gas and Electric Corp., Consolidated Edison Co. of NY, Inc., and Niagara Mohawk Power Corp. (3 vols.)

Quirk, Lawler, and Matusky, Engineers. 1970. Hudson River water quality and waste assimilative capacity study. Prepared for N.Y.S. Dept. of Environmental Conservation.

Quirk, Lawler, and Matusky, Engineers. 1971a. Circulation in the Hudson estuary—a technical bulletin.

Quirk, Lawler, and Matusky, Engineers. 1971b. Effect of winter-time operation of Indian Point Units 1 and 2 on Hudson River temperature distribution. Prepared for Consolidated Edison Co. of NY, Inc.

Quirk, Lawler, and Matusky, Engineers. 1971c. Environmental effects on Hudson River, Albany Steam Station discharge. Prepared for Niagara Mohawk Power Corp.

Quirk, Lawler, and Matusky, Engineers. 1971d. Effects of Cornwall Pumped Storage Plant on Hudson River saltwater intrusion. Prepared for Consolidated Edison Co. of NY, Inc.

Quirk, Lawler, and Matusky, Engineers. 1971e. Environmental effects of Bowline Point Generating Station on Hudson River. Prepared for Orange and Rockland Utilities, Inc. and Consolidated Edison Co. of NY, Inc.

Quirk, Lawler, and Matusky, Engineers. 1971f. Environmental effects on Hudson River Lovett Plant Unit 5 submerged discharge. Prepared for Orange and Rockland Utilities, Inc.

Quirk, Lawler, and Matusky, Engineers. 1972a. Environmental impact evaluation of proposed steam-electric generating station, town of Ulster, NY. Prepared for Central Hudson Gas and Electric Corp.

Quirk, Lawler, and Matusky, Engineers. 1972b. Supplemental study of effect of submerged discharge of Indian Point cooling water on Hudson River temperature distribution. Prepared for Consolidated Edison Co. of NY, Inc.

Quirk, Lawler, and Matusky, Engineers. 1973a. A study of impinged organisms at the Astoria Generating Station. Prepared for Consolidated Edison Co. of NY, Inc.

Quirk, Lawler, and Matusky, Engineers. 1973b. Aquatic ecology studies, Kingston, NY, 1971–1972. Prepared for Central Hudson Gas and Electric Corp.

Quirk, Lawler, and Matusky, Engineers. 1973c. Interim report on larval fish studies in the vicinity of Bowline Generating Station, 1971–1972. Prepared for Orange and Rockland Utilities, Inc.

Quirk, Lawler, and Matusky, Engineers. 1973d. Interim report on plankton studies in the vicinity of Bowline Generating Station, 1971–1972. Prepared for Orange and Rockland Utilities, Inc.

Quirk, Lawler, and Matusky, Engineers. 1973e. Interim report on impingement studies in the vicinity of Bowline Generating Station, 1973. Prepared for Orange and Rockland Utilities, Inc.

Quirk, Lawler, and Matusky, Engineers. 1973f. Interim report on Hudson River water quality. Bowline Generating Station, 1971–1972. Prepared for Orange and Rockland Utilities, Inc.

Quirk, Lawler, and Matusky, Engineers. 1973g. Interim report on benthos in the vicinity of Bowline Generating Station, 1971–1972. Prepared for Orange and Rockland Utilities, Inc.

Quirk, Lawler, and Matusky, Engineers. 1973h. Statistical analysis of fish impingement data at Indian Point Generating Station. Prepared for Consolidated Edison Co. of NY, Inc.

Quirk, Lawler, and Matusky, Engineers. 1973i. The biological effects of entrainment at Indian Point. Prepared for Consolidated Edison Co. of NY, Inc.

Quirk, Lawler, and Matusky, Engineers. 1973j. Effect of Cornwall pumped storage plant operation on Hudson River salinity distribution. Prepared for Consolidated Edison Co. of NY, Inc.

Quirk, Lawler, and Matusky, Engineers. 1973k. Effect of three-unit operation at Indian Point using deep submerged discharge on Hudson River temperature distribution. Prepared for Consolidated Edison Co. of NY, Inc.

Quirk, Lawler, and Matusky, Engineers. 1973l. Preliminary evaluation of aquatic environmental effects of a proposed 125 Mw combined-cycle generating unit at Poughkeepsie. Prepared for Central Hudson Gas and Electric Corp.

Quirk, Lawler, and Matusky, Engineers. 1973m. Roseton/Danskammer Point Generating Stations, aquatic ecology studies, 1971–1972. Prepared for Central Hudson Gas and Electric Corp.

Quirk, Lawler, and Matusky, Engineers. 1974a. Cornwall gear evaluation study. Prepared for Consolidated Edison Co. of NY, Inc.

Quirk, Lawler, and Matusky, Engineers. 1974b. Danskammer Point Generating Station. 1972 fish impingement study. Prepared for Central Hudson Gas and Electric Corp.

Quirk, Lawler, and Matusky, Engineers. 1974c. 1972 Hudson River aquatic ecology studies at Bowline Unit 1. Pre-operational studies. Prepared for Orange and Rockland Utilities, Inc.

Quirk, Lawler, and Matusky, Engineers. 1974d. Hudson River hydrology study. Report to the North Atlantic Division, U.S. Army Corps of Engineers.

Quirk, Lawler, and Matusky, Engineers. 1974e. Hudson River study: Aquatic factors governing the siting of power plants. Empire State Electric Energy Research Corp.

Quirk, Lawler, and Matusky, Engineers. 1974f. 1973 Hudson River aquatic ecology study at Bowline and Lovett Generating Stations, Vols. I–V. Prepared for Orange and Rockland Utilities, Inc.

Raytheon Co., Marine Research Laboratory. 1970. Data report for June-Dec. 1969. Ecology of thermal additions: Lower Hudson River cooperative fishery study, vicinity of Indian Point, Buchanan, NY. Prepared for Consolidated Edison Co. of NY, Inc.

Raytheon Co. 1971a. Indian Point ecological survey: Report 1, 1969, Appendix P to Supplement No. 1 to environmental report for Indian Point Unit 2. Prepared for Consolidated Edison Co. of NY, Inc.

Raytheon Co. 1971b. Ecology of thermal additions, lower Hudson River cooperative fishery study vicinity of Indian Point, 1969–1971. Final report to Consolidated Edison Co. of NY, Inc.

Stevens, R.E. 1973. Final technical report, Marine Protein Corp. Culture of Hudson River striped bass in Florida. 24 pp. Prepared for Consolidated Edison Co. of NY, Inc.

Stone and Webster Engineering Corp. 1975. First progress report, Indian Point flume study. Prepared for Consolidated Edison Co. of NY, Inc.

Stone and Webster Engineering Corp. 1976. Alternative intake design with fish handling capabilities. Bowline Point Generating Stations Units 1 and 2. Prepared for Orange and Rockland Utilities, Inc.

Texas Instruments, Inc. 1972a. Airborn infrared survey of the Hudson River in the vicinity of the Indian Point Nuclear Power Station. Prepared for Consolidated Edison Co. of NY, Inc.

Texas Instruments, Inc. 1972b. Hudson River ecological study in the area of Indian Point. First semiannual report. Vol. I (Biological Sampling) and Vol. II (Standard Procedures). Prepared for Consolidated Edison Co. of NY, Inc.

Texas Instruments, Inc. 1972c. Monthly status report no. 4. Ossining environmental study for the month of Aug. 1972. Prepared for Consolidated Edison Co. of NY, Inc.

Texas Instruments, Inc. 1973a. An atlas of chemical, physical and meteorological parameters in the area of Indian Point, Hudson River, NY. Prepared for Consolidated Edison Co. of NY, Inc.

Texas Instruments, Inc. 1973b. Hudson River ecological study in the area of Indian Point. First annual report. Prepared for Consolidated Edison Co. of NY, Inc.

Texas Instruments, Inc. 1973c. Hudson River ecological study in the area of Indian Point. Second semiannual report. Prepared for Consolidated Edison Co. of NY, Inc.

Texas Instruments, Inc. 1973d. Indian Point Unit 3 environmental report. Prepared for Consolidated Edison Co. of NY, Inc.

Texas Instruments, Inc. 1973e. Hudson River environmental study in the area of Ossining. Final report, Vol. II. Temperature distribution. Prepared for Consolidated Edison Co. of NY, Inc.

Texas Instruments, Inc. 1973f. Hudson River program fisheries data summary, Vols. I and II. Prepared for Consolidated Edison Co. of NY, Inc.

Texas Instruments, Inc. 1973g. The biological effects of chemical discharges at Indian Point. Prepared for Consolidated Edison Co. of NY, Inc.

Texas Instruments, Inc. 1974a. Acute and chronic effects of evaporative cooling tower blowdown and power plant chemical discharges on white perch (*Morone americana*) and striped bass (*Morone saxatilis*). Prepared for Consolidated Edison Co. of NY, Inc.

Texas Instruments, Inc. 1974b. Feasibility study and design development striped bass fish hatchery, Hudson River, NY. Prepared for Consolidated Edison Co. of NY, Inc.

Texas Instruments, Inc. 1974c. Feasibility of culturing and stocking Hudson River striped bass. 1973 annual report. Prepared for Consolidated Edison Co. of NY, Inc.

Texas Instruments, Inc. 1974d. First annual progress report on fisheries investigation of the Hudson River as related to the Cornwall pumped-storage Hydroelectric Plant, Cornwall, NY. Prepared for Consolidated Edison Co. of NY, Inc.

Texas Instruments, Inc. 1974e. Fisheries survey of the Hudson River Mar.–July 1973. Vol. III. Prepared for Consolidated Edison Co. of NY, Inc.

Texas Instruments, Inc. 1974f. Hudson River ecological study in the area of Indian Point. 1973 annual report. Prepared for Consolidated Edison Co. of NY, Inc.

Texas Instruments, Inc. 1974g. Indian Point impingement study report for the period June 15, 1972 through Dec. 31, 1973. Prepared for Consolidated Edison Co. of NY, Inc.

Texas Instruments, Inc. 1974h. Second semi-annual report related to the feasibility study for spawning, hatching and stocking striped bass in the Hudson River. Prepared for Consolidated Edison Co. of NY, Inc.

Texas Instruments, Inc. 1975a. A preliminary evaluation of the potential impact on the fisheries of the Hudson River by the Cornwall pumped-storage project. Prepared for Consolidated Edison Co. of NY, Inc.

Texas Instruments, Inc. 1975b. Benthic land-fill studies. Cornwall final report. Prepared for Consolidated Edison Co. of NY, Inc.

Texas Instruments, Inc. 1975c. Feasibility of culturing and stocking Hudson River striped bass. 1974 annual report. Prepared for Consolidated Edison Co. of NY, Inc.

Texas Instruments, Inc. 1975d. First annual report for the multiplant impact study of the Hudson River estuary. Vol. I (text) and Vol. II (appendices). Prepared for Consolidated Edison Co. of NY, Inc.

Texas Instruments, Inc. 1975e. Hudson River ecological study in the area of Indian Point. 1974 annual report. Prepared for Consolidated Edison Co. of NY, Inc.

Texas Instruments, Inc. 1975f. Hudson River environmental study in the area of Ossining. Final report. Prepared for Consolidated Edison Co. of NY, Inc.

Texas Instruments, Inc. 1975g. 1974 Hudson River program fisheries data display, Vols. I and III. Prepared for Consolidated Edison Co. of NY, Inc.

Texas Instruments, Inc. 1975h. 1974 Hudson River program ichthyoplankton data summary. Prepared for Consolidated Edison Co. of NY, Inc.

Texas Instruments, Inc. 1975i. Indian Point impingement study report for the period Jan. 1, 1974 through Dec. 31, 1974. Prepared for Consolidated Edison Co. of NY, Inc.

Texas Instruments, Inc. 1975j. Semiannual progress report for Hudson River ecological study in the area of Indian Point, Jan. 1 to June 30, 1974. Prepared for Consolidated Edison Co. of NY, Inc.

Texas Instruments, Inc. 1975k. Semiannual progress report for the multiplant impact study of the Hudson River estuary, May to Nov. 1974. Prepared for Consolidated Edison Co. of NY, Inc.

Texas Instruments, Inc. 1975l. The final report of the synoptic subpopulations analysis. Phase I: A report on the feasibility of using innate tags to identify striped bass (*Morone saxatilis*) from various spawning rivers. Prepared for Consolidated Edison Co. of NY, Inc.

Texas Instruments, Inc. 1976a. A synthesis of available data pertaining to major physicochemical variables within the Hudson River estuary emphasizing the period from 1972 through 1975. Prepared for Consolidated Edison Co. of NY, Inc.

Texas Instruments, Inc. 1976b. 1975 fisheries data display. Vols. I–III. Prepared for Consolidated Edison Co. of NY, Inc.

Texas Instruments, Inc. 1976c. 1975 Cornwall transect data display. Prepared for Consolidated Edison Co. of NY, Inc.

Texas Instruments, Inc. 1976d. Fisheries survey of the Hudson River, March to Dec. 1973. Vol. IV (revised ed.). Prepared for Consolidated Edison Co. of NY, Inc.

Texas Instruments, Inc. 1976e. Hudson River ecological study in the area of Indian Point. 1975 annual report. Prepared for Consolidated Edison Co. of NY, Inc.

Texas Instruments, Inc. 1976f. Hudson River ecological study in the area of Indian Point. Thermal effects report. Prepared for Consolidated Edison Co. of NY, Inc.

Texas Instruments, Inc. 1976g. 1974 Hudson River program fisheries data display. Vol. II. Prepared for Consolidated Edison Co. of NY, Inc.

Texas Instruments, Inc. 1976h. 1975 ichthyoplankton data display. Prepared for Consolidated Edison Co. of NY, Inc.

Texas Instruments, Inc. 1976i. Indian Point impingement study report for the period Jan. 1, 1975 through Dec. 31, 1975. Prepared for Consolidated Edison Co. of NY, Inc.

Texas Instruments, Inc. 1976j. Liberty State Park ecological study. Final report. Prepared for the Port Authority of New York and New Jersey.

Texas Instruments, Inc. 1976k. Predation by bluefish in the lower Hudson River. Prepared for Consolidated Edison Co. of NY, Inc.

Texas Instruments, Inc. 1976l. Report on relative contributions of Hudson River striped bass to the Atlantic coastal fishery. Prepared for Consolidated Edison Co. of NY, Inc.

Texas Instruments, Inc. 1977a. 1972–1973 benthic data display. Prepared for Consolidated Edison Co. of NY, Inc.

Texas Instruments, Inc. 1977b. 1974 benthic data display. Prepared for Consolidated Edison Co. of NY, Inc.

Texas Instruments, Inc. 1977c. 1975 biological characteristics data display. Prepared for Consolidated Edison Co. of NY, Inc.

Texas Instruments, Inc. 1977d. 1976 biological characteristics data display. Prepared for Consolidated Edison Co. of NY, Inc.

Texas Instruments, Inc. 1977e. Feasibility of culturing and stocking Hudson River striped bass. 1975 annual report. Prepared for Consolidated Edison Co. of NY, Inc.

Texas Instruments, Inc. 1977f. Feasibility of culturing and stocking Hudson River striped bass. An overview, 1973–1975. Prepared for Consolidated Edison Co. of NY, Inc.

Texas Instruments, Inc. 1977g. 1972 fisheries data display. Prepared for Consolidated Edison Co. of NY, Inc.

Texas Instruments, Inc. 1977h. Gear evaluation data display. Prepared for Consolidated Edison Co. of NY, Inc.

Texas Instruments, Inc. 1977i. 1974 and 1975 gear evaluation studies. Prepared for Consolidated Edison Co. of NY, Inc.

Texas Instruments, Inc. 1977j. Hudson River ecological study in the area of Indian Point. 1976 annual report. Prepared for Consolidated Edison Co. of NY, Inc.

Texas Instruments, Inc. 1977k. 1973 ichthyoplankton data display. Prepared for Consolidated Edison Co. of NY, Inc.

Texas Instruments, Inc. 1977l. 1976 ichthyoplankton data display. Prepared for Consolidated Edison Co. of NY, Inc.

Texas Instruments, Inc. 1977m. 1972 impingement data display. Prepared for Consolidated Edison Co. of NY, Inc.

Texas Instruments, Inc. 1977n. 1973 impingement data display. Prepared for Consolidated Edison Co. of NY, Inc.

Texas Instruments, Inc. 1977o. 1974 impingement data display. Prepared for Consolidated Edison Co. of NY, Inc.

Texas Instruments, Inc. 1977p. 1975 impingement data display. Prepared for Consolidated Edison Co. of NY, Inc.

Texas Instruments, Inc. 1977q. 1976 impingement data display. Prepared for Consolidated Edison Co. of NY, Inc.

Texas Instruments, Inc. 1977r. Production of striped bass for experimental purposes. 1976 hatchery report. Prepared for Consolidated Edison Co. of NY, Inc.

Texas Instruments, Inc. 1977s. Production of striped bass for power plant entrainment studies. 1977 hatchery report. Prepared for Consolidated Edison Co. of NY, Inc.

Texas Instruments, Inc. 1977t. 1975 relative contribution data display. Prepared for Consolidated Edison Co. of NY, Inc.

Texas Instruments, Inc. 1977u. 1974 synoptic subpopulation data display. Prepared for Consolidated Edison Co. of NY, Inc.

Texas Instruments, Inc. 1977v. 1972 water quality data display. Prepared for Consolidated Edison Co. of NY, Inc.

Texas Instruments, Inc. 1977w. 1973 water quality data display. Prepared for Consolidated Edison Co. of NY, Inc.

Texas Instruments, Inc. 1977x. 1974 water quality data display. Prepared for Consolidated Edison Co. of NY, Inc.

Texas Instruments, Inc. 1977y. 1975 water quality data display. Prepared for Consolidated Edison Co. of NY, Inc.

Texas Instruments, Inc. 1977z. 1976 water quality data display. Prepared for Consolidated Edison Co. of NY, Inc.

Texas Instruments, Inc. 1977aa. 1974 year-class report for the multiplant impact study of the Hudson River estuary, Vol. I. Prepared for Consolidated Edison Co. of NY, Inc.

Texas Instruments, Inc. 1977bb. 1974 year-class report for the multiplant impact study of the Hudson River estuary, Vol. II (appendices). Prepared for Consolidated Edison Co. of NY, Inc.

Texas Instruments, Inc. 1977cc. 1974 year-class report for the multiplant impact study of the Hudson River estuary, Vol. III (lower estuary study). Prepared for Consolidated Edison Co. of NY, Inc., Orange and Rockland Utilities, Inc., and Central Hudson Gas and Electric Corp.

Texas Instruments, Inc. 1978a. 1977 biocharacteristics data display. Prepared for Consolidated Edison Co. of NY, Inc.

Texas Instruments, Inc. 1978b. Catch efficiency of 100 ft. (30 m) beach seines for estimating density of young-of-the-year striped bass and white perch in the shore zone of the Hudson River estuary. Prepared for Consolidated Edison Co. of NY, Inc.

Texas Instruments, Inc. 1978c. Evaluation of a submerged weir to reduce fish impingement at Indian Point for the period May 25 to July 26, 1977. Prepared for Consolidated Edison Co. of NY, Inc.

Texas Instruments, Inc. 1978d. 1973 fisheries data display. Prepared for Consolidated Edison Co. of NY, Inc.

Texas Instruments, Inc. 1978e. 1976 fisheries data display. Prepared for Consolidated Edison Co. of NY, Inc.

Texas Instruments, Inc. 1978f. 1977 fisheries data display. Prepared for Consolidated Edison Co. of NY, Inc.

Texas Instruments, Inc. 1978g. 1977 ichthyoplankton data display. Prepared for Consolidated Edison Co. of NY, Inc.

Texas Instruments, Inc. 1978h. 1977 impingement data display. Prepared for Consolidated Edison Co. of NY, Inc.

Texas Instruments, Inc. 1978i. Initial and extended survival of fish collected from a fine mesh continuously operating traveling screen at the Indian Point Generating Station for the period June 15 to Dec. 22, 1977. Prepared for Consolidated Edison Co. of NY, Inc.

Texas Instruments, Inc. 1978j. Relationship between population density and growth for juvenile striped bass in the Hudson River. Prepared for Consolidated Edison Co. of NY, Inc. Submitted in the EPA-utilities Adjudicatory Hearing No. C/II-WP-77-01, Exb. UT-49.

Texas Instruments, Inc. 1978k. 1977 water quality data display. Prepared for Consolidated Edison Co. of NY, Inc.

Texas Instruments, Inc. 1978l. 1975 year-class report for the multiplant study of the Hudson River estuary. Prepared for Consolidated Edison Co. of NY, Inc.

Texas Instruments, Inc. 1979a. A holistic perspective of the Hudson River ecosystem with major focus on the fish populations. Prepared for Consolidated Edison Co. of NY, Inc. (draft.)

Texas Instruments, Inc. 1979b. Data on shortnose sturgeon (*Acipenser brevirostrum*) collected incidentally from 1969 through June 1979 in sampling programs conducted for the Hudson River ecological study. Prepared for Consolidated Edison Co. of NY, Inc.

Texas Instruments, Inc. 1979c. Efficiency of a 100-ft beach seine for estimating shore zone densities at night of juvenile striped bass, juvenile white perch and yearling and older (150mm) white perch. Prepared for Consolidated Edison Co. of NY, Inc.

Texas Instruments, Inc. 1979d. Efficiency of the $1.0m^2$ epibenthic sled for collecting young-of-the-year striped bass, white perch and Atlantic tomcod in the Hudson River estuary, 1978. Prepared for Consolidated Edison Co. of NY, Inc.

Texas Instruments, Inc. 1979e. 1978 fisheries data display. Prepared for Consolidated Edison Co. of NY, Inc.

Texas Instruments, Inc. 1979f. Hudson River ecological study in the area of Indian Point. 1977 annual report. Prepared for Consolidated Edison Co. of NY, Inc.

Texas Instruments, Inc. 1979g. Hatchery letter report. Sept. 1979. Prepared for Consolidated Edison Co. of NY, Inc.

Texas Instruments, Inc. 1979h. Collection efficiency and survival estimates of fish impinged on a fine mesh continuously operating traveling screen at the Indian Point Generating Station for the period Aug. 8 to Nov. 10, 1978. Prepared for Consolidated Edison Co. of NY, Inc.

Texas Instruments, Inc. 1979i. 1978 ichthyoplankton data display. Prepared for Consolidated Edison Co. of NY, Inc.

Texas Instruments, Inc. 1979j. 1978 impingement data display. Prepared for Consolidated Edison Co. of NY, Inc.

Texas Instruments, Inc. 1979k. Production of striped bass for experimental purposes. 1978 hatchery report. Prepared for Consolidated Edison Co. of NY, Inc.

Texas Instruments, Inc. 1979l. 1978 water quality data display. Prepared for Consolidated Edison Co. of NY, Inc.

Texas Instruments, Inc. 1979m. 1976 year-class report for the multiplant impact study of the Hudson River estuary. Prepared for Consolidated Edison Co. of NY, Inc.

Texas Instruments, Inc. 1980a. 1979 fisheries data display. Prepared for Consolidated Edison Co. of NY, Inc.

Texas Instruments, Inc. 1980b. Hudson River ecological study in the area of Indian Point. 1978 annual report. Prepared for Consolidated Edison Co. of NY, Inc.

Texas Instruments, Inc. 1980c. Hudson River ecological study in the area of Indian Point. 1979 annual report. Prepared for Consolidated Edison Co. of NY, Inc.

Texas Instruments, Inc. 1980d. 1979 ichthyoplankton data display. Prepared for Consolidated Edison Co. of NY, Inc.

Texas Instruments, Inc. 1980e. 1979 impingement data display. Prepared for Consolidated Edison Co. of NY, Inc.

Texas Instruments, Inc. 1980f. Report on 1978–1979 studies to evaluate catch efficiency of the $1.0m^2$ epibenthic sled. Prepared for Consolidated Edison Co. of NY, Inc.

Texas Instruments, Inc. 1980g. 1979 water quality data display. Prepared for Consolidated Edison Co. of NY, Inc.

Texas Instruments, Inc. 1980h. 1977 year-class report for the multiplant impact study of the Hudson River estuary. Prepared for Consolidated Edison Co. of NY, Inc.

Texas Instruments, Inc. 1980i. 1978 year-class report for the multiplant impact study of the Hudson River estuary. Prepared for Consolidated Edison Co. of NY, Inc.

Texas Instruments, Inc. 1981a. 1979 year-class report for the multiplant impact study of the Hudson River estuary. Prepared for Consolidated Edison Co. of NY, Inc.

Texas Instruments, Inc. 1981b. 1979 bottom trawl comparability study for the interregional trawl survey. Prepared for Consolidated Edison Co. of NY, Inc.

Bibliography B: Open Literature

I. Organisms

A. Phytoplankton

Bates, S.S. and J.S. Craigie. 1978. Chloroplast pigments of a green phytoplankter from the Hudson Estuary, USA. Phycologie 17(1): 79–84.

Brook, A.J. and A.L. Baker. 1972. Chlorination at power plants: Impact on phytoplankton productivity. Science 176:1414–1415.

Burkholder, P.R. 1932. Plankton studies in some lakes of the upper Hudson watershed, pp. 239–263 *In* A Biological Survey of the Upper Hudson Watershed. Supplement to 22nd Annual Report. N.Y.S. Conservation Dept., Albany, NY.

Ducklow, H.W. and D.L. Kirchman. 1983. Bacterial dynamics and distribution during a spring diatom bloom in the Hudson River plume, USA. J. Plankton Research 5(3):333–356.

Feldman, A.E. 1973. A microfaunal and microflora population investigation of the Hudson River in the vicinity of Roseton, NY. Paper 16 In Hudson River Ecology, 3rd Symp. Hudson River Environ. Soc., Bear Mountain, NY.

*Fredrick, S.W., R.L. Heffner, A.T. Packard, P.M. Eldridge, J.C. Eldridge, G.J. Schumacher, K.L. Eichorn, J.H. Currie, J.N. Richards, and O.C. Broody. 1976. Notes on phytoplankton distribution in the Hudson River estuary, In Hudson River Ecology, 4th Symp. Hudson River Environ. Soc., Bronx, NY.

Gold, K. 1969. Plankton as indicators of pollution in New York waters, pp. 281–292 In Hudson River Ecology, 2nd Symp. NYS Dept. of Environmental Conservation, Albany, NY.

Heffner, R.L. 1973. Phytoplankton community dynamics in the Hudson River estuary between mile points 39 and 77. Paper 17 In Hudson River Ecology, 3rd Symp. Hudson River Environ. Soc., Bear Mountain, NY.

Heffner, R.L., G.P. Howells, G.J. Lauer, and H.I. Hirshfield. 1971. Effects of power plant operation on Hudson River estuary micro-biota. Third Nat. Symp. on Radioecology. Oak Ridge, TN. USAEC Conf.-710501-PI.

*Heffner, R.L., A.T. Packard, and S.W. Fredrick. 1976. A review of some phytoplankton methodologies. Paper 18 In Hudson River Ecology, 4th Symp. Hudson River Environ. Soc., Bronx, NY.

*Howells, G.P. and S. Weaver. 1969. Studies on phytoplankton at Indian Point, pp. 231–261. In Hudson River Ecology, 2nd Symp. N.Y.S. Dept. of Environmental Conservation, Albany, NY.

Kingsbury, J.M. 1968. Review of algal literature for New York State, In Algae, Man and the Environment (D.J. Jackson, ed.). Syracuse University Press, Syracuse, NY.

Lee, J.A. 1951. A study of the marine algae of New York Harbor, Long Island and adjacent waters with particular reference to the location of weed beds, In The Hydrology of the New York Area (J.C. Ayers et al., eds.). Status Report 4. Contract N6 ONR 264. Task 15. Cornell University, Ithaca, NY.

Malone, T.C. 1974. Phytoplankton productivity, nutrient recycling and energy flow in the inner New York Bight, Sept. 1973 to Feb. 1974. Lamont-Doherty Geological Observatory, Columbia University, Palisades, NY. 7 pp.

Malone, T.C. 1976. Phytoplankton productivity in the apex of the New York Bight: Environmental regulation of productivity/chlorophyll a, pp. 260–272 In Middle Atlantic Coastal Shelf and the New York Bight, (M.G. Gross, ed.). Am. Soc. Limn. Ocean., Inc., Lawrence, KA. 441 pp.

Malone, T.C. 1976. Phytoplankton productivity in the apex of the New York Bight: Sept. 1973 to Aug. 1974. NOAA, Boulder, CO. NOAA-TM-ERL-MESA-5. 106 pp. NTIS - PB-256-245.

*Malone, T.C. 1977. Environmental regulation of phytoplankton productivity in the lower Hudson estuary. Est. Coastal Mar. Sci. 5:157–171.

Malone, T.C. 1977. Light-saturated photosynthesis by phytoplankton size fractions in the New York Bight, USA. Mar. Biol. 42(4):281–292.

Malone, T.C. and M. Chervin. 1979. The production and fate of phytoplankton

---

*References included in summary tables in Chapter 7.

size fractions in the plume of the Hudson River, New York Bight. Limnol. Oceanogr. 24:683-696.

Malone, T.C., P.J. Neale, and D.C. Boardman. 1979. Influences of estuarine circulation on the distribution and biomass of phytoplankton size fractions, *In* Estuarine Perspectives (V.S. Kennedy, ed.). Academic Press, New York, NY.

Malone, T.C., C. Garside, and P.J. Neale. 1980. Effects of silicate depletion on photosynthesis by diatoms in the plumes of the Hudson River. Mar. Biol. 58:197-204.

Malone, T.C., P.G. Falkowski, T.S. Hopkins, G.T. Rowe, and T.E. Whitledge. 1983. Mesoscale response of diatom populations to a wind event in the plume of the Hudson River, USA. Sea Res. Part A: Oceangr. Research Papers 30(2):147-170.

*Neale, P.J., T.C. Malone, and D.C. Boardman. 1981. Effects of fresh water flow on salinity and phytoplankton biomass in the lower Hudson estuary, pp. 168-184 *In* Proc. of the Nat. Symp. on Freshwater Inflow to Estuaries (R. Cross and D. Williams, eds.). U.S. Fish and Wildl. Serv. Office of Biological Services, FWS/OBS-81/04. (2 vols.)

*O'Reilly, J., J.P. Thomas, and C. Evans. 1976. Annual primary production (nannoplankton, netplankton, dissolved organic matter) in the lower New York Bay. Paper 19 *In* Hudson River Ecology, 4th Symp. Hudson River Environ. Soc., Bronx, NY.

Pierce, M.E. and D. Alessandrello. 1972. A study of the organisms living in the heated effluent of a power plant. Proc. Northeast Weed Soc. 26:167-172.

Powers, C.D., G.M. Nau-Ritter, R.G. Rowland, and C.F. Wurster. 1982. Field and laboratory studies of the toxicity to phytoplankton of polychlorinated biphenyls (PCBs) desorbed from fine clays and natural suspended particulates. J. Great Lakes Res. 8(2):350-357.

Sirois, D.L. 1973. Community metabolism and water quality in the lower Hudson River estuary. Paper 15 *In* Hudson River Ecology, 3rd Symp. Hudson River Environ. Soc., Bear Mountain, NY.

Sirois, D.L. and S.W. Fredrick. 1978. Phytoplankton and primary production in the lower Hudson River estuary. Est. Coastal Mar. Sci. 7:413-423.

Storm, P.C. 1979. A description and analysis of phytoplankton production in the Hudson River estuary during 1974 and 1975. Ph.D. Thesis. New York University, New York, NY. 109 pp.

*Storm, P.C. and P.L. Heffner. 1976. A comparison of phytoplankton abundance, chlorophyll *a* and water quality factors in the Hudson River and its tributaries. Paper 17 *In* Hudson River Ecology, 4th Symp. Hudson River Environ. Soc., Bronx, NY.

Weaver, S.S. 1970. Phytoplankton in the Hudson River at Indian Point: 1968-1969. Thesis. New York University, New York, NY. 111 pp.

Weiss, D., K. Geitzenauer, and F.J. Shaw. 1976. Foraminifera, diatoms, and mollusks as potential Holocene paleoecologic indicators in the Hudson Estuary (abstract). Geol. Soc. Am. Abstr. Prog. 8(2):296-297.

*Weiss, D., K. Geitzenauer, and F.C. Shaw. 1978. Foraminifera, diatom and bivalve distribution in recent sediments of the Hudson estuary. Est. Coastal Mar. Sci. 7(4):393-400.

*Williams, L.G. and C. Scott. 1962. Principal diatoms of major waterways of the United States. Limnol. Oceanogr. 7:365-379.

B. Macrophytes

Bard, R.L. 1959. Battle report on water chestnut. Conservationist 19(4):13-15.

Bard, R.L. 1965. Eurasian water milfoil. A new threat to our waterways. Conservationist 14(6):10-11.

Bard, R.L. 1965. Eradication of water chestnut. Conservationist 19(4):13-15.

Bard, R.L. 1976. Region 3 water chestnut eradication summary, 1976. N.Y.S. Dept. of Environmental Conservation, New Paltz, NY.

*Buckley, E.H. and S.S. Ristich. 1976. Distribution of rooted vegetation in the brackish marshes and shallows of the Hudson River estuary. Paper 20 In Hudson River Ecology, 4th Symp. Hudson River Environ. Soc., Bronx, NY.

Foley, D. 1948. Hudson River marsh surveys, pp. 215-254. N.Y.S. Conservation Dept., Albany, NY.

Hook, S.M. 1978. Control of water chestnut in New York State, pp. 55-59 In Proceedings Research Planning Control Program, Oct. 19-22, 1976. Atlantic Beach, FL.

*Kiviat, E. 1976. Goldenclub, a threatened plant in the tidal Hudson River. Paper 21 In Hudson River Ecology, 4th Symp. Hudson River Environ. Soc., Bronx, NY.

Lehr, J.H. 1967. The marshes at Piermont, NY. A field report. Sarracenia 11:31-34.

Lehr, J.H. 1967. The plants of Iona Island. A field report. Sarracenia 11:35-38.

McKeon, W. 1974. An appraisal of the current status of marshes or wetland areas along the Hudson River. Hudson River Environ. Soc., Bronx, NY. 7 pp. (unpublished.)

*McVaugh, R. 1936. Some aspects of plant distribution in the Hudson River estuary. Bartonia 17:13-16.

Menzie, C.A. 1979. Growth of the aquatic plant *Myriophyllum spicatum* in a littoral area of the Hudson River estuary. Aquat. Bot. 6(4):365-375.

Menzie, C.A. unpublished. Observations of the distribution of aquatic macrophytes in the Hudson estuary with respect to salinity. City College of NY Biology Dept., New York, NY.

Muenscher, W.C. 1932. Aquatic vegetation of the upper Hudson watershed, pp. 216-238 In A Biological Survey of the Upper Hudson Watershed. Supplement to 22nd Annual Report. N.Y.S. Conservation Dept., Albany, NY.

Muenscher, W.C. 1935. Aquatic vegetation of the Mohawk-Hudson watershed, pp. 228-249 In Supplement to 24th Annual Report. N.Y.S. Conservation Dept., Biol. Surv. 9, Albany, NY.

Muenscher, W.C. 1937. Aquatic vegetation of the lower Hudson area, pp. 231-248 In Supplement to 26th Annual Report. N.Y.S. Conservation Dept., Biol. Surv. 11, Albany, NY.

Schuyler, A.E. 1975. *Scirpus cylindricus*: An ecologically restricted eastern North American tuberous bullrush. Bartonia (43):29-37.

Svenson, H.K. 1935. Plants from the estuary of the Hudson River. Torreya 35(5):117-125.

Torrey, R.H. 1931. Trip of August 2. Torreya 31:153-154.

U.S. Army Corps of Engineers. 1971. Aquatic plant control program, Hudson and Mohawk Rivers, New York. (Final environmental impact statement.) U.S. Dept. of the Army, Corps of Engineers, New York District, NY. 17 pp. NTIS-PB-200-003-F.

C. Invertebrates

Ayers, J.C. 1951. The average rate of fouling of surface objects and of submerged objects in the waters adjacent to New York Harbor, In The Hydrography of the New York Area (J.C. Ayers et al., eds.). Status Rep. No. 6, Contract N6 ORN 264. Cornell University, Ithaca, NY.

Blackford, E.G. 1887. Report of Commission of Fisheries (of New York State) in charge of the oyster investigation and of survey of oyster territory for the years 1885-1886.

Boesch, D.F. 1982. Ecosystem consequences of alterations of benthic community structure and function in the New York Bight region, pp. 543-568 In Ecological Stress and the New York Bight: Science and Management (G.F. Mayer, ed.). Estuarine Research Federation, Columbia, SC.

*Burbanck, W.D. 1962. Further observations on the biotype of the estuarine isopod, Cyathura polita. Ecology 43(4):719-722.

Chervin, M.B., T.C. Malone, and P.J. Neale. 1981. Interactions between suspended organic matter and copepod grazing in the plume of the Hudson River. Est. Coast. and Shelf Sci. 13(2):169-184.

Crandall, M.E. 1976. The distribution of bivalve mollusks in the Hudson River estuary. M.S. Thesis. Long Island University, C.W. Post Campus, Brentwood, NY.

*Crandall, M.E. 1977. Epibenthic invertebrates of Croton Bay in the Hudson River. N.Y. Fish Game J. 24(2):178-186.

Crocker, D.W. 1957. The crayfishes of New York State. N.Y.S. Mus. Bull. 355. 97 pp.

Duch, T.M. 1976. Aspects of the feeding habits of Viviparus georgianus. Nautilus 90(1):7-10.

Feigl, F.J. 1955. The northernmost limit of living foraminifera in the Hudson River. M.S. Thesis. New York University, New York, NY. 28 pp.

Feldman, A.E. 1973. A microfaunal and microfloral population investigation of the Hudson River in the vicinity of Roseton, NY. Paper 16 In Hudson River Ecology, 3rd Symp. Hudson River Environ. Soc., Bear Mountain, NY.

Fitzpatrick, J.F., Jr. and J.F. Pickett, Sr. 1980. A new crayfish of the genus Orconectes from eastern New York (Decapoda: Cambaridae). Proc. of the Biol. Soc. of Wash. 93:373-382.

Fowler, H. 1912. The crustacea of New Jersey. Ann. Rep. N.J. State Museum, Trenton, NJ.

Franz, D.R. 1982. An historical perspective on molluscs in lower New York harbor, with emphasis on oysters, pp. 181-197 In Ecological Stress and the New York Bight: Science and Management (G.F. Mayer, ed.). Estuarine Research Federation, Columbia, SC.

Ginn, T.C. 1977. An ecological investigation of Hudson River macrozooplankton in the vicinity of a nuclear power plant. Ph.D. Dissertation. New York University, New York, NY.

*Ginn, T.C., W.T. Waller, and G.J. Lauer. 1976. Survival and reproduction of Gammarus spp. (Amphipoda) following short-term exposure to elevated temperatures. Chesapeake Sci. 17(1):8-14.

Ginn, T.C. and J.M. O'Connor. 1978. Response of the estuarine amphipod Gammarus diabert to chlorinated power plant effluent. Est. Coastal Mar. Sci. 6(5):459-469.

Gosner, K.L. 1969. Biological fouling on navigational buoys in the Hudson River estuary. Bull. N.J. Acad. Sci. 14(1–2):49.

Gould, J.D. 1969. Conservation interests in the Hudson River, pp. 30–39 *In* Hudson River Ecology, 2nd Symp. N.Y.S. Dept. of Environmental Conservation, Albany, NY.

Grabe, S.A. and J. Alber. 1977. The occurrence of *Chiridotea almyra* Bowman, 1955 (Isopoda, Valvifera) in the limnetic sector of the lower Hudson River. Crustaceana 33(1):103–104.

*Hirschfield, H.I., J.W. Rachlin, and E. Leff. 1966. A survey of the invertebrates from selected sites of the lower Hudson, pp. 220–257 *In* Hudson River Ecology, 1st Symp. Hudson River Valley Comm.

Hogan, T.M., B.S. Williams, and Z. Zo. 1974. The ecology of the estuarine isopod, *Cyathura polita* (Stimpson), in the Hudson River. Paper 19 *In* Hudson River Ecology, 3rd Symp. Hudson River Environ. Soc., Bear Mountain, NY.

*Howells, G.P., E. Musnick, and H.I. Hirschfield. 1969. Invertebrates of the Hudson River, pp. 262–280 *In* Hudson River Ecology, 2nd Symp. N.Y.S. Dept. of Environmental Conservation, Albany, NY.

Lauer, G.J., W.T. Waller, D.W. Bath, W. Meeks, R. Heffner, T. Ginn, L. Zubarik, P. Bibko, and P.C. Storm. 1974. Entrainment studies on Hudson River organisms, *In* Proceedings of the Second Workshop on Entrainment and Intake Screening (L.D. Jensen, ed.). Johns Hopkins University, Baltimore, MD. Feb. 5–9, 1973.

Means, E.A. 1898. A study of the vertebrate fauna of the Hudson Highlands, with observations on the Mollusca, Crustacea, Lepidoptera and the flora of the region. Bull. Am. Mus. Nat. Hist. 10:303–352.

Menzie, C.A. 1978. Productivity of chironomid larvae in a littoral area of the Hudson River estuary. Ph.D. Thesis. City University of NY, New York, NY.

Menzie, C.A. 1980. The chironomid (Insecta: Diptera) and other fauna of a *Myriophyllum spicatum* L. plant bed in the lower Hudson River. Estuaries 3(1):38–54.

*Menzie, C.A., B. Woodard, and R. Hyman. 1976. Investigation of the chironomid insect fauna of Haverstraw Bay. Paper 22 *In* Hudson River Ecology, 4th Symp. Hudson River Environ. Soc., Bronx, NY.

Musnick, E. 1969. Hudson River invertebrates. Masters Thesis. New York University, New York, NY.

Neuderfer, G.N. and A.L. Cooper. 1972. A macro-invertebrate study of Moodna Creek. N.Y.S. Dept. of Environmental Conservation, New Paltz, NY. 32 pp. (unpublished.)

*Occhiogrosso, T., W. Waller, and G. Lauer. 1976. Effects of heavy metals on benthic macroinvertebrate densities in Foundry Cove on the Hudson River. Paper 9 *In* Hudson River Ecology, 4th Symp. Hudson River Environ. Soc., Bronx, NY.

Pate, V.S.L. 1932. Studies of fish food in selected areas, pp. 130–149 *In* A Biological Survey of the Upper Hudson Watershed. Supplement to 22nd Annual Report. N.Y.S. Conservation Dept., Albany, NY.

Paulmier, F.C. 1905. Higher crustacea of New York City. N.Y.S. Mus. Bull. 12(81). 189 pp.

*Pearce, J.B. 1974. Invertebrates of the Hudson River Estuary, pp. 137–143 *In* Hudson River Colloquium (O.A. Roels, ed.). Ann. N.Y. Acad. Sci. 250.

*Ristich, S.S., M. Crandall, and J. Fortier. 1977. Benthic and epibenthic macroinvertebrates of the Hudson River. Est. Coastal Mar. Sci. 5:255–266.

Schupak, B. 1934. Some foraminifera from western Long Island Sound and New York harbor. Am. Mus. Noviates No. 737. 12 pp.

Shaw, F.C. 1975. Distribution of shelled macroinvertebrate benthos in the Hudson estuary, *In* Programs and Abstracts of the 7th Long Island Sound Conference, Jan. 11, 1975, (abstract).

*Simpson, K.W. 1976. A water quality evaluation of the Hudson River, based on the collection and analysis of macroinvertebrate communities. Paper 24 *In* Hudson River Ecology, 4th Symp. Hudson River Environ. Soc., Bronx, NY.

Simpson, K.W., T.B. Lyons, and S.P. Allen. 1974. Macroinvertebrate survey of the upper Hudson River, New York—1972. N.Y.S. Dept. of Health, Environmental Health Rep. No. 2. 52 pp.

Smith, D.G. 1979. New locality records of crayfishes from the middle Hudson River system. Ohio J. Sci. 79(3):133–135.

Smith, D.G. 1981. A note on the morphological variability of *Orconectes kinderhookensis* (Decapoda: Cambaridae) from the Hudson River system in New York. J. Crustacean Biology 1(3):389–391.

Stepien, J.C., T.C. Malone, and M.B. Chervin. 1981. Copepod communities in the estuary and coastal plume of the Hudson River. Est. Coast. and Shelf Sci. 13(2):185–196.

Thomas, J.P., W. Phoel, J.E. O'Reilly, and C. Evans. 1976. Seabed oxygen consumption in the lower Hudson estuary. NOAA, NMFS, Sandy Hook Lab., Highlands, NJ. 21 pp.

Townes, H.K., Jr. 1937. Studies on the food organisms of fish, pp. 217–229 *In* Supplement to 26 Annual Report. N.Y.S. Conservation Dept., Biol. Survey No. 11, Albany, NY.

Udell, H.F. 1962. Oysters—a delicacy from the cradle of the sea. Conservationist 17(2):9–11.

Weiss, D. 1976. Distribution of benthic foraminifera in the Hudson estuary, pp. 119–128 *In* First International Symposium on Benthic Foraminifera of Continental Margins, Part A, Ecology and Biology. Maritime Sediments Spec. Publ. L.

Weiss, D., K. Geitzenauer, and F.J. Shaw. 1976. Foraminifera, diatoms, and mollusks as potential Holocene paleoecologic indicators in the Hudson estuary (abstract). Geol. Soc. Am. Abstr. Prog. 8(2):296–297.

*Weiss, D., K. Geitzenauer, and F.C. Shaw. 1978. Foraminifera, diatom and bivalve distribution in recent sediments of the Hudson estuary. Est. Coastal Mar. Sci. 7(4):393–400.

Williams, B.S., T. Hogan, and Z. Zo. 1974. Indian Point benthic studies—1972. Paper 19 *In* Hudson River Ecology, 3rd Symp. Hudson River Environ. Soc., Bear Mountain, NY.

*Williams, B.S., T. Hogan, and Z. Zo. 1975. The benthic environment of the Hudson River in the vicinity of Ossining, NY, during 1972 and 1973. N.Y. Fish Game J. 22(1):25–31.

*Zubarik, L. 1976. Entrainment effects of microzooplankton. Paper 23 *In* Hudson River Ecology, 4th Symp. Hudson River Environ. Soc., Bronx, NY.

D. Birds, Reptiles, and Amphibians

Craig, R.J., M.W. Klemens, and S.S. Craig. 1980. The northeastern range limit of the eastern mud turtle *Kinosteron s. subrubrum* (Lacepode). J. Herpetology 14(3):295-297.

Deed, R.F. 1959. Birds of Rockland County and the Hudson Highlands. Rockland Audubon Soc., West Nyack, NY. 44 pp.

Deed, R.F. 1968. Supplement to Birds of Rockland County and the Hudson Highlands. Rockland Audubon Soc., West Nyack, NY. 27 pp.

Deed, R.F. 1976. Rockland County's on-again, off-again resting places for shorebirds. Linnaean News-Letter 29(7):1-3.

Deed, R.F. 1981. The endless change in a local checklist. Linnean News-Letter 34(7-8):1-2.

Drennan, S.R. 1981. Where to find birds in New York State; the top 500 sites. Syracuse University Press, Syracuse, NY. 499 pp.

Foley, D.D. and R.W. Taber. 1951. Lower Hudson waterfowl investigation. Pittman-Robertson Project 47-R. N.Y.S. Conservation Dept. Albany, NY. 796 pp.

Griscom, L. 1933. The birds of Dutchess County, New York. Trans. Linnean Soc. of NY 3. 184 pp.

Kiviat, E. 1980. A Hudson River tidemarsh snapping turtle population. Trans. Northeast Section Wildl. Soc. 37:158-159.

Landry, J.L. 1977. Status of the red-bellied turtle in New York (*Pseudemys rubriventris*) for the Dept. of the Interior, Fish and Wildlife Service endangered and threatened wildlife and plants; Review of the status of twelve species of turtles. Fed. Reg. June 6, 1977, 42 FR 28903.

Nicholas, G.L. 1900. The swallow-tailed kite at Piermont, NY. Auk 17:386.

*Stone, W.B., E. Kiviat, S.A. Butkas. 1980. Toxicants in snapping turtles. N.Y. Fish Game J. 27(1):39.

E. Fish

1. Striped Bass

Alexander, J.E., J. Foerenbach, S. Fisher, and D. Sullivan. 1973. Mercury in striped bass and bluefish. N.Y. Fish and Game J. 20(2):147-151.

Alperin, I.M. 1965. Recent records of pugheaded striped bass from New York. N.Y. Fish and Game J. 12:114-115.

Alperin, I.M. 1966. Dispersal, migration and origins of striped bass from Great South Bay, Long Island. N.Y. Fish and Game J. 13(1):79-112.

Alperin, I.M. 1966. Occurrence of yearling striped bass along the South shore of Long Island. N.Y. Fish and Game J. 13(1):113-120.

*Austin, H.M. and O. Custer. 1977. Seasonal migration of striped bass in Long Island Sound. N.Y. Fish and Game J. 24(l):53-68.

Barnthouse, L.W., S.W. Christensen, B.L. Kirk, K.D. Kumar, W. Van Winkle, and D.S. Vaughan. 1980. Methods to assess impacts on Hudson River striped bass: Report for the period Oct. 1, 1977 to Sept. 30, 1979. ORNL/NUREG/TM-374. Oak Ridge National Laboratory, Oak Ridge, TN.

Baumann, P.C. and R.J. Klauda. 1977. Fluctuations in annual abundance of young striped bass in the Hudson River estuary. Paper Delivered at 1977 N.E. Fish and Wildl. Conf., Boston, MA.

Bean, T.H. 1903. Fishes of New York, pp. 524-527, 744 *In* Catalogue of the Fishes of N.Y. N.Y.S. Museum, University of the State of N.Y. Bulletin 278.

Berggren, T.J. and J.T. Liberman. 1978. Relative contribution of Hudson, Chesapeake, and Roanoke striped bass, *Morone saxatilis*, stocks to the Atlantic coast fishery. Fish. Bull. 76(2):335–345.

Boreman, J. 1983. Simulation of striped bass egg and larva development based on temperature. Trans. Am. Fish. Soc. 112:286–292.

Campbell, K.P., R.J. Klauda, and D. McKenzie. 1975. Movement of striped bass tagged in the Hudson River. Presented at the Atlantic Estuarine Res. Soc. Meeting, Annapolis, MD.

Carlson, F.T. and J.A. McCann. 1969. Hudson River fisheries investigations, 1965–1968: Evaluations of a proposed pumped storage project at Cornwall, NY in relation to fish in the Hudson River, pp. 320–372 *In* Hudson River Ecology, 2nd Symp. N.Y.S. Dept. of Environmental Conservation, Albany, NY.

Christensen, S.W., D.S. Vaughan, W. Van Winkle, L.W. Barnthouse, D.L. DeAngelis, K.D. Kumar, and R.M. Yoshiyama. 1982. Methods to assess impacts on Hudson River striped bass: Final report. Environmental Sciences Division Publication No. 1978. NUREG/CR-2674, ORNL/TM 8309. Oak Ridge National Laboratory, Oak Ridge, TN.

*Clark, J.R. 1968. Seasonal movements of striped bass contingents of Long Island Sound and the New York Bight. Trans. Am. Fish. Soc. 97(4):320–343.

Cole, J.N. 1978. Striper, a story of fish and man. Little, Brown and Co., Boston, MA. 271 pp.

*Dahlberg, M.D. 1979. A review of survival rates of fish eggs and larvae in relation to impact assessments. Mar. Fish. Rev. 41(3):1–12.

Dew, C.B. 1980. Biological characteristics of commercially caught Hudson River striped bass, 1973–1975. Hudson River Ecology, 5th Symp. Hudson River Environ. Soc., Vassar College, Poughkeepsie, NY.

Dey, W.P. 1981. Mortality and growth of young-of-the-year striped bass in the Hudson River estuary. Trans. Am. Fish. Soc. 110:151–157.

Englert, T.L. and D. Sugerman. 1986. Patterns of larval movement of striped bass and white perch in the Hudson River estuary. *In* Fisheries Research in the Hudson River (C.L. Smith, ed.). Proceedings of a workshop sponsored by the Hudson River Environmental Society, September, 1981. State University of N.Y. Press, Albany, NY. (in press.)

Eraslan, A.H., W. Van Winkle, R.D. Sharp, S.W. Christensen, C.P. Goodyear, R.M. Rush, and W. Fulkerson. 1976. A computer simulation model for the striped bass young-of-the-year population in the Hudson River. ORNL-NUREG-8 (Special), Oak Ridge National Laboratory, Oak Ridge, TN. 208 pp.

Gomber, J. 1977. Hudson River striped bass. Underwater Naturalist 10(3):27.

Grove, T.L. and T.J. Berggren. 1975. A multivariate approach to determining the relative contribution of the Hudson River striped bass population to the Atlantic fishery. Presented at the 105th Annual Meeting, Am. Fish. Soc. Las Vegas, NV.

*Hickey, C.R., Jr., B.H. Young, and R.D. Bishop. 1977. Skeletal abnormalities in striped bass. N.Y. Fish Game J. 24(1):69–85.

Hitron, J.W. 1974. Serum transferrin phenotypes in striped bass, *Morone saxatilis*, from the Hudson River. Chesapeake Sci. 15(4):246–247.

Hjorth, D.A. 1986. Feeding selectivity of larval striped bass (*Morone saxatilis*) and white perch (*Morone americana*) in the Peekskill region of the Hudson River. *In* Fisheries Research in the Hudson River (C.L. Smith, ed.). Proceedings of a workshop sponsored by the Hudson River Environmental Society, September, 1981. State University of N.Y. Press, Albany, NY. (in press.)

Klauda, R.J., K.P. Campbell, and W.E. Cooper. 1975. The striped bass commercial fishery of the Hudson River. Presented at the Hudson River Ecology Meeting. Saratoga, NY. June 11, 1975. 12 pp.

*Klauda, R.J., W.P. Dey, T.B. Hoff, J.B. McLaren, and Q.E. Ross. 1980. Biology of Hudson River juvenile striped bass, pp. 101-123 *In* Proceedings of the Fifth Annual Marine Recreational Fisheries Symp., (H. Clepper, ed.). Sport Fishing Institute, Washington, D.C.

*Koo, T.S. 1970. The striped bass fishery in the Atlantic states. Chesapeake Sci. 11(2):73–93.

Kumar, K.D. and W. Van Winkle. 1978. Estimates of relative stock composition of the Atlantic coast striped bass population based on the 1975 Texas Instruments data set. Paper presented Feb. 27, 1978 at Northeast Fish and Wildlife Conference. The Greenbrier, WV. Sponsored by the Am. Fish Soc., N.E. Division.

Lawler, J.P., R.A. Norris, G. Goldwyn, K.A. Abood, and T.L. Englert. 1974. Hudson River striped bass life cycle model, pp. 83–94 *In* Second Workshop on Entrainment and Intake Screening. (L.D. Jensen, ed.). Elec. Power Res. Inst. Palo Alto, CA. Publ. 74-049-00-5.

Lewis, R. 1957. Comparative study of populations of the striped bass. Special Sci. Report Fish. No. 204. M.S. Thesis. Cornell University, Ithaca, NY. 54 pp.

Luna, W.A. 1957. Morphometric study of the striped bass *Roccus saxatilis*. Special Sci. Report Fish. No. 216. 24 pp.

*McCann, J.A. and F.T. Carlson. 1969. Effects of a proposed pumped storage hydroelectric generating plant on fish life in the Hudson River, *In* Hudson River Ecology, 2nd Symp. N.Y.S. Dept. of Environmental Conservation, Albany, NY.

McLaren, J.B., R.J. Klauda, T. Hoff, and M.N. Gardinier. 1986. Commercial fishery for striped bass in the Hudson River. *In* Fisheries Research in the Hudson River (C.L. Smith, ed.). Proceedings of a workshop sponsored by the Hudson River Environmental Society, September, 1981. State University of N.Y., Albany, NY. (in press.)

McLaren, J.B., J.C. Cooper, T.B. Hoff, and V. Lander. 1981. Movements of Hudson River striped bass. Trans. Am. Fish. Soc. 110:158–167.

Merriman, D. 1938. A report of progress on the striped bass investigations along the Atlantic coast. North Am. Wildl. Conf. 3:476–485.

*Merriman, D. 1941. Studies on the striped bass (*Roccus saxatilis*) of the coast. U.S. Fish Wildl. Serv., Fish. Bull. 50:1–77.

Murawski, W.S. 1958. Comparative study of populations of the striped bass, *Roccus saxatilis* (Walbaum) based on lateral line scale counts. M.S. Thesis. Cornell University, Ithaca, NY. 80 pp.

Murawski, W.S. 1966. Fish and wildlife resources and their management in the Hudson-Mohawk corridor, *In* The Hudson Fish and Wildlife. Hudson River Valley Commission. 45 pp.

National Oceanic and Atmospheric Administration. 1979. Action plan, emergency striped bass study. Natl. Mar. Fish. Serv.

O'Connor, J.M. and S.A. Sachaeffer. 1977. The effects of sampling gear on the survival of striped bass ichthyoplankton. Chesapeake Sci. 18 (3):312–315.

Pakkala, I.S., W.H. Gutenmann, D.J. Lisk, G.E. Burdick, and E.J. Harris. 1972. A survey of the selenium content of fish from 49 New York State waters. Pesticides Monitoring J. 6(2):107–114.

Pizza, J.C. and J.M. O'Connor. 1983. PCB dynamics in Hudson River striped bass. II. Accumulation from dietary sources. Aquatic Toxicol. 3(4):313–328.

*Rachlin, J.W., A.P. Beck, and J.M. O'Connor. 1978. Karyotypic analysis of the Hudson River striped bass *Morone saxatilis*. Copeia 2:343–345.

Raney, E.C. 1952. The life history of the striped bass, *Roccus saxatilis* (Walbaum). Bull. Bingham. Oceanogr. Collection 14:5–97.

*Raney, E.C. and D.P. deSylva. 1953. Racial investigations of the striped bass (*Roccus saxatilis* Walbaum). J. Wildlife Management 17:495–509.

Raney, E.C. 1954. The striped bass in New York waters. Conservationist 8(4):14–16.

Raney, E.C., W.S. Woolcott, and A.G. Nehring. 1954. Migratory patterns and racial structure of Atlantic coast striped bass. Trans. 19th North Am. Wildl. Conf., pp. 376–396.

*Rathjen, W.F. and L.C. Miller. 1957. Aspects of the early life history of the striped bass (*Roccus saxatilis*) in the Hudson River. N.Y. Fish and Game J. 4(1):43–60.

Rehwoldt, R., L.W. Menapace, B. Nerrie, and D. Alessandrello. 1972. The effect of increased temperature upon the acute toxicity of some heavy metal ions. Bull. Environ. Contam. Toxicol. 8(2):91–96.

Retzsch, W.C. 1975. A legislative and management plan for the recreational and commercial striped bass fisheries of New York State. Rep. to the N.Y.S. Assembly, Pub. Service Legisl. Studies Prog. 128 pp.

Ricker, W.E. 1979. Notes on certain of the testimonial documents that pertain to the effects of power plants on striped bass in the lower Hudson River and estuary, Appendix I *In* Appraisal of Certain Arguments, Analyses, Forecasts, and Precedents Contained in the Utilities Evidentiary Studies on Power Plant Insult to Fish Stocks of the Hudson River Estuary (R.I. Fletcher, and R.B. Deriso, eds.). Submitted in the EPA-Utilities Adjudicatory Hearings. No. C/II-WP-77-01, Exb. 218.

Rogers, B.A. 1978. Temperature and rate of early development of striped bass *Morone saxatilis* (Walbaum). Ph.D. Dissertation. University of Rhode Island, Providence, RI. 196 pp.

Rogers, B.A., D.T. Westin, and S.B. Saila. 1977. Life stage duration studies of Hudson River striped bass. Univ. Rhode Island Tech. Rep. 31. 111 pp.

Saila, S.B. and E. Lorda. 1977. Sensitivity analysis applied to a matrix model of the Hudson River striped bass population, pp. 311–332 *In* Proceedings of the Conference on Assessing the Effects of Power-Plant-Induced Mortality on Fish Populations (W. Van Winkle, ed.). Pergamon Press, New York, NY.

Schaefer, R.H. 1968. Size, age composition and migration of striped bass from the surf waters of Long Island. N.Y. Fish and Game J. 15:1–51.

Sykes, J.E., R.J. Mansueti, and A.H. Swartz. 1961. Striped bass research on the Atlantic coast. Minutes of the 20th Annual Meeting. Atlantic States Mar. Fish. Commission. Appendix 7(1–3). 170 pp.

Van Winkle, W., B.W. Rust, C.P. Goodyear, S.R. Blum, and P. Thall. 1974. A striped-bass populations model and computer programs. Oak Ridge National Laboratory Rep. ORNL-TM-4578. 200 pp.

*Van Winkle, W., B.L. Kirk, and B.W. Rust. 1979. Periodicities in Atlantic coast striped bass (*Morone saxatilis*) commercial fisheries data. J. Fish. Res. Board Can. 36:54–62.

Wallace, D.N. 1975. A critical analysis of the biological assumptions of Hudson River striped bass models and field survey data. Trans. Am. Fish. Soc. 104:710–717.

*Wallace, D.N. 1978. Two anomalies of fish larval transport and their importance in environmental assessment. N.Y. Fish and Game J. 25(1):59–71.

*Whipple, J.A. 1982. The impact of estuarine degradation and chronic pollution on populations of andromous striped bass (*Morone saxitilis*) in the San Francisco Bay Delta, California. Tiburon Lab. Southwest Fisheries Center. NOAA. Tiburon, CA.

Young, J.R. and T.B. Hoff. 1986. Age-specific variation in reproductive effort in female Hudson River striped bass. *In* Fisheries Research in the Hudson River. Proceedings of a workshop sponsored by the Hudson River Environmental Society, September, 1981. State University of N.Y. Press, Albany, NY. (in press.)

2. White Perch

Barnthouse, L.W., B.L. Kirk, K.D. Kumar, W. Van Winkle, and D.S. Vaughn. 1980. Methods to assess impacts on Hudson River white perch: Report for the period Oct. 1, 1978 to Sept. 30, 1979. Oak Ridge National Laboratory. ORNL/NUREG/TM-373.

Barnthouse, L.W., W. Van Winkle, and D.S. Vaughan. 1983. Impingement losses of white perch at Hudson River power plants: magnitude and biological significance. Environmental Management 7(4):355–364.

*Dorfman, D. 1973. Serum proteins of white perch. N.Y. Fish and Game J. 20(1):62–67.

Englert, T.L. and D. Sugerman. 1986. Patterns of larval movement of striped bass and white perch in the Hudson River estuary. *In* Fisheries Research in the Hudson River (C.L. Smith, ed.). Proceedings of a workshop sponsored by the Hudson River Environmental Society, September, 1981. State University of N.Y. Press, Albany, NY. (in press.)

Hjorth, D.A. 1986. Feeding selectivity of larval striped bass (*Morone saxatilis*) and white perch (*Morone americana*) in the Peekskill region of the Hudson River. *In* Fisheries Research in the Hudson River (C.L. Smith, ed.). Proceedings of a workshop sponsored by the Hudson River Environmental Society, September, 1981. State University of N.Y. Press, Albany, NY. (in press.)

*Holsapple, J.C. and L.E. Foster. 1975. Reproduction of white perch in the lower Hudson River. N.Y. Fish and Game J. 22(2):122–127.

Van Winkle, W., L.W. Barnthouse, B.L. Kirk, and D.S. Vaughan. 1980. Evaluation of impingement losses of white perch at the Indian Point Nuclear Station and other Hudson River power plants. ORNL/NUREG/TM-361. Oak Ridge National Laboratory, Oak Ridge, TN.

Vaughan, D.S. and W. Van Winkle. 1982. Corrected analysis of the ability to detect reductions in year-class strength of the Hudson River white perch (*Morone americana*) population. Can. J. Fish. Aquatic Sci. 39(5):782–785.

*Woolcott, W.S. 1962. Infraspecific variation in the white perch, *Roccus americanus* (Gmelin). Chesapeake Sci. 3(2):94–113.

3. Atlantic Tomcod

Booth, R.A. 1967. A description of the larval stages of the Tomcod, *Microgadus tomcod*, with comments on its spawning ecology. PhD. Dissertation. University of Connecticut, Storrs, CT. 53 pp.

Campbell, K.P., T.H. Peck, J.B. McLaren, and M. Nittel. 1975. Biology of the Atlantic tomcod *Microgradus tomcod* in the Hudson River. Presented at the 105th Annual Meeting, Am. Fish. Soc., Las Vegas, NV.

*Dew, C.B. and J.H. Hecht. 1976. Observations on the population dynamics of Atlantic tomcod (*Microgadus tomcod*) in the Hudson River estuary. Paper 25 *In* Hudson River Ecology, 4th Symp. Hudson River Environ. Soc., Bronx, NY.

Fikslin, T. and J. Golumbek. 1979. Analysis of Atlantic tomcod in the Hudson River. Summary and conclusions. Submitted in the EPA-Utilities Adjudicatory Hearing. Exhibit 220.

*Grabe, S.A. 1978. Food and feeding habits of juvenile Atlantic tomcod, *Microgadus tomcod*, from Haverstraw Bay, Hudson River. Fish. Bull. 76(1):89–94.

Grabe, S.A. 1980. Food of age 1 and 2 Atlantic tomcod, *Microgadus tomcod*, from Haverstraw Bay, Hudson River, NY. Fish. Bull. 77(4):1003–1006.

Klauda, R.J., R.E. Moos, and R.E. Schmidt. 1986. Life history of Atlantic tomcod, *Microgadus tomcod* in the Hudson River estuary, New York, with emphasis on spatio-temporal distributions and movements. *In* Fisheries Research in the Hudson River (C.L. Smith, ed.). Proceedings of a workshop sponsored by the Hudson River Environmental Society, September, 1981. State University of N.Y. Press, Albany, NY. (in press.)

*Nittel, M. 1976. Food habits of Atlantic tomcod in the Hudson River. Paper 26 *in* Hudson River Ecology, 4th Symp. Hudson River Environ. Soc., Bronx, NY.

Ruppert, G.F. 1964. The tomcod. N.J. Outdoors, Feb. 1964, pp. 12–17.

*Smith, C.E., T.H. Peck, R.J. Klauda, and J.B. McLaren. 1979. Hepatomas in Atlantic tomcod *Microgadus tomcod* (Walbaum) collected in the Hudson River estuary in New York. J. Fish Diseases 2:313–319.

4. Shad and Blueback Herring

Baskous, A.A. and P. Lanahan. 1973. Hudson River shad commands top price. Conservationist 28(2):7–9.

*Burdick, G.E. 1954. An analysis of factors, including pollution, having possible influence on the abundance of shad in the Hudson River. N.Y. Fish and Game J. 1(2):188–205.

Cheney, A.N. 1896. Shad of the Hudson River, pp. 125–134 *In* First Ann. Rep. of N.Y. Fisheries, Game, and Forest Comm. Albany, NY.

Cooper, J.C. (editor). 1984. Workshop on critical data needs for shad research on Atlantic Coast of North America. February 8–9, 1984. Proceedings. Hudson River Foundation for Science and Environmental Research, Inc. New York, NY. 70 pp.

*Davis, W.S. 1957. Ova production of American shad in Atlantic coast rivers. U.S. Fish Wildl. Serv. Res. Rep. 49. 5 pp.

*Fischler, K.J. 1959. Contributions of Hudson and Connecticut Rivers to New York-New Jersey shad catch of 1956. U.S. Fish Wildl. Serv., Fish. Bull. 60:161–174.

Grabe, S.A. and R.E. Schmidt. 1978. Overlap and diel variation in feeding habits of *Alosa* spp. juveniles from the lower Hudson River estuary. Paper presented at Annual Meeting, N.Y. Chapter of Am. Fish. Soc., Feb. 3-4, 1978. Marcy, NY. 28 pp.

Grim, J.S. 1954. Hudson River juvenile shad investigation progress report: 1948-1953. N.Y.S. Conservation Dept. (PASNY 1975:IV 3R-4). 33 pp. (Leaflet.)

Hill, D.R. 1959. Some uses of statistical analysis in classifying races of American shad (*Alosa sapidissima*). U.S. Fish and Wild. Serv., Fish. Bull. 59(147):269-286.

Johnson, F.F. 1938. Marketing of shad on the Atlantic coast. Investigational Report No. 38, Vol. II. Bur. Fish., U.S. Dept. of Commerce.

Klauda, R.J., M. Nittel, and K.P. Campbell. 1977. The commercial fishery for American shad in the Hudson River: Fishing effort and stock abundance trends, pp. 107-134 *In*: Proceedings of a Workshop on American Shad. U.S. Government Printing Office, Washington, D.C.

*Leggett, W.C. and R.R. Whitney. 1972. Water temperature and the migration of American shad. Fish. Bull. 70(3):659-670.

*Lehman, B.A. 1953. Fecundity of Hudson River shad. Fish. and Wildlife Serv., Res. Report 33. 8 pp.

Mansueti, R.J. and H. Kolb. 1953. A historical review of the shad fisheries of North America. State of Maryland, Board of Nat. Res., Dept. of Res. and Ed., Chesapeake Biol. Lab. Solomons, MD. Publ. No. 97.

Medeiros, W.H. 1974. Legal mechanisms to rehabilitate the Hudson River shad fishery. N.Y. State Assembly Scientific Staff and N.Y. State Sea Grant Program, Albany, NY. 65+ pp.

*Medeiros, W.H. 1975. The Hudson River shad fishery: Background, management problems and recommendation. N.Y. Sea Grant Institute, Albany, NY. 54 pp.

*Nichols, P.R. 1958. Effects of New Jersey-New York pound-net catches on shad runs of Hudson and Connecticut. U.S. Fish and Wildl. Serv., Fish. Bull. 58(143):491-500.

*Nichols, P.R. 1966. Comparative study of juvenile American shad populations by fin ray and scute counts. U.S. Fish and Wildl. Serv. Spec. Sci. Rep. Fish. 525. 10 pp.

Schmidt, R.E., R.J. Klauda, and J.M. Bartells. 1986. Distributions and movements of the early life stages of three *Alosa* spp. in the Hudson River estuary, with comments on the evidence for mechanisms that may reduce interspecific competition. *In* Fisheries Research in the Hudson River (C.L. Smith, ed.). Proceedings of a workshop sponsored by the Hudson River Environmental Society, September, 1981. State University of N.Y. Press, Albany, NY. (in press.)

*Sykes, J.E. and G.B. Talbot. 1958. Progress in Atlantic coast shad investigations-migrations, pp. 82-90 *In* Proc. of the Gulf and Caribbean Fish. Inst., 11th Annual Session.

*Talbot, G.B. 1954. Factors associated with fluctuations in abundance of Hudson River shad. U.S. Fish and Wildl. Serv., Fish. Bull. 56(101):373-414.

Talbot, G.B. 1954. Shad in the Hudson. Conservationist 8(5):17-19.

*Talbot, G.B. 1956. Conservation of an east coast shad fishery. Proc. of the Gulf and Caribbean Fish. Inst., 8th Annual Session, pp. 92-99.

*Talbot, G.B. and J.E. Sykes. 1958. Atlantic coast migrations of American shad. U.S. Fish and Wildl. Serv., Fish. Bull. 58(142):473-488.

*Walburg, C. 1957. Observations on the food and growth of juvenile American shad, *Alosa sapidissimn*. Trans. Am. Fish. Soc. 86:302–306.

*Walburg, C.H. and P.R. Nichols. 1967. Biology and management of the American shad and status of the fisheries, Atlantic coast of the United States, 1960. U.S. Fish and Wildl. Serv. Spec. Sci. Rep. Fish. 550. 105 pp.

5. *Sturgeon*

Atz, J. and C.L. Smith. 1976. Hermaphroditism and gonadal teratoma-like growths in sturgeon (Acipenser). Bull. South. Calif. Acad. Sci. 75(2):119–126.

Bean, T.H. 1908. Fish culture in New York. Forest, Fish and Game Comm. 243 pp.

*Crandall, M.E. 1978. Dinosaurs of the Hudson. Conservationist 32(4):17–19.

*Dovel, W.L. 1978. Biology and management of shortnose and Atlantic sturgeon of the Hudson River. Project No. AFS-9-R-3. N.Y.S. Dept. of Environmental Conservation, Albany, NY.

Dovel, W.L. 1979. The biology and management of shortnose sturgeon and Atlantic sturgeon of the Hudson River. Final Report. Prepared for the N.Y.S. Dept. of Environmental Conservation and the Boyce Thompson Institute for Plant Research, Inc. 54 pp.

Dovel, W.L. 1979. The endangered shortnose sturgeon of the Hudson River: Its life history and vulnerability to the activities of man. Periodic Progress Report No. 1, Apr. 1–30. Prepared for Federal Energy Regulatory Commission. 10 pp.

Dovel, W.L. 1979. The endangered shortnose sturgeon of the Hudson River: Its life history and vulnerability to the activities of man. Periodic Progress Report No. 2, May 1–31. Prepared for Federal Energy Regulatory Commission. 5 pp.

Dovel, W.L. 1979. The endangered shortnose sturgeon of the Hudson River: Its life history and vulnerability to the activities of man. Periodic Progress Report No. 3, June 1–30. Prepared for Federal Energy Regulatory Commission. 6 pp.

Dovel, W.L. 1980. Life history aspects of the Atlantic sturgeon, *Acipenser oxyrhynchus* (Mitchill) of the Hudson River. 36th Ann. Northeast Fish and Wildl. Conf., Fish. Abstrs., Apr. 27–30. Ellenville, NY.

Dovel, W.L. 1981. The endangered shortnose sturgeon of the Hudson estuary: Its life history and vulnerability to the activities of man. Final Report. Prepared for the Federal Energy Regulatory Commission, Contract No. DEAC 39-79 RC-10074. 139 pp.

*Hoff, T.B. and R.J. Klauda. 1979. Data on shortnose sturgeon (*Acipenser brevirostrum*) collected incidentally from 1969 through June 1979 in sampling programs conducted for the Hudson River ecological study. Prepared for Meeting of Shortnose Recovery Team, Nov. 30. Financed by Consolidated Edison Co. of NY, Inc.

Hoff, T.B., R.J. Klauda, and J.R. Young. 1986. Contributions to the biology of shortnose sturgeon (*Acipenser brevirostrum*) in the Hudson River estuary. *In* Fisheries Research in the Hudson River (C.L. Smith, ed.). Proceedings of a workshop sponsored by the Hudson River Environmental Society, September, 1981. State University of N.Y. Press, Albany, NY. (in press.)

Koski, R.T., E.C. Kelley, and B.E. Turnbaugh. 1971. A record sized sturgeon from the Hudson River. N.Y. Fish and Game J. 18(1):75.

Murawski, S.A. and A.L. Pacheco. 1977. Biological and fisheries data on Atlantic sturgeon, *Acipenser oxyrhynchus* (Mitchill). NOAA, NMFS, NEFC, Sandy Hook Lab., Technical Report 10. 69 pp.

Pekovitch, A.W. 1980. Distribution and some life history aspects of the shortnose sturgeon (*Acipenser brevirostrum*) in the upper Hudson River estuary. 36th Annual Northeast Fish and Wildl. Conf., Fisheries Abstrs., Apr. 27–30. Ellenville, NY.

Ryder, J.A. 1888. The sturgeon and sturgeon industries of the eastern coast of the United States, with an account of experiments bearing upon sturgeon culture. Bull. U.S. Fish Comm.

State of New York Conservation Commission. 1925. Hudson River sturgeon situation. 15th Annual Report.

*Vladykov, V.D. and J.R. Greeley. 1963. Order Acipenseroidei, pp. 24–60 *In* Fishes of the Western North Atlantic. Part 3. Memoirs Sears Foundation Mar. Resources 1(3):630.

Young, J.R., T.B. Hoff, W.P. Dey, and J.G. Hoff. 1986. Management recommendations for a Hudson River sturgeon fishery based on an age-structured population model. *In* Fisheries Research in the Hudson River (C.L. Smith, ed.). Proceedings of a workshop sponsored by the Hudson River Environmental Society, September, 1981. State University of N.Y. Press, Albany, NY. (in press.)

6. *Fish—General*

Adams, C.C., T.L. Hankinson, and W.C. Kendall. 1919. A preliminary report on a fish cultural policy for the Palisades Interstate Park. Trans. Am. Fish. Soc. 48:193–204.

*Aleveras, R.A. 1973. Occurrence of a lookdown in the Hudson River. N.Y. Fish and Game J. 20(1):75.

Arcement, R.J. and J. Rachlin. 1976. A study of the karyotype of a population of banded killifish (*Fundulus diaphanus*) from the Hudson River. J. Fish. Biology 8:119–125.

Atlantic States Marine Fisheries Commission. 1958. Important fisheries of the Atlantic coast. Supplement to 16th Annual Report, Atl. States Mar. Fish. Comm., Mt. Vernon, NY. 52 pp.

Atlantic States Marine Fisheries Commission. 1958. Recent advances in marine fishery research along the Atlantic coast. Atl. States Mar. Fish. Comm., Mt. Vernon, NY. 34 pp.

*Bath, D., C.A. Beebe, C.B. Dew, R.H. Reider, and J.H. Hecht. 1976. A list of common and scientific names of fishes collected from the Hudson River. Paper 33 *In* Hudson River Ecology, 4th Symp. Hudson River Environ. Soc., Bronx, NY.

Beebe, A. 1978. Ichthyoplankton in the lower Hudson River system. Presented at the Am. Fish. Soc. Nat. Meeting, University of Rhode Island, Aug. 21–26.

Boyle, R.H. Notes on the fishes of the lower Hudson. Bull. Am. Littoral Soc. 5(2):32–33, 40.

Brant, R. Unpublished information on Hudson River goldfish decline, and skin infections perhaps related to high PCB residues in goldfish. N.Y.S. Dept. of Environmental Conservation, New Paltz, NY.

Breder, C.H., Jr. 1938. The species of fish in New York Harbor. Bull. N.Y. Zool. Soc. 41:23–29.

Breder, C.M., Jr. 1948. Field book of marine fishes of the Atlantic coast from Labrador to Texas. G.P. Putnam and Sons. New York, NY. 332 pp.

Carlson, F.T. and J.A. McCann. 1969. Evaluation of a proposed pumped-storage project at Cornwall, NY in relation to fish in the Hudson River, *In* Hudson River Fisheries Investigations, 1965–1968. Hudson River Policy Comm., N.Y.S. Conservation Dept., Albany, NY. 50 pp.

Carlson, F.T. and J.A. McCann. 1969. Report of the biological findings of the Hudson River fisheries investigation, 1965–1968, *In* Hudson River Fisheries Investigations, 1965–1968. Hudson River Policy Comm., N.Y.S. Conservation Dept., Albany, NY. 50 pp.

Casne, S.R. and T.C. Cannon. 1976. Occurrence of fish in thermal discharges from power plants on the Hudson River. Paper 30 *In* Hudson River Ecology, 4th Symp. Hudson River Environ. Soc., Bronx, NY.

Clark, J.R. 1969. Ecology of anadromous fish of the Hudson River, pp. 11–26. *In* Hearings on the Impact of the Hudson River Expressway Proposal on Fish and Wildlife Resources of the Hudson River and Atlantic Coastal Fisheries. House of Rep., Subcomm. on Fish. and Wildl. Conservation, Comm. on Merchant Marine and Fish. Serial No. 91–10.

*Clark, J.R. and S. Smith. 1969. Migratory fish of the Hudson estuary, pp. 293–319 *In* Hudson River Ecology, 2nd Symp. N.Y.S. Dept. of Environmental Conservation, Albany, NY.

Committee on Merchant Marine and Fisheries. 1965. Hearings before the Subcommittee on Fisheries and Wildlife Conservation of the Committee on Merchant Marine and Fisheries. House of Representatives, Anadromous Fish, 1965. No. 89-9. U.S. Government Printing Office, Washington, D.C.

Curran, H.W. and D.T. Ries. 1936. Fisheries investigations in the lower Hudson River, *In* A Biological Survey of the Lower Hudson Watershed. Supplement to 26th Annual Report. N.Y.S. Conservation Dept., Albany, NY.

Dew, C.B. 1974. Comments on the recent incidence of gizzard shad, *Dorosoma cepedianum*, in the lower Hudson River. Paper 20 *In* Hudson River Ecology, 3rd Symp. Hudson River Environ. Soc., Bear Mountain, NY.

Dovel, W.L. 1974. Summary report: An investigation of the population dynamics of Hudson River fishes found in the general area of Tappan Zee and Haverstraw Bays with particular application to the Rockwood Hall property. Submitted to N.Y.S. Office of Parks and Recreation and Taconic State Park and Recreation Commission by Boyce Thompson Institute for Plant Research. 80 pp.

Dovel, W.L. 1975. Early developmental stages of estuarine-dependent fishes of the lower Hudson River, New York. Report to the Rockefeller Foundation. 82 pp.

*Dovel, W.L. 1981. Ichthyoplankton of the lower Hudson estuary. N.Y. Fish and Game J. 28(1):21–39.

*Englert, T.L., F.N. Aydin, J.C. Huang, and G. Vachtsevanos. 1976. Computer simulation of spatial and temporal distributions of early fish life stages in the Hudson River estuary. Paper 27 *In* Hudson River Ecology, 4th Symp. Hudson River Environ. Soc., Bronx, NY.

*Esser, S.C. 1982. Long term changes in some finfishes of the Hudson-Raritan estuary, *In* Ecological Stress and the New York Bight: Science and Management (G.F. Mayer, ed.). Estuarine Research Federation, Columbia, SC.

Farrell, M.A. 1932. Pollution studies of the upper Hudson watershed, pp. 208–215 *In* A Biological Survey of the Upper Hudson Watershed. Supplement to 22nd Annual Report. N.Y.S. Conservation Dept., Albany, NY.

Foster, N.R. 1967. Comparative studies on the biology of killifishes (Pisces, Cyprinodontidae). Ph.D. Dissertation. Cornell University, Ithaca, NY.
Francis, J.D. and L. Busch. 1973. New York State's commercial fisheries: Industry and manpower projections. N.Y.S. Food and Life Sci. Bull. No. 28.
Friedman, B.R. and C.T. Hamilton. 1980. The fish life of Upper New York Bay. Underwater Nat. 12(2):18–21.
Goode, G.B. 1887. The fisheries and fishery industries of the United States. U.S. Government Printing Office, Washington, D.C. 787 pp.
Greeley, J.R. 1935. Annotated list of fishes occurring in the watershed, pp. 88–101 In A Biological Survey of the Mohawk-Hudson Watershed. Supplement to 24th Annual Report. N.Y.S. Conservation Dept., Albany, NY.
*Greeley, J.R. 1936. Fishes of the area with annotated list, pp. 45–103 In A Biological Survey of the Lower Hudson Watershed. Supplement to 26th Annual Report. N.Y.S. Conservation Dept., Albany, NY.
*Greeley, J.R. and S.C. Bishop. 1932. Fishes of the upper Hudson watershed with annotated list, pp. 64–125 In A Biological Survey of the Upper Hudson Watershed. Supplement to 22nd Annual Report. N.Y.S. Conservation Dept., Albany, NY.
*Grim, J.S. 1966. Summarized progress report of the Hudson River fisheries investigation, 1965–1966, pp. 201–219 In Hudson River Ecology, 1st Symp. Hudson River Valley Comm., Tuxedo, NY.
Hall, C.A.S. 1977. Models and the decision making process: The Hudson River power plant case, In Ecosystem Modeling in Theory and Practice (C.A.S. Hall and J.W. Day, eds.). John Wiley & Sons, New York, NY.
*Heller, R.F. and H. Hermo, Jr. 1969. Fluctuations of fish populations, Part 2. Observed by 24 hour sampling, pp. 373–389 In Hudson River Ecology, 2nd Symp. N.Y.S. Dept. of Environmental Conservation, Albany, NY.
Hudson River Policy Committee. 1969. Hudson River fisheries investigations, 1965–1968. Evaluations of a proposed pumped storage project at Cornwall, NY in relation to fish in the Hudson River. N.Y.S. Conservation Dept. Presented to Consolidated Edison Co. of NY, Inc.
Hudson River Valley Commission. 1966. The Hudson, fish and wildlife. A report on fish and wildlife resources in the Hudson River valley. N.Y.S. Conservation Dept., Fish and Game Division, Albany, NY.
Jensen, A.C. (ed.) 1969. Hudson River fisheries investigations, 1965–1968. Hudson River Policy Committee, Albany, NY.
Jensen, A.C. 1970. Thermal loading in the marine district. N.Y. Fish Game J. 17:65–80.
Jensen, A.C. 1977. New York's marine fisheries: Changing needs in a changing environment. N.Y. Fish Game J. 24(2):99–128.
Kiviat, E. 1977. Dutchess County, NY zoological survey, 1975–1976. Summary of data. Bard College. 10 pp. (mimeo.)
Kiviat, E. 1978. Hudson River east bank natural areas, Clermont to Norrie. The Nature Conservancy, Arlington, VA. 115 pp.
Klauda, R.J., P.H. Muessig, and J.A. Matousek. 1986. Fisheries data sets compiled by utility-sponsored research in the Hudson River estuary. In Fisheries Research in the Hudson River (C.L. Smith, ed.). Proceedings of a workshop sponsored by the Hudson River Environmental Society, September, 1981. State University of N.Y. Press, Albany, NY. (in press.)

*Koski, R.T. 1974. Life history and ecology of the hogchoker, *Trinectes maculatus*, in its northern range. Ph.D. Thesis. University of Connecticut, Storrs, CT. 111 pp.

Koski, R.T. 1974. Sinistrality and albinism among hogchokers in the Hudson River. N.Y. Fish and Game J. 21(2):186–187.

*Koski, R.T. 1978. Age, growth, and maturity of the hogchoker, *Trinectes maculatus*, in the Hudson River, New York. Trans. Am. Fish. Soc. 107(3):449–453.

Lanahan, P. 1973. Returning the Hudson to shad, bass. N.Y.S. Environment. Feb. 1, 1973, p. 7.

McCann, J.A. and F.T. Carlson. 1969. Effects of a proposed pumped storage hydroelectric generating plant on fish life in the Hudson River pp. 320–356 *In* Hudson River Ecology, 2nd Symp. N.Y.S. Dept. of Environmental Conservation, Albany, NY.

McDonald, M. 1887. The fisheries of the Hudson River, pp. 658–659 *In* The Fisheries and Fishery Industries of the United States (G.B. Goode, ed.). U.S. Government Printing Office, Washington, D.C.

McHugh, J.L. 1972. Marine fisheries of New York State. Fish. Bull. 70(3).

McHugh, J.L. 1977. Limiting factors affecting commercial fisheries in the Middle Atlantic estuarine areas, pp. 149–169 *In* Estuarine Pollution Control and Assessment. Proc. of a Conf., Vol. 1, U.S. Environmental Protection Agency, Washington, D.C.

McHugh, J.L. and J.J.C. Ginter. 1978. Fisheries, MESA New York Bight Atlas. Monograph 16. New York Sea Grant Institute, Albany, NY. 129 pp.

Mearns, E.A. 1898. A study of the vertebrate fauna of the Hudson Highlands, with observations on the mollusca, crustacea, lepidoptera and the flora of the region. Bull. Am. Mus. Nat. Hist. 10:302–352.

*Moore, E. 1936. Introduction, *In* A Biological Survey of the Lower Hudson Watershed. Supplement to 26th Annual Report. N.Y.S. Conservation Dept., Albany, NY.

Myers, G.S. 1930. Introduction of the European bitterling (*Rhodeus*) in New York and of the rudd (*Scardinus*) in New Jersey. Copeia 140:20–21.

Mylod, J. 1980. Commercial fishing in the Hudson River, *In* Hudson River Ecology, 5th Symp. Hudson River Environ. Soc., Vassar College, Poughkeepsie, NY.

New York State Dept. of Environmental Conservation. 1980. Hudson Estuary fisheries development program. Report to State Legislature. 12 pp.

New York State Dept. of Environmental Conservation and U.S. Fish and Wildlife Service Division of Ecological Services. 1978. Hudson River fish and wildlife report, Hudson River Level B Study. N.Y.S. Dept. of Environmental Conservation, Albany, NY.

Oceanic Society. 1980. Hudson River fishery management program study. Report to the N.Y.S. Dept. of Environmental Conservation, Stanford, CT.

Perlmutter, A., E. Leff, and E.E. Schmidt. 1967. Distribution and abundance of fishes along the shores of the lower Hudson River during the summer of 1965. N.Y. Fish and Game J. 14(1):47–75.

*Perlmutter, A., R.F. Heller, and H. Hermo, Jr. 1969. Fluctuating of fish populations, Part I. As a monitor of environmental changes in the Hudson River, pp. 357–372 *In* Hudson River Ecology, 2nd Symp. N.Y.S. Dept. of Environmental Conservation, Albany, NY.

*Perlmutter, A., E. Leff, E.E. Schmidt, R. Heller, and M. Siciliano. 1966. Distribution and abundances of fishes along the shores of the lower Hudson during the summer of 1966, pp. 147–200 *In* Hudson River Ecology, 1st Symp. Hudson River Valley Comm., Tuxedo, NY.

Reider, R.H. 1979. Occurrence of a silver lamprey in the Hudson River. N.Y. Fish Game J. 26(l):93.

Reider, R.H. 1979. Occurrence of a kokanee in the Hudson River. N.Y. Fish Game J. 26(l):94.

Sheppard, J.D. 1976. Valuation of the Hudson River fishery resources: Past, present and future. N.Y.S. Dept. of Environmental Conservation Report, Albany, NY.

*Smith, C.L. 1976. The Hudson River fish fauna. Paper 32 *In* Hudson River Ecology, 4th Symp. Hudson River Environ. Soc., Bronx, NY.

Smith, C.L. 1985. The Inland Fishes of New York State. N.Y.S. Dept. of Environmental Conservation, Albany, NY. 523 pp.

Snelson, F.F., Jr. 1968. Systematics of the Cyprinid fish *Notropis amoenus*, with comments on the subgenus *Notropis*. Copeia (4):776–802.

Squires, D.F. 1981. The bight of the big apple. The New York Sea Grant Institute of the State University of N.Y. and Cornell University.

*Tabery, M.A., A.P. Ricciardi, and T.J. Chambers. 1978. Occurrence of larval inshore lizardfish in the Hudson River estuary. N.Y. Fish and Game J. 25(l):87–89.

Townes, H.K. 1936. Studies on the food organisms of fish, pp. 217–229 *In* A Biological Survey of the Lower Hudson Watershed. Supplement to 26th Annual Report. N.Y.S. Conservation Dept. Albany, NY.

Young, B.H., I.H. Morrow, and S.R. Wanner. 1982. First record of the bluespotted cornetfish in the Hudson River. N.Y. Fish Game J. 29(l):106.

F. Diseases and Parasites

Alperin, I.M. 1965. Recent records of pugheaded striped bass from New York. N.Y. Fish and Game J. 12:114–115.

Alperin, I.M. 1966. A new parasite of striped bass. N.Y. Fish and Game J. 13(1).

*Bowman, T.E., S.A. Grabe, and J.H. Hecht. 1977. Range extension and new hosts for the cymothoid isopod *Anilocra acuta*. Ches. Sci. 18(4):390–393.

Goodchild, C.G. 1939. *Cerearidia conica* n. sp. from the clam *Pisidium abditum* Haldeman. Trans. Am. Microscopical Soc. 58(2):179–184.

*Hickey, C.R., B.H. Young, and R.D. Bishop. 1977. Skeletal abnormalities in striped bass. N.Y. Fish and Game J. 24(l):69–85.

*Hogan, I.M. and B.S. Williams. 1976. Occurrence of the gill parasite *Ergasilus labraris* on striped bass, white perch and tomcod in the Hudson River. N.Y. Fish and Game J. 23(1):97.

Hunter, G.W. 1936. Parasitism of fishes in the lower Hudson area, pp. 264–273 *In* A Biological Survey of the Lower Hudson Watershed. Supplement to 26th Annual Report, N.Y.S. Conservation Dept. Albany, NY.

Kaczynski, V.W. and T. Cannon. 1974. Incidence and effect of the parasitic isopod *Lironeca ovalis*, on bluefish and white perch in the lower Hudson estuary. Paper 21 *In* Hudson River Ecology, 3rd Symp. Hudson River Environ. Soc., Bear Mountain, NY.

Levine, N.E. 1982. Scheduling power plant operations to minimize impacts on entrainable fish in the Hudson River. M.S. Thesis. Cornell University, Ithaca, NY. 86 pp.

Liguori, V.M. 1979. The gill parasites of the white perch, *Morone americana*, from the Hudson River at Vincent's Point. M.S. Thesis. Adelphi Univ., Garden City, NY. 86 pp.

Mahoney, J.B., F.H. Midlige, and D.G. Deuel. 1973. A fin rot disease of marine and euryhaline fishes in the New York Bight. Trans. Am. Fish. Soc. 102(3):596–605.

*Smith, C.E., T.H. Peck, R.J. Klauda, and J.B. Mclaren. 1979. Hepatomas in Atlantic tomcod *Microgadus tomcod* (Walbaum) collected in the Hudson River estuary in New York. J. Fish Diseases 2:313–319.

G. Entrainment and Impingement

Alevras, R.A. 1974. Status of air-bubble fish protection system at Indian Point Station of the Hudson River, *In* Proceedings of the Second Entrainment and Intake Screening Workshop (L.D. Jensen, ed.). Johns Hopkins University and EPRI Publ. No. 74-040-00-5.

Barnthouse, L.W. 1979. Density-dependent growth: A critique of the utilities' "empirical evidence." Oak Ridge National Laboratory, Oak Ridge, TN. Prepared for the U.S. EPA-Utilities Adjudicatory Hearing No. C/II-WP-77- 01, Exb. 210.

Barnthouse, L.W. and W. Van Winkle. 1980. Impact of impingement on the Hudson River white perch population. National Workshop on Entrainment and Impingement, San Francisco, CA. 19 pp. NTIS PC A02/MF A01.

Barnthouse, L.W., J.B. Cannon, and S.G. Christensen. 1977. A selective analysis of power plant operation on the Hudson River with emphasis on the Bowline Point Generating Station. Oak Ridge National Laboratory, Oak Ridge, TN. ORNL/TM 5877/V1 and V2.

Barnthouse, L.W., D.L. DeAngelis, and S.W. Christensen. 1979. An empirical model of impingement impact. Oak Ridge National Laboratory. Environmental Sciences Division Publication No. 1289. NUREG/CR-0639, ORNL/NUREG/TM-290. 20 pp.

Barnthouse, L.W., B.L. Kirk, K.D. Kumar, W. Van Winkle, and D.S. Vaughan. 1980. Methods to assess impacts on Hudson River white perch: Report for the period Oct. 1, 1978 to Sept. 30, 1979. Oak Ridge National Laboratory, Oak Ridge, TN. ORNL/NUREG/TM-373.

Barnthouse, L.W., W. Van Winkle, and D.S. Vaughan. 1983. Impingement losses of white perch at Hudson River power plants: magnitude and biological significance. Environmental Management 7(4):355–364.

Baslow, M.H. 1980. Empirical evidence of control mechanisms in the Cape Fear, Chesapeake, and Hudson River estuaries. Submitted to the U.S. Environmental Protection Agency in the EPA-Utilities Adjudicatory Hearing.

Beck, A.P., G.V. Poje, and W.T. Willet. 1975. A laboratory study on the effects of the exposure of some entrainable Hudson River biota to hydrostatic pressure regime calculated for the proposed Cornwall storage plant, *In* Fisheries and Energy Production. A Symposium. (S.B. Saila, ed.). Lexington Books, Lexington, MA. 300 pp.

Boreman, J., L.W. Barnthouse, D.S. Vaughan, C.P. Goodyear, S.W. Christensen, W. Van Winkle, J. Golumbek, G.F. Cada, and B.L. Kirk. 1982. The impact of entrainment and impingement on fish populations in the Hudson River estuary. Oak Ridge National Laboratory, Oak Ridge, TN. ORNL/NUREG/TM-385/V1-V3.

Campbell, K.P., I.R. Savidge, W.P. Dey, and J.B. McLaren. 1977. Impacts of recent power plants on the Hudson River striped bass (*Morone saxatilis*) population, p. 184 In Proceedings of the Conference on Assessing the Effects of Power-Plant-Induced Mortality on Fish Populations (W. Van Winkle, ed.). Gatlinburg, TN. May 3-6, 1977. Pergamon Press, New York, NY.

Cannon, T.C., S.M. Jinks, L.R. King, and G.J. Lauer. 1978. Survival of entrained ichthyoplankton and macroinvertebrates at Hudson River power plants, pp. 71-89 In Fourth National Workshop on Entrainment and Impingement. Dec. 5, 1977. Chicago, IL.

Carter, H.H., R.E. Wilson, and G. Carroll. 1979. An assessment of the thermal effects on striped bass larvae entrained in the heated discharge of the Indian Point Generating Facilities Units 2 and 3. SR-24, REF-79-7. S.U.N.Y. at Stony Brook, Marine Sciences Research Center, Stony Brook, NY. 33 pp.

*Casne, S.R. and T.C. Cannon. 1976. Occurrence of fish in thermal discharges from power plants on the Hudson River—preliminary investigations. Paper 30 In Hudson River Ecology, 4th Symp. Hudson River Environ. Soc., Bronx, NY.

Christensen, S.W., C.P. Goodyear, and B.L. Kirk. 1979. Analysis of the validity of the utilities' stock recruitment curve-fitting exercise. Oak Ridge National Laboratory, Oak Ridge, TN. 316 pp.

Christensen, S.W., D.S. Vaughan, W. Van Winkle, L.W. Barnthouse, D.L. DeAngelis, K.D. Kumar, and R.M. Yoshiyama. 1982. Methods to assess impacts on Hudson River striped bass: Final report. Oak Ridge National Laboratory. Environmental Sciences Division Publication No. 1978. NUREG/CR-2674 ORNL/TM-8309.

Fletcher, R.I. and R.B. Deriso. 1979. Appraisal of certain arguments, analyses, forecasts, and precedents contained in the utilities' evidentiary studies on power plant insult to fish stocks of the Hudson River estuary. University of Washington, Seattle, WA. Prepared for the U.S. EPA. Submitted in the EPA-Utilities Adjudicatory Hearings No. C/II-WP-77-01, Exb. 218.

Ginn, T.C. and J.M. O'Connor. 1978. Response of the estuarine amphipod *Gammarus diabert* to chlorinated power plant effluent. Est. Coastal Mar. Sci. 6(5):459-469.

Ginn, T.C., W.T. Waller, and G.J. Lauer. 1974. The effects of power plant condenser cooling water entrainment on the Amphipod, *Gammarus* sp. Water Research 8(ll):937-945.

Heffner, R.L., G. Howells, G. Lauer, and H. Hirshfield. 1971. The effects of power plant entrainment and heat discharge on Hudson River biota. Proc. of 3rd Nat. Symp. on Radioecology. Oak Ridge, TN. May 10-12, 1971. Conf.-710501.

Hillegas, J., R. Alevras, D. Logan, E. Pikitch, and R. Armco. 1979. Ecosystem effects of phytoplankton and zooplankton entrainment. EPRI interim report. EPRI EA-1038. Electric Power Research Institute, Palo Alto, CA.

Jinks, S.M., T. Cannon, D. Latimer, and L. Claflin. 1978. An approach for the analysis of striped bass entrainment survival at Hudson River power plants, pp. 343-350 In Fourth National Workshop on Entrainment and Impingement. Dec. 5, 1977. Chicago, IL.

Kellogg, R.L., J.J. Salerno, and D.L. Latimer. 1978. Effects of acute and chronic thermal exposures on the eggs of three Hudson River anadromous fishes, pp. 714–725 In Energy and Environmental Stress in Aquatic Systems. Technical Info. Center, U.S. DOE, Oak Ridge, TN.

King, L.R., J.B. Hutchinson, Jr., and T.G. Higgins. 1978. Impingement survival studies on white perch, striped bass, and Atlantic tomcod at three Hudson River power plants, In Fourth National Workshop on Entrainment and Impingement (L.D. Jensen, ed.). EA Communications, Melville, NY.

Lanza, G.R., G.J. Lauer, T.C. Ginn, P.C. Storm, and L. Zubarik. 1975. Biological effects of simulated discharge plume entrainment at Indian Point Nuclear Power Station, Hudson River estuary, USA, In The International Atomic Energy Association Symposium on the Combined Effects of Radioactive, Chemical and Thermal Releases to the Environment. Stockholm, Sweden.

Lauer, G.J., W.T. Waller, D.W. Bath, W. Meeks, R. Heffner, T. Ginn, L. Zubarik, P. Bibko, and P.C. Storm. 1974. Entrainment studies on Hudson River organisms, pp. 37–82, In Entrainment and Intake Screening: Proceedings of the Second Entrainment and Screening Workshop (L.D. Jensen, ed.). Johns Hopkins University and Edison Electric Institute. Report 15-1-347.

Lawler, J.P. 1976. Physical measurements—their significance in the prediction of entrainment effects, In Proceedings of the Third National Workshop on Entrainment and Impingement, Feb. 1976 (L.D. Jensen, ed.). Ecological Analysts, Melville, NY.

Lawler, J.P., T.L. Englert, R.A. Norris, and C.B. Dew. 1977. Modeling of compensatory response to power plant impact, In Proceedings of the Conference on Assessing the Effects of Power-Plant-Induced Mortality on Fish Populations (W. Van Winkle, ed.). Gatlinburg, TN. May 3–6, 1977. Pergamon Press, New York, NY.

Lawler, Matusky, and Skelly, Engineers. 1979. Methodology for assessing population and ecosystem level effects related to the intake of cooling waters. EPRI Final Report. Electric Power Research Institute, Palo Alto, CA. EPRI EA-1238.

Lawler, Matusky, and Skelly, Engineers. 1980. Methodology for assessing population and ecosystem level effects related to the intake of cooling waters. Handbook of Methods, Vol. 1: Population Level Techniques. EPRI Final Report. Electric Power Institute, Palo Alto, CA. EPRI EA-1402, Vol. 1.

Lawler, Matusky, and Skelly, Engineers. 1980. Methdology for assessing population and ecosystem level effects to intake of coding waters. Handbook of Methods, Vol. 2: Community Analysis Techniques. EPRI Final Report Electric Power Institute, Palo Alto, CA. EPRI EA-1402, Vol. 2.

Levin, S.A. 1979. The concept of compensatory mortality in relation to impacts of power plants on fish populations. Cornell University, Ithaca, NY. Prepared for the U.S. EPA. Submitted in the EPA-Utilities Adjudicatory Hearings. No. C/II-WP-77-01, Exb. 212.

Muessig, P.H., D.R. Mayercek, and T.L. Grove. 1975. The interaction of natural history and environmental variables as influencing factors in the impingement of fish. Presented at the 105th Annual Meeting. Am. Fish. Soc., Las Vegas, NV. (abstract only.)

*Muessig, P.H. and D.R. Mayercek. 1976. Diel distribution patterns of impingement at Indian Point Nuclear Generating Stations. Paper 26 *In* Hudson River Ecology, 4th Symp. Hudson River Environ. Soc., Bronx, NY.

Ricker, W.E. 1979. Notes on certain of the testimonial documents that pertain to the effects of power plants on striped bass in the lower Hudson River and estuary, Appendix I *In* Fletcher, R.I. and R.B. Deriso. 1979. Appraisal of certain arguments, analyses, forecasts, and precedents contained in the utilities' evidentiary studies on power plant insult to fish stocks of the Hudson River estuary. Submitted in the EPA-Utilities Adjudicatory Hearings. No. C/II-WP-77-01, Exb. 218.

Rohlf, F.J. 1979. Analysis of (1) population density and growth and (2) striped bass stock recruitment models. S.U.N.Y. at Stony Brook, NY. Prepared for the U.S. EPA. Submitted in the EPA-Utilities Adjudicatory Hearings. No. C/II-WP-77-01, Exb. 208.

Schubel, J.R., H.H. Carter, and J.M. O'Connor. 1979. Effects of increasing T on power plant entrainment mortality at Indian Point, NY. Marine Sciences Research Center, S.U.N.Y. at Stony Brook. Special Report 19. 16 pp.

Van Winkle, W., S.W. Christensen, and G. Kauffman. 1976. Critique and sensitivity analysis of the compensation function used in the LMS Hudson River striped bass models. Environmental Sciences Division, Publ. 944. Oak Ridge National Laboratory, Oak Ridge, TN. 107 pp.

*Wallace, D.N. 1978. Two anomalies of fish larval transport and their importance in environmental assessment. N.Y. Fish and Game J. 25(1):59–71.

Zubarik, L. 1976. Entrainment effects on microzooplankton. Paper 23 *In* Hudson River Ecology, 4th Symp. Hudson River Environ. Soc., Bronx, NY.

II. Physical and Chemical Characteristics

A. Geology, Hydrology

*Abood, K.A. 1974. Circulation in the Hudson estuary, pp. 39–111 *In* Hudson River Colloquium (D.A. Roels, ed.). Annals of the N.Y. Acad. Sci. 250.

Abood, K.A. 1977. Evaluation of circulation in partially stratified estuaries as typified by the Hudson River. Ph.D. Dissertation. Rutgers University. New Brunswick, NJ.

Aggarwal, Y. and L.R. Sykes. 1978. Earthquakes, faults, and nuclear power plants in southern New York and northern New Jersey. Science 200:425–429.

Alden Research Laboratories. 1966. Hudson River model study in the area of the Cornwall pumped storage project for Consolidated Edison Co. of New York, Inc. Preliminary Data Report. Worcester Polytechnic Institute. 24 pp.

Ayers, J.C. *et al.* (eds.) 1951. The Hydrography of the New York Area. Status Reports. Contract N6 ORN 264. Prepared for U.S. Office of Naval Research. Cornell University, Ithaca, NY.

Ayers, J.C. 1951. An examination of the factor used in correcting for the effect of bottom friction on the velocity of flow in tidal stream, and the manner in which it is affected by stage of tide. *In* The Hydrography of the New York Area. Status Report No. 3 (J.C. Ayers *et al.*, eds.). Contract N6 ORN 264. Cornell University, Ithaca, NY.

Ayers, J.C. 1951. The steady-state condition and the unbalanced condition in tidal estuaries, with special reference to the unbalanced condition of New York Harbor in Mar.-Apr. 1950, *In* The Hydrography of the New York Area. Status

Report No. 2 (J.C. Ayers *et al.*, eds.). Contract N6 ORN 264. Cornell University, Ithaca, NY.

Ayers, J.C. 1951. The "normal" flushing time and half-life of New York Harbor, *In* The Hydrography of the New York Area. Status Report No. 1 (J.C. Ayers *et al.*, eds.). Contract N6 ORN 264. Cornell University, Ithaca, NY.

Ayers, J.C. 1951. The transparency (by white secchi disc) of the waters in and about New York Harbor. *In* The Hydrography of the New York Area. Status Report No. 7 (J.C. Ayers *et al.*, eds.). Contract N6 ORN 264. Cornell University, Ithaca, NY.

Ayers, J.C. 1951. Tabulated temperature and salinity data of the New York area, *In* The Hydrography of the New York Area. Interim Report No. 1 (J.C. Ayers *et al.*, eds.). Contract N6 ORN 263. Cornell University, Ithaca, NY.

Ayers, J.C. and R.H. Backus. 1952. First annual report, *In* The Hydrography of the New York Area. Progress Report No. 1, (J.C. Ayers *et al.*, eds.). Contract N6 ORN 264. Cornell University, Ithaca, NY.

Backus, R.H. 1951. The distribution of dissolved oxygen in the New York area, *In* The Hydrography of the New York Area. Status Report No. 8 (J.C. Ayers *et al.*, eds.). Contract N6 ORN 264. Cornell University, Ithaca, NY.

Bacon, E.M. 1907. The Hudson River: From Ocean to Source. Putnam Press.

Banino, G.M. and W.E. Cutcliffe. 1979. Economic geology of the Hudson River Valley, pp. 292–309 *In* Guidebook Joint Annual Meeting of New York State Geological Association (51st Annual Meeting) and New England Intercollegiate Geological Conference (71st Annual Meeting) (G.M. Friedman, ed.). Troy, NY.

Blumstead, J. 1977. Comprehensive public water supply studies. Water Use Information for Hudson Basin Counties. N.Y.S. Dept. of Health. Albany, NY.

*Busby, M.W. 1966. Flow quality and salinity in the Hudson River estuary, pp. 135–146 *In* Hudson River Ecology, 1st Symp. Hudson River Valley Comm., Tuxedo, NY.

*Busby, M.W. and K.I. Darmer. 1970. A look at the Hudson estuary. Water Resources Bull. 6(5):802–812.

Carmody, D.J. 1972. The distribution of five heavy metals in the sediments of the New York Bight. Ph.D. Thesis. Columbia University, New York, NY. 121 pp.

Coch, N.K. 1976. Temporal and areal variations in some Hudson River estuary sediments, Mar.-Oct. 1974. Geol. Soc. Am. (abstract.) 8(2):153.

Coch, N.K., J. Helfond, and C. Kasdorf. 1979. Hudson River sedimentation in the vicinity of Kingston, NY. Geol. Soc. Am. (abstract.) 11(1):7.

Crawford, R.W., C.F. Powers, and R.H. Backus. 1951. The depths of unconsolidated sediments in the New York Harbor area and its approaches, *In* The Hydrography of the New York Area. Status Report No. 9 (J.C. Ayers *et al.*, eds.). Contract N6 ORN 264. Cornell University, Ithaca, NY.

*Darmer, K.I. 1969. Hydrologic characteristics of the Hudson River estuary, pp. 40–55 *In* Hudson River Ecology, 2nd Symp. N.Y.S. Dept. of Environmental Conservation, Albany, NY.

Dunn, B. and G.C. Gravlee. 1978. Dye-dispersion study at proposed pumped-storage project on Hudson River at Cornwall-on-the-Hudson, NY. Geological Survey, Water Resources Div. Open-file Report 78-589. Albany, NY. 40+ pp.

Eissler, B.B. 1978. Low-flow data and frequency analysis of streams in New York excluding New York City and Long Island. N.Y.S. Dept. of Environmental Conservation, Albany, NY.

Eissler, B.B. and T.J. Zembrzuski, Jr. 1974. Summary and evaluation of crest-stage-gage data in New York. U.S. Dept. of the Interior, Albany, NY.

*Giese, G.L. and J.W. Barr. 1967. The Hudson River estuary, a preliminary investigation of flow and water characteristics. N.Y.S. Conservation Dept., Water Resources Comm. Bull. 61. 30 pp.

Gravlee, G.C., Jr. and B. Dunn. 1978. Dye used to study mixing of waters of Hudson River estuary. (abstract only.) U.S. Geological Survey Prof. Paper (XGPPA9) No. 1100. U.S. Dept. of the Interior, Washington, DC.

Gross, M.G. 1970. New York metropolitan region—a major sediment source. Water Resources Res. 6(3):927–931.

*Gross, M.G. 1974. Sediment and waste deposition in New York Harbor, pp. 112–128 In Proceedings of the Hudson Estuary Colloquium (E.O. Roels, ed.). Ann. N.Y. Acad. Sci. 250.

Hammond, D.E., H.J. Simpson, and G. Mathieu. 1976. Distribution of Radon-222 in the Delaware and Hudson estuaries as an indicator of migration rates of dissolved species across the sediment-water interface. Eos: Am. Geophys. Union Trans. 57(3):151.

Heath, R.C. 1966. Work of the U.S. Geological Survey on Hudson River estuary, pp. 131–134 In Hudson River Ecology, 1st Symp. Hudson River Valley Comm., Tuxedo, NY.

Helsinger, M.H. and G.M. Friedman. 1975. The effects of industrialization and urbanization on the upper Hudson River basin, New York. Geol. Soc. Am. Abstr. Prog. 7(1):73.

*Hohman, M.S. and D.B. Parke. 1966. Hudson River dye studies: 150 miles of red water, pp. 60–81 In Hudson River Ecology, 1st Symp. Hudson River Valley Comm., Tuxedo, NY.

*Hooper, L.J. 1966. Model studies at the Alden Laboratories. River model study of the Cornwall pumped storage projects, pp. 114–130 In Hudson River Ecology, 1st Symp. Hudson River Valley Comm., Tuxedo, NY.

Howells, G.P. 1972. The estuary of the Hudson River, USA. Proc. Roy. Soc., London 180(1061):521–534.

*Johnson, J.H. 1966. Geology and mineral resources of the Hudson estuary, pp. 8–40 In Hudson River Ecology, 1st Symp. Hudson River Valley Comm., Tuxedo, NY.

Keenan, E.M. 1979. Hydrocarbon distributions in sediments from the Hudson estuary. Geol. Soc. Am. Abstr. Prog. 11(1):19.

Keenan, E.M. 1980. Sources of fatty acids in sediments from the Hudson estuary. Geol. Soc. Am. Abstr. Prog. 12(2):44–45.

Knebel, H.J. and S.A. Wood. 1978. Hudson River: Evidence for extensive migration on the continental shelf during the Pleistocene. Geol. Soc. Am. Abstr. Prog. 10(7):436.

Knebel, H.J. and S.A. Wood. 1979. Hudson River. Evidence for extensive migration on the exposed continental shelf. Geology 7(5):254–258.

Lawler, Matusky, and Skelly, Engineers. 1978. Progress report on Hudson River bottom sediments analysis in the vicinity of the West Side of Manhattan. Preliminary chemistry and bioassay results. (draft.) Prepared for Parsons, Brinckerhoff, Quade, and Douglas, Inc., NY.

Lilley, W.D. and C. Assini. 1976. Engineering and environmental geology of the Hudson Valley power sites (Trip B-10), In Field Guide Book. New York State Geological Association 48th Annual Meeting (J.H. Johnson, ed.). Vassar College, Poughkeepsie, NY. Oct. 15–16, 1976.

Lovegreen, J.R. 1974. Paleodrainage history of the Hudson River estuary. Geol. Soc. Am. Abstr. Prog. 6(1):49–50.

McCrone, A.W. 1966. The Hudson River estuary: hydrology, sediments, and pollution. Geogr. Rev. 56:175–189.

*McCrone, A.W. and R.C. Koch. 1966. Some geochemical properties of Hudson River sediments: Kingston to Manhattan, pp. 41–59 In Hudson River Ecology, 1st Symp. Hudson River Valley Comm., Tuxedo, NY.

McCrone, A.W. and C. Shafer. 1966. Geochemical and sedimentary environments of foraminifera in the Hudson River estuary, New York. Micropaleontology 12:505–509.

*Neal, L.A. 1969. Modeling of Hudson River, pp. 56–77 In Hudson River Ecology, 2nd Symp. N.Y.S. Dept. of Environmental Conservation, Albany, NY.

Newman, W.S., D.H. Thurber, H.S. Zeiss, A. Rokach, and L. Musich. 1969. Late quaternary geology of the Hudson River estuary: A preliminary report. Trans. N.Y. Acad. Sci., Ser. 2, 31(5):548–570.

Normandeau Associates, Inc. 1977. Hudson River survey 1976–1977 with cross-section and planimetric maps. Bedford, NH. 351 pp.

Olsen, C.R., H.J. Simpson, and R.M. Trier. 1977. Anthropogenic radionuclides as tracers for recent sediment deposition in the Hudson estuary. Eos: Am. Geophys. Union Trans. 58(6):406.

*Olsen, C.R., H.J. Simpson, S.C. Williams, T.H. Peng, and B.L. Deck. 1978. A geochemical analysis of the sediments and sedimentation in the Hudson estuary. J. Sedimentary Petrology 48:401–418.

Olsen, C.R., H.J. Simpson, R.F. Bopp, R.M. Trier, and S.C. Williams. 1978. Pollution history and sediment accumulation patterns in the Hudson estuary. Lamont-Doherty Geological Observatory, Palisades, NY. 26 pp.

*Panuzio, F.L. 1966. The Vicksburg model. The Hudson River model, pp. 82–113 In Hudson River Ecology, 1st Symp. Hudson River Valley Comm., Tuxedo, NY.

Postmentier, E.S. and J.M. Raymont. 1979. Variations of longitudinal diffusivity in the Hudson estuary. Est. Coastal Mar. Sci. 8(6):555–564.

Quirk, Lawler, and Matusky, Engineers. 1974. Hudson River hydrology study. U.S. Army Corps of Engineers North Atlantic Division.

*Sanders, J.E. 1974. Geomorphology of the Hudson estuary, pp. 4–38 In Hudson River Colloquium (O.A. Roels, ed.). Annals of the N.Y. Acad. Sci. 250.

*Schureman, P. 1934. Tides and currents in Hudson River. U.S. Coast and Geodetic Survey Spec. Publ. 180.

Shindel, H.L. 1969. Time-of-Travel study upper Hudson River. Report of Investigation RI-10. N.Y.S. Conservation Dept., Water Resources Comm., Albany, NY.

Simmons, H.B. and W.H. Bobb. 1965. Hudson River channel, New York and New Jersey plans to reduce shoaling in Hudson River channels and adjacent pier slips, hydraulic model investigation. U.S. Army Corps of Engineers. Vicksburg, MS. NTIS-AD-720-971.

Simpson, H.J., R. Bopp, and D. Thurber. 1973. Salt movement patterns in the

Hudson, *In* Hudson River Ecology, 3rd Symp. Hudson River Environ. Soc., Bear Mountain, NY.

*Stakhiv, E.Z., A.G. Sklarew, and R.W. Reinhardt. 1976. An approach to the environmental evaluation of the Hudson River high-flow skimming project, pp. 2–27 *In* Hudson River Ecology, 4th Symp. Hudson River Environ. Soc., Bronx, NY.

Stedfast, D.A. 1980. Cross sections of the Hudson River estuary from Troy to New York City, NY. Water Resources Investigations, U.S. Geological Survey, Albany, NY. 76 pp. NTIS-PB-81 202 384. (WRI 80-24.)

Stedfast, D.A. 1982. Flow model of the Hudson River estuary from Albany to New Hamburg, NY. U.S. Geological Survey, Water Resources Investigation Report, 81–55, Albany, NY. 69 pp.

Stefansson, K. 1974. Chemical, mineralogic, and palynologic character of the upper Wisconsin-lower Holocene fill in parts of the Hudson, Delaware, and Chesapeake estuaries. J. Sediment. Petrol. 44(2):390–408.

*Stewart, H.B. 1958. Upstream bottom currents in New York Harbor. Science 127:1113–1115.

Thompson, H.D. 1935. Geomorphology of the Hudson Gorge in the Highlands, New York. Ph.D. Thesis. New York University, New York, NY. 100 pp.

Tucker, W. 1977. "Environmentalism and the leisure class." Harper's Magazine 225(1531):49–62. NTIS-AN79-02/895.

U.S. Army Corps of Engineers. 1977. Water supply for southeastern New York. Prepared by North Atlantic Division.

U.S. Army Corps of Engineers. 1977. Northeastern United States water supply study (NEWS). Summary report. Appendix—Part 1. Hudson River Project Main Report. Prepared by the North Atlantic Division.

U.S. Dept. of the Interior. 1972. River basins of the United States: the Hudson. U.S. Geological Survey. U.S. Government Printing Office, Washington, D.C.

U.S. Dept. of the Interior. 1978. Minimum average 7-day, 10-year flows in the Hudson River basin, New York, with release-flow data on Roundout and Ashokan Reservoirs. Open-File Report 78-333. Geological Survey.

U.S. Geological Survey. 1982. Water resources data for New York. Water year 1981. U.S. Geological Survey Water Data Report NY-81-1. (Streamflow and water chemistry data are available for water year 1946 to the present for monitoring stations within the Hudson basin.)

U.S. Dept. of the Interior. 1978. Discharge, gage weight, and elevation of 100-year floods in the Hudson River basin. Open-File Report 78-322. Geological Survey.

U.S. Dept. of the Interior. 1978. Ground water data on the Hudson River basin, New York. Open File Report 78-710. Geological Survey.

Weiss, D. 1974. Late pleistocene stratigraphy and paleoecology of the lower Hudson River estuary. Geol. Soc. Am. Bull. 85:1561–1570.

Weiss, D., J.W. Rachlin, and N.K. Coch. 1975. The Hudson estuary, pp. 30–54 *In* 67th New England Intercollegiate Geol. Conf., New York, 1975. Guidebook for field trips in western Massachusetts, northern Connecticut, and adjacent areas of New York. City College of NY, New York, NY.

Weiss, D., K. Geitzenauer, and F.J. Shaw. 1976. Foraminifera, diatoms, and mollusks as potential Holocene paleoecologic indicators in the Hudson Estuary (abstract). Geol. Soc. Am. Abstr. Prog. 8(2):296–297.

B. Temperature

*Abood, K.A., E.A. Maikish, and R.R. Kimmel. 1976. Field and analytical investigations of ambient temperature distribution in the Hudson River. Paper 6 *In* Hudson River Ecology, 4th Symp. The Hudson River Environ. Soc., Bronx, NY.

Boreman, J. 1983. Simulation of striped bass egg and larva development based on temperature. Trans. Am. Fish. Soc. 112:286-292.

Hammond, D.E. 1975. Dissolved gases and kinetic processes in the Hudson River Estuary. Ph.D. Dissertation. Columbia University, New York, NY. 161 pp.

Howells, G.P., T.J. Kneip, and M. Eisenbud. 1970. Water quality in industrial areas: profile of a river. Environ. Sci. Tech. 4(1):26-35.

*Hudson River Research Council. 1980. Results of Hudson River field weeks, Apr. 1977 and Aug. 1978. R.E. Henshaw, Chairman, Bronx, NY.

*Jensen, A.C. 1970. Thermal loading in the marine district. N.Y. Fish Game J. 17:65-80.

Jensen, A.C. 1970. Thermal pollution in the marine environment. Conservationist 25(2):8-13.

*Lawler, J.P. and K.A. Abood. 1969. Thermal state of the Hudson River and potential changes, pp. 407-456 *In* Hudson River Ecology, 2nd Symp. N.Y.S. Dept. of Environmental Conservation, Albany, NY.

New York State Atomic and Space Development Authority. 1968. A thermal profile of the waters of New York State. N.Y.S. ASDA, New York, NY. 27 pp.

New York State Atomic and Space Development Authority. 1974. Survey of New York State water temperatures. Aerial infrared surveys of thermal discharges from electric generating stations into New York State waters. Final Report. 65 pp. NTIS-PB-244 998.

New York State Energy Research and Development Authority. 1974. A survey of New York surface water temperatures. Aerial infrared surveys of thermal discharges from electric generating stations into New York State waters. N.Y.S. ERDA. 56 pp. NTIS-PB-244998.

Posmentier, E.S. and J.W. Rachlin. 1976. Distribution of salinity and temperature in the Hudson estuary. J. Phys. Oceanogr. 6(5):775-777.

Powers, C.F. and R.H. Backus. 1951. The distribution of temperature in New York Harbor and its approaches, *In* The Hydrography of the New York Area. Status Report No. 10 (J.C. Ayers *et al.*, eds.). Contract N6 ORN 264. Task 15. Cornell University, Ithaca, NY.

*Sorge, E.V. 1969. The status of thermal discharges east of the Mississippi River. Chesapeake Sci. 10(3+4):131-138.

*Wrobel, W.E. 1974. Thermal balance in the Hudson estuary, pp. 157-168 *In* Hudson River Colloquium, (O.A. Roels, ed.). Ann. N.Y. Acad. Sci. 250.

Zamanis, A. 1960. Thermodynamics of steady and unsteady processes, involving heat transfer for the Consolidated Edison Nuclear Power Plant at Indian Point, NY. M.S. Thesis. City University of NY, New York, NY.

C. Water Chemistry

Bates, S.S. 1976. Effects of light and ammonium on nitrate uptake by two species of estuarine phytoplankton. Limnol. Oceanogr. 21(2):212-218.

Carpenter, J.H., D.W. Pritchard, and R.C. Whaley. 1969. Observations of eutrophication and nutrient cycles in some coastal plain estuaries, pp. 210-221 *In*

Proceedings of the International Symposium on Eutrophication: Causes, Consequences, Correctives. Nat. Acad. of Sci., Washington, D.C.

Chervin, M.B., T.C. Malone, and P.J. Neale. 1981. Interactions between suspended organic matter and copepod grazing in the plume of the Hudson River. Est. Coast. and Shelf Sci. 13(2):169–184.

Crawford, R.W. 1952. The distribution of oxygen in the waters of the New York Bight, Block Island Sound, and Newport Bight; Cruise STIRNI I, July-Sept. 1951, In The Hydrography of the New York Area. Status Report No. 17 (J.C. Ayers et al., eds.). Contract N6 ONR 264. Cornell University, Ithaca, NY.

Deck, B. 1981. Nutrient element distribution in the Hudson estuary, New York. Ph.D. Dissertation. Columbia University, New York, NY.

Faigenbaum, H.M. 1932. Chemical investigation of the Upper Hudson watershed, pp. 157–171 In A Biological Survey of the Upper Hudson Watershed. Supplement to 22nd Annual Report. N.Y.S. Conservation Dept., Albany, NY.

Faigenbaum, H.M. 1936. Chemical investigation of the lower Hudson area, pp. 146–216 In A Biological Survey of the Lower Hudson Watershed. Supplement to 26th Annual Report. N.Y.S. Conservation Dept., Albany, NY.

Hammond, D.E. 1975. Dissolved gases and kinetic processes in the Hudson River estuary. Ph.D. Dissertation. Columbia University, New York, NY. 161 pp.

*Hudson River Research Council. 1980. Results of Hudson River field weeks: Apr. 1977 and Aug. 1978. R.E. Henshaw, Chairman, Bronx, NY.

Ingram, R.J. 1979. Selected phytoplankton nutrients in the lower New York estuary. Ph.D. Dissertation. Fordham University, Bronx, NY. 131 pp.

New York State Bureau of Public Water Supply. 1977. A study of chemicals in water from selected community water systems with major emphasis in the Mohawk and Hudson River basins. N.Y.S. Dept. of Health, Albany, NY. 64 pp.

New York University Medical Center, Laboratory for Environmental Studies. 1970. Water chemistry of the lower Hudson River, pp. 53–75 In Water Pollution in the Greater New York Area Symposium. Gordon and Breach, New York, NY.

Schaffer, S.A. 1978. Concentrations of organic carbon and protein in the Hudson estuary near Indian Point. M.S. Thesis. New York University, New York, NY. 60 pp.

*Simpson, H.D., D.E. Hammond, B.L. Deck, and S.C. Williams. 1975. Nutrient budgets in the Hudson River estuary, pp. 616–635 In Marine Chemistry in the Coastal Environment (T.M. Church, ed.). ACS Symp. Series No. 18.

Simpson, H.J., S. Williams, B. Deck, and D. Hammond. 1975. Nutrient budgets in the Hudson estuary. American Chemical Society 169th National Meeting. Special Symposium on Marine Chemistry in the Coastal Environment. (abstract only.)

Thomas, J.P., W. Phoel, J.E. O'Reilly, and C. Evans. 1976. Seabed oxygen consumption in the lower Hudson estuary. Paper 12 In Hudson River Ecology, 4th Symp. Hudson River Environ. Soc., Bronx, NY.

*U.S. Geological Survey. 1982. Water resources data for New York. Water year 1981. U.S. Geological Survey Water Data Report NY-81-1. (Streamflow and water chemistry data are available for water year 1946 to the present for monitoring stations within the Hudson basin.)

D. Salinity

Buckley, E.H. and W.E. Hopkins. (in press.) Salinity profiles in the estuary of the Hudson River: An approach toward a 3-dimensional portrayal. Boyce Thompson Institute.

*Busby, M.W. 1966. Flow, quality, and salinity in the Hudson River estuary, pp. 135–146 In Hudson River Ecology, 1st Symp. Hudson River Valley Comm., Tuxedo, NY.

*Hunkins, K. 1981. Salt transport in the Hudson estuary. J. Phys. Oceanogr. 11(5):729–938.

Malone, T.C. 1979. Influence of estuarine circulation on the distribution and biomass of phytoplankton size fractions. 5th Biennial Intern. Estuarine Research Conf. Abstr. Estuarine Res. Fed.

Menzie, C.A. 1974. Observations of the distribution of aquatic macrophytes in the Hudson River estuary with respect to salinity. City College of NY, Biol. Dept. (unpublished.), New York, NY.

*Neale, P.J., T.C. Malone, and D.C. Boardman. 1981. Effects of fresh water flow on salinity and phytoplankton biomass in the lower Hudson estuary, pp. 168–184 In Proc. of the Nat'l. Symp. on Freshwater Inflow to Estuaries (R. Cross and D. Williams, eds.). U.S. Fish and Wildl. Serv. Office of Biol. Serv. FWS/OBS-81/04. (2 vols.)

Posmentier, E.S. and J.W. Rachlin. 1976. Distribution of salinity and temperature in the Hudson estuary. J. Phys. Oceanogr. 6(5):775–777.

Schaefer, P. 1969. The upper Hudson: time for decision. Conservationist 23(3):2–5.

*Simpson, H.J., R. Bopp, and D. Thurber. 1973. Salt movement patterns in the Hudson. Paper 9 In Hudson River Ecology, 3rd Symp. The Hudson River Environ. Soc., Bear Mountain, NY.

U.S. Dept. of the Interior, Office of Saline Water. 1966. Saline water conversion report.

E. Water Quality

Alpine Geophysical Associates, Inc. 1974. West Side highway project technical report on water quality. Part 2. Water quality sampling program, biological populations and inshore area studies. Prepared for N.Y.S. Dept. of Transportation.

Eisenbud, M. 1970. Water quality in industrial areas: Profile of a river. Environ. Sci. Technol. 4(1):26–35.

*Flemming, A., K.A. Abood, H.F. Mulligan, C.B. Dew, C.A. Menzie, W. Sydor, and W. Su. 1976. The environmental impact of PL 92-500 on the Hudson River estuary. Paper 16 In Hudson River Ecology, 4th Symp. Hudson River Environ. Soc., Bronx, NY.

*Giese, G.L. and J.W. Barr. 1967. The Hudson River estuary: A preliminary investigation of flow and water-quality characteristics. N.Y.S. Conservation Dept., Water Resources Comm., Bull. 61. 30 pp.

Giese, G.L. and W.A. Hobbs, Jr. 1970. Water resources of the Champlain-Upper Hudson basins in New York State. N.Y.S. Water Resources Commission and Office of Planning Coordination Report.

*Howells, G.P., T.J. Kneip, and M. Eisenbud. 1970. Water quality in industrial areas: Profile of a river. Environ. Sci. Tech. 4(1):26–35.

Hudson River Valley Commission. 1966. The Hudson: Water resources. The Hudson River as a resource. N.Y.S. Conservation Dept., Albany, NY.

Interstate Sanitation Commission. 1971. 1971 Hudson River Survey.

Jay, D.A. and M.J. Bowman. 1975. The physical oceanography and water quality of New York Harbor and western Long Island Sound. Mar. Sci. Res. Center. S.U.N.Y. at Stony Brook, NY. Tech. Rep. 23, Ref. No. 75-7. 71 pp.

Kneip, T.J. 1971. Sampling and analytical problems in a complex natural system: The Hudson River estuary. Am. Lab. 3(12):47–52.

Lawler, Matusky, and Skelly, Engineers. 1974. West Side highway project. Technical Report on Water Quality. Part I. Water quality and hydrodynamic effects. Prepared for N.Y. State.

Lawler, Matusky, and Skelly, Engineers. 1975. Final report on the water quantity and quality and environmental impact assessment studies of the Hudson River. Prepared for the National Commission on Water Quality.

Lawler, Matusky, and Skelly Engineers. 1978. Industrial water use study, Hudson River basin. Prepared for N.Y. District, U.S. Army Corps of Engineers.

Lawler, Matusky, and Skelly, Engineers. 1980. Biological and water quality data collected in the Hudson River near the proposed Westway Project during 1979–1980. Prepared for the N.Y.S. Dept. of Transportation and System Design Concepts, Inc.

Male, C.T. and Associates. 1971. Regional water supply and wastewater disposal plan and program, Capital District Region. Capital District Regional Planning Commission, Elnora, NY.

Manhattan College and College of Mount St. Vincent. 1974. Interface: Where the river meets a city. A water quality and land use study, Oct. 1973 to Mar. 1974. Environmental Action Council, Manhattan College, Bronx, NY. 49 pp.

Martin, R.M. 1973. Hudson River, pp. 149–174 *In* Our Environment: The Outlook for 1980, Part 1. Our Environment: Water (A.J. Van Tassel, ed.). Lexington Books, Lexington, MA.

*Maylath, R.E. 1969. Water quality surveillance of the Hudson River, pp. 147–161 *In* Hudson River Ecology, 2nd Symp. N.Y.S. Dept. of Environmental Conservation, Albany, NY.

McCarthy, G.T. and N.L. Barbarossa. 1968. Surface water resources planning in Hudson Basin. J. Hydraul. Div. 94(HY2):375–389.

Mitchell, R.D. 1968. Hudson River as a water source for New York City. J. Sanit. Eng. Div. 94(SA3):447–453.

*Mt. Pleasant, R.C. and H. Bagley. 1975. Towards purer waters. Progress Report. Environ. Quality News. Conservationist 30(2):ii–iii.

*Mytelka, A.I. 1976. A steady state water quality model of the New York Harbor complex. Paper 15 *In* Hudson River Ecology, 4th Symp. Hudson River Environ. Soc., Bronx, NY.

National Commission on Water Quality. 1976. Environmental impact assessment. Water quality analysis: Hudson River. NTIS-PB-251 099. 533 pp.

New York State Dept. of Environmental Conservation. 1974–1978. Water quality management plans (303e) for the Hudson River basin. Office of Program Development, Planning and Research and the Division of Pure Waters, Albany, NY.

*New York State Dept. of Environmental Conservation. 1983. Water quality data, lower Hudson River. (Unpublished computerized data base of twenty-

one sites from 1947 to the present.) N.Y.S. Dept. of Environmental Conservation, Bureau of Environmental Protection, Albany, NY.
*New York State Dept. of Health. 1952. Mohawk River drainage basin survey. Series Report No. 2. (Except Sanquoit Creek, West Canada Creek, East Canada Creek, and Schoharie Creek.) Water Pollution Control Board.
*New York State Dept. of Health. 1955. Upper Hudson River drainage survey. Series Report No. 1. Hoosic River Drainage Basin, Water Pollution Control Board.
*New York State Dept. of Health. 1960. Mohawk River drainage basin survey. Series Report No. 3. (West Canada Creek Drainage Basin.) Water Pollution Control Board.
*New York State Dept. of Health. 1960. Mohawk River drainage basin survey. Series Report No. 4. (Schoharie Creek Drainage Basin.) Water Pollution Control Board.
*New York State Dept. of Health. 1960. Mohawk River drainage basin survey. Series Report No. 5. (East Canada Creek Drainage Basin.) Water Pollution Control Board.
*New York State Dept. of Health. 1960. Lower Hudson River. (Drainage basins of streams: entering the Hudson River in Orange, Ulster, Dutchess, Putnam counties.) Lower Hudson River drainage basin survey. Series Report No. 8. Water Pollution Control Board.
*New York State Dept. of Health. 1960. Lower Hudson River (from mouth to northern Westchester-Rockland County lines). Lower Hudson River drainage basin survey. Series Report No. 9. Water Pollution Control Board.
*New York State Dept. of Health. 1962. Lower Hudson River drainage basin survey. Series Report No. 11. (Drainage basins of streams entering the Hudson River in Albany, Columbia, Greene, Rensselaer counties.) Water Resources Commission.
*New York State Dept. of Health. 1963. Lower Hudson River drainage basin survey. Series Report No. 10. (Hackensack River-Saddle River and Greenwood Lake-Sterling Lake Drainage Basins.) Water Resources Commission.
*New York State Dept. of Health. 1963. Upper Hudson River drainage survey. Series Report No. 2. (Upper Hudson River Drainage except Hoosic River Drainage Basin.) Water Pollution Control Board.
Quirk, Lawler, and Matusky, Engineers. 1970. Hudson River water quality and waste assimilative capacity study. Final report submitted to N.Y.S. Dept. of Environmental Conservation, Albany, NY. 198+ pp.
*Rehe, M.E. and M.S. Kohn. 1976. The water quality index—significance, advantages and disadvantages. Paper 13 In Hudson River Ecology, 4th Symp. Hudson River Environ. Soc., Bronx, NY.
Schroeder, R.A. and C.R. Barnes. 1983. Polychlorinated biphenyl concentrations in Hudson River water and treated drinking water at Waterford, NY. U.S. Geological Survey, Water-Resources Investigations Report. 83-4188. Albany, NY. 13 pp.
*Simpson, K.W. 1976. A water quality evaluation of the Hudson River, based on the collection and analysis of macroinvertebrate communities. Paper 24 In Hudson River Ecology, 4th Symp. Hudson River Environ. Soc., Bronx, NY.
*Storm, P.C. and L. Heffner. 1976. A comparison of phytoplankton abundance, chlorophyll "a" and water quality factors in the Hudson River and its tributar-

ies. Paper 17 *In* Hudson River Ecology, 4th Symp. Hudson River Environ. Soc., Bronx, NY.
Swanson, R.L. and M. Devine. 1982. Ocean dumping policy. Environment 24:14–20.
U.S. Army Corps of Engineers. 1963. Pollution studies for Interstate Sanitation Commission. New York Harbor model. Hydraulic model investigation. Waterways Experiment Station Misc. Pap. No. 2-558. 110 pp.
U.S. Dept. of Commerce. 1981. Northeast monitoring program. Annual NEMP report on the health of the northeast coastal waters of the United States, 1980. NOAA Technical Memorandum NMFS-F/NEC-10. NEMP IV 81 A-H 0043.
*U.S. Dept. of the Interior. Federal Water Pollution Control Administration. 1969. Conference in the matter of pollution of the interstate waters of the Hudson River and its tributaries—New York and New Jersey. (2 vols.)
U.S. Dept. of the Interior. Federal Water Pollution Control Administration. 1969. An evaluation of the significance of combined sewer overflows in the Hudson River conference area. Hudson-Delaware Basins Office, Edison, NJ. Rep. CWT-10-11.
U.S. Environmental Protection Agency. 1973. Water quality criteria, 1972. A report on the Committee on Water Quality Criteria prepared by Environmental Studies Board, National Academy of Sciences/National Academy of Engineering, Washington, D.C.

III. Pollutants

A. Sewage
Anderson, A.R. and J.A. Mueller. 1976. Estimate of New York Bight future (2000) contaminant inputs. Environ. Engineering and Science Program. Manhattan College, Bronx, NY. 31 pp.
Bowman, M.J. 1978. Spreading and mixing of the Hudson River effluent into the New York Bight, pp. 373–386 *In* Hydrodynamics of Estuaries and Fjords. (J.C.J., ed.) Elsevier, Amsterdam.
Farrell, M.A. 1932. Pollution studies of the upper Hudson watershed, pp. 208–215 *In* A Biological Survey of the Upper Hudson Watershed. Supplement to the 22nd Annual Report. N.Y.S. Conservation Dept., Albany, NY.
*Garside, C., T.C. Malone, O.A. Roels, and B.A. Sharfstein. 1976. An evaluation of sewage-derived nutrients and their influence on the Hudson estuary and New York Bight. Est. Coastal Mar. Sci. 4:281–286.
Gross, G.M. 1970. Preliminary analyses of urban wastes, New York Metropolitan Region. Marine Sciences Research Center, S.U.N.Y. at Stony Brook, NTIS-AD-746 959.
Hetling, L.J. 1974. An analysis of past, present and future Hudson River wastewater loadings, *In* Hudson River Ecology, 3rd Symp. Hudson River Environ. Soc., Bear Mountain, NY.
*Hetling, L. 1976. Trends in waste water loading, 1900–1976. Paper 14 *In* Hudson River Ecology, 4th Symp. Hudson River Environ. Soc., Bronx, NY.
*Ketchum, B.H. 1974. Population, resources, and pollution, and their impact on the Hudson estuary, pp. 144–156 *In* Hudson River Colloquium (O.A. Roels, ed.). Ann. N.Y. Acad. Sci., Vol. 250.
*Lang, M. 1974. Problems on pollution and water resources in the New York City

metropolitan area, pp. 178–181 *In* Hudson River Colloquium (O.A. Roels, ed.). Ann. NY Acad. Sci., Vol. 250.

Leslie, J.A., K.A. Abood, E.A. Maikish, and P.J. Keeser. 1986. Recent dissolved oxygen trends in the Hudson River. *In* Fisheries Research in the Hudson River (C.L. Smith, ed.). Proceedings of a workshop sponsored by the Hudson River Environmental Society, September, 1981. State University of N.Y. Press, Albany, NY. (in press.)

*Molof, A.H. 1969. Nutrient loads in the Hudson and their control, pp. 390–406 *In* Hudson River Ecology, 2nd Symp. N.Y.S. Dept. of Environmental Conservation, Albany, NY.

*Mt. Pleasant, R.C. 1966. Industrial and domestic waste discharges into the Hudson River from the Troy Lock to the George Washington Bridge, pp. 258–267 *In* Hudson River Ecology, 1st Symp. Hudson River Valley Comm., Tuxedo, NY.

Mueller, J.A., J.S. Jeris, A. Anderson, and C. Hughs. 1976. Contaminant inputs to the New York Bight. NOAA Tech. Memorandum ERL-MESA-6.

Mueller, J.A., T.A. Gerrish, and M.C. Casey. 1982. Contaminant inputs to the Hudson-Raritan estuary. NOAA Tech. Memorandum OMPA-21. Boulder, CO.

Mytelka, A.I., L.P. Cagliostro, D.J. Deutsch, and C.A. Haupt. 1973. Combined sewer overflow for the Hudson River Conference. Environmental Protection Agency Tech. Ser. Rep. EPA-R2-73-152. NTIS-PB-227 341. 287 pp.

*O'Connor, J.D. and J.P. St. John. 1967. Pollution analysis of the upper Hudson River estuary, pp. 222–233 *In* The Fresh Water of New York State: Its Conservation and Use. Proc. of a Symp. held at S.U.N.Y. at Buffalo, June 13–17, 1966.

Pringle, L. 1971. The upper Hudson, whitewater or washwater. Audubon 73(2):88–100.

Schindler, D.W. 1974. Biological and chemical mechanisms in eutrophication of fresh water lakes, *In* Hudson River Colloquium (O.A. Roels, ed.). Ann. N.Y. Acad. Sci. 250:129–135.

*Tofflemire, T.J. and L.J. Hetling. 1969. Pollution sources and loads in the lower Hudson River, pp. 78–146 *In* Hudson River Ecology, 2nd Symp. N.Y.S. Dept. of Environmental Conservation, Albany, NY.

U.S. Dept. of Health, Education, and Welfare. 1965. Report on pollution of the Hudson River and its tributaries. Division of Water Supply and Pollution Control, Public Health Service, Region II, NY.

B. *Polychlorinated Biphenyls (J.M. and E.H. Buckey)*

Ahmed, A.K. 1976. A review of federal and state government roles in controlling impacts of PCB's on the environment, pp. 412–427 *In* Proceedings of the National Conference on Polychlorinated Biphenyls, Nov. 19–21, 1975. Chicago, IL.

Ahmed, A.K. 1976. PCBs in the environment: the accumulation continues. Environment 18(2):6–11.

Ahmed, A.K., S. Chasis, and J.K. Thornton. 1982. Comments of the Natural Resources Defense Council regarding the Hudson River PCB reclamation project addressed to the Region II administrator of the Environmental Protection Agency, Mar. 11, 1981. Northeastern Environ. Sci. 1(1):69–74.

Alben, K. and E. Shpirt. 1983. Distribution profiles of chloroform, weak organic

acids and PCBs on granular activated carbon columns from Waterford, NY. Environ. Sci. Technol. 17(4):187–192.
Apicella, G. 1984. Trends in PCB transport in the upper Hudson River. Northeastern Environ. Sci. 3(3/4):197-202.
Apicella, G.A. and T.F. Zimmie. 1978. Sediment and PCB transport model of the Hudson River, pp. 645–653 In Proceedings of the 26th Annual Hydraulics Division Specialty Conference. University of Maryland, College Park, MD. Aug. 9–11.
Armstrong, R.W. and R.J. Sloan. 1980. Trends in levels of several known chemical contaminants in fish from New York State waters. Technical Report 80-2, Bur. Environ. Protection, Div. Fish and Wildl., N.Y.S. Dept. of Environmental Conservation, Albany, NY. 77 pp.
Armstrong, R.W. and R.J. Sloan. 1986. PCB patterns in Hudson River fish. I. Resident/freshwater species, In Fisheries Research in the Hudson River (C.L. Smith, ed.). Proceedings of a workshop sponsored by the Hudson River Environmental Society, September, 1981. State University of N.Y. Press, Albany, NY. (in press.)
*Baker, F.D., C.F. Tumasonis, W.B. Stone, and B. Bush. 1976. Levels of PCB and trace metals in waterfowl in New York State. N.Y. Fish Game J. 23(1):82–91.
Blasland, W.V., Jr., W.H. Bouck, E.R. Lynch, and R.K. Goldman. 1981. The Fort Miller site: remedial program for securement of an inactive disposal site containing PCB's, pp. 215–222 In Proceedings of the National Conference for Management of Uncontrolled Hazardous Waste Sites. Silver Spring, MD.
Bopp, R. 1979. The geochemistry of polychlorinated biphenyls in the Hudson River. Ph.D. Dissertation. Columbia University, New York, NY.
Bopp, R.F., H.J. Simpson, and C.R. Olsen. 1977. PCBs and Cs-137 in sediment of the Hudson estuary. Eos: Trans. Am. Geophys. Union 58(6):407.
Bopp, R., H.J. Simpson, and C. Olsen. 1978. PCB analysis in the sediments of the lower Hudson: Final report. Lamont-Doherty Geological Observatory, Palisades, NY. 35 pp.
Bopp, R.F., H.J. Simpson, and C.R. Olsen. 1978. PCBs and the sedimentary record of the Hudson River, New York. Lamont-Doherty Geological Observatory, Palisades, NY. 12 pp.
Bopp, R.F., H.J. Simpson, C.R. Olsen, and N. Kostyk. 1979. PCBs in Hudson River sediments. Lamont-Doherty Geological Observatory, Palisades, NY. 10 pp.
Bopp, R.F., H.J. Simpson, C.R. Olsen, and N. Kostyk. 1981. Polychlorinated biphenyls in sediments of the tidal Hudson River, New York. Environ. Sci. Technol. 15(2):210–215.
Bopp, R.F., H.J. Simpson, C.R. Olsen, R.M. Trier, and N. Kostyk. 1982. Chlorinated hydrocarbons and radionuclide chronologies in sediments of the Hudson River and estuary, New York. Environ. Sci. Technol. 16(10):666–676.
Bopp, R.F., H.J. Simpson, B.L. Deck, and N. Kostyk. 1984. The persistence of PCB components in sediments of the lower Hudson. Northeastern Environ. Sci. 3(3/4):180-184.
Boyle, R.H. 1975. Of PCB ppms from GE and a SNAFU from EPA and DEC. Audubon 77(6):127–133.

Boyle, R.H. 1975. "Poisoned fish, troubled waters." Sports Illustrated, Sept. 1, pp. 14–17.

Boyle, R.H. 1978. Update: getting rid of PCBs. Audubon 80(6):150–151.

Boyle, R.H. and J.H. Highland. 1979. The persistence of PCBs. Environment 21(5):6–37.

Brant, R. Unpublished information on Hudson River goldfish decline, and skin infections perhaps related to high PCB residues in goldfish. N.Y.S. DEC, Region III. New Paltz, NY.

Brinkman, M., K. Fogelman, J. Hoeflein, T. Lindh, M. Pastel, W.C. Trench, and D.A. Aikens. 1978. Levels of polychlorinated biphenyls in the Ft. Edward water supply ecosystem: a baseline approach. Rensselaer Polytechnic Institute, Troy, NY.

Brinkman, M., K. Fogelman, J. Hoeflein, T. Lindh, M. Pastel, W.C. Trench, and D.A. Aikens. 1980. Distribution of polychlorinated biphenyls in the Ft. Edward, NY water system. Environ. Management 4(6):511–520.

Brown, J.F., Jr., R.E. Wagner, D.L. Bedard, M.J. Brennan, J.C. Carnahan, R.J. May, and T.J. Tofflemire. 1984. PCB transformations in upper Hudson sediments. Northeastern Environ. Sci. 3(3/4):167-179.

Brown, M.P. 1981. PCB desorption from river sediments suspended during dredging: an analytical framework. Technical Paper No. 65, N.Y.S. Dept. of Environmental Conservation, Albany, NY.

Brown, M.P. and G.S. Kleppel. 1980. Summary of findings: an evaluation of the lower food chain kinetics of PCB's in the Hudson River ecosystem. The Louis Calder Conservation and Ecology Study Center of Fordham University, Armonk, NY.

Brown, M.P., J. McLaughlin, J.M. O'Connor, and K. Wyman. 1982. A mathematical model of PCB bioaccumulation in plankton. Ecol. Modell. 15(1):29–47.

Brown, M.P., M.B. Werner, R.J. Sloan, and K.W. Simpson. 1985. Polychlorinated biphenyls in the Hudson River. Environ. Sci. Technol. 19(8):656–661.

Buckley, E.H. 1980. PCBs in vegetation. Draft report to the N.Y.S. Dept. of Environmental Conservation, Albany, NY.

Buckley, E.H. 1982. Accumulation of airborne polychlorinated biphenyls in foliage. Science 216:520–522.

Buckley, E.H. 1983. Decline of background PCB concentrations in vegetation in New York State. Northeastern Environ. Sci. 2(3/4):181–187.

Buckley, E.H. and T.J. Tofflemire. 1983. Uptake of airborne PCBs by terrestrial plants near the tailwater of a dam, pp. 662–669 In Proceedings of the 1983 National Conference on Environmental Engineering. ASCE Specialty Conference, July 6–8, 1983, Boulder, CO.

Califano, R.J., J.M. O'Connor, and L.S. Peters. 1980. Uptake, retention, and elimination of PCB (Aroclor 1254) by larval striped bass (*Morone saxatilis*). Bull. Environ. Contam. Toxicol. 24(3):467–472.

Califano, R.J., J.M. O'Connor, and J.A. Hernandez. 1982. Polychlorinated biphenyl dynamics in Hudson River striped bass. I. Accumulation in early life history stages. Aquat. Toxicol. 2(4):187–204.

Carcich, I.G. and T.J. Tofflemire. 1982. Distribution and concentration of PCB in the Hudson River and associated management problems. Environ. International 7:73–85.

Carlson, G.A. and R.L. Collins (eds.). 1978. Toxic materials in the environment:

Collected papers from a workshop held in Albany, NY on Feb. 8–10, 1978. Technical Paper No. 52, N.Y.S. Dept. of Environmental Conservation, Albany, NY.

Cross, R.F. 1978. PCB's in the Hudson River. Conservationist 33(3):EQN2-5.

Daly, C.J. 1978. An assessment of the performance of laboratories providing PCB data for the PCB settlement. Environmental Health Center, Division of Laboratories and Research, N.Y.S. Dept. of Health, Albany, NY.

Economic Development Board. 1975. Economic impacts of regulating the use of PCBs in New York State. N.Y.S. Executive Dept., Albany, NY. 27 pp.

Finlay, D.J., F.H. Siff, and V.J. DeCarlo. 1976. Review of PCB levels in the environment. U.S. EPA, Office of Toxic Substances. Report EPA-560/7-76-001. Washington, D.C. NTIS-PB-253 735. 137 pp.

Gahagan and Bryant Associates. 1980. Dredging system decision report, dredging of PCB contaminated hot spots, upper Hudson River, New York, *In* Program Report No. 2, Appendix A. Malcolm Pirnie, Inc., White Plains, NY.

General Electric Co. 1977. Research on removal or treatment of PCB in liquid or sediments dredged from the Hudson River. Progress report. General Electric Co., Phys. Chem. Lab., Corporate Research and Development, Fairfield, CT.

Griffen, P.M. and C.M. McFarland. 1977. Research on removal or treatment of PCB in liquid or sediments dredged from the Hudson River: proposed study. General Electric Co., Corporate Research and Development, Schenectady, NY. 8 pp.

Griffen, P.M., C.M. McFarland, and A.R. Sears. 1978. Research on removal or treatment of PCB in liquid or sediments dredged from the Hudson River: Semi-annual progress report. General Electric Co., Corporate Research and Development, Schenectady, NY.

Hall, A.J. 1978. The Hudson: "That river's alive." National Geographic 153(1):62–89.

Henningson, J.C. and R.F. Thomas. 1981. Hudson River cleanup, pp. 205–214 *In* Proceedings of the Environmental Decontamination Workshop (G.A. Cristy and H.C. Jernigan, eds.). 1979.

Henningson, J.C., J.S. Reed, and S.P. Maslansky. 1982. Upland site screening for the disposal of hazardous dredged material: two Hudson River studies, *In* Proceedings of the Conference on the Impact of Waste Storage and Disposal on Ground-Water Resources. Cornell University Center for Environmental Research, June 28 to July 1, 1982, Ithaca, NY.

Hetling, L.J. 1978. Current problems in New York State. A. Water, pp. 81–95 *In* Toxic Materials in the Environment: Collected Papers from a Workshop Held in Albany, NY on Feb. 8–10, 1978. Technical Paper No. 52, N.Y.S. Dept. of Environmental Conservation, Albany, NY.

Hetling, L.J. and E.G. Horn. 1977. Hudson River PCB study description and detailed work plan. N.Y.S. Dept. of Environmental Conservation, Albany, NY. 62 pp.

Hetling, L.J., E. Horn, and J. Tofflemire. 1978. Summary of Hudson River PCB study results. Technical Paper 51. Bureau of Water Research, N.Y.S. Dept. of Environmental Conservation, Albany, NY. 88 pp.

Hetling, L.J., T.J. Tofflemire, E.G. Horn, R. Thomas, and R. Mt. Pleasant. 1979. The Hudson River PCB problem: management alternatives. Ann. N.Y. Acad. Sci. 320:630–650.

Highland, J.H., M.E. Fine, R.H. Harris, J.M. Warren, R.J. Rauch, A. Johnson, and R.H. Boyle. 1979. Malignant neglect. Alfred A. Knopf Publishers, New York, NY. 275 pp.

Horn, E.G., L.J. Hetling and T.J. Tofflemire. 1979. The problem of PCB's in the Hudson River system. Ann. N.Y. Acad. Sci. 320:591–609.

Horn, E., R. Sloan, and M. Brown. 1983. PCB in Hudson River striped bass, 1983. N.Y.S. Dept. of Environmental Conservation, Albany, NY. 8 pp. (Mss.)

Horstman, K.H. 1977. Evaluation of non-dredging alternatives to the removal of PCB contamination in the Hudson River. Comp. Ed. 18, Thesis. Union College, Schenectady, NY.

HRF. 1984. Proceedings of a PCB workshop held January 19, 1984. Hudson River Foundation for Science and Environmental Research, Inc., New York, NY.

Hullar, T., R.C. Mt. Pleasant, S. Pagano, J. Spagnoli, and W. Stasiuk. 1976. PCB data in Hudson River fish, sediments, water and wastewater. N.Y.S. Dept. of Environmental Conservation, Albany, NY. 26 pp.

Hydroscience, Inc. 1978. Estimation of PCB reduction by remedial action on the Hudson River ecosystem. Report to the N.Y.S. Dept. of Environmental Conservation. Hydroscience, Inc., Westwood, NJ. 107 pp.

Hydroscience, Inc. 1979. Analysis of the fate of PCB's in the ecosystem of the Hudson estuary. Report to the N.Y.S. Dept. of Environmental Conservation. Hydroscience, Inc., Westwood, NJ.

Johnson, B.J. 1981. PCB's in Hudson River sediments. 10/7/81–10/22/81. U.S. EPA Surveillance and Monitoring Branch, Region II, Edison, NJ.

Klauda, R.J., T.H. Peck, and G.K. Rice. 1981. Accumulation of polychlorinated biphenyls in Atlantic tomcod (*Microgadus tomcod*) collected from the Hudson River estuary, New York. Bull. Environ. Contam. Toxicol. 27(6):829–835.

Knabel, R.R. 1976. A failure of government, pp. 367–369 *In* Proceedings of the National Conference on Polychlorinated Biphenyls, Nov. 19–21, 1975. Chicago, Ill.

Kuzia, E. and H. Dean. 1978. Environmental effects of toxic materials. A. Fish, *In* Toxic Materials in the Environment: Collected Papers from a Workshop Held in Albany, NY on Feb. 8–10, 1978. Technical Paper No. 52. N.Y.S. Dept. of Environmental Conservation, Albany, NY.

Lawler, Matusky, and Skelly, Engineers. 1978. Upper Hudson River PCB no action alternative study: Final report to N.Y.S. Dept. of Environmental Conservation. Lawler, Matusky, and Skelly, Engineers, Pearl River, NY. 190 pp.

Lawler, Matusky, and Skelly, Engineers. 1979. Upper Hudson River PCB transport modeling study. Final report to N.Y.S. Dept. of Environmental Conservation. Lawler, Matusky, and Skelly, Engineers, Pearl River, NY.

Limburg, K.E. 1984. Environmental impact assessment of the PCB problem: A review. Northeastern Environ. Sci. 3(3/4):124–137.

Malcolm Pirnie, Inc. 1975. Engineering report, investigation of conditions associated with the removal of Ft. Edward dam, Ft. Edward, NY. Prepared for the N.Y.S. Dept. of Environmental Conservation. Malcolm Pirnie, Inc., White Plains, NY. 118 pp.

Malcolm Pirnie, Inc. 1976. Engineering report, preliminary appraisal sediment transport relations, upper Hudson River. Malcolm Pirnie, Inc., White Plains, NY. 41 pp.

Malcolm Pirnie, Inc. 1977a. Environmental assessment of maintenance dredging at Ft. Edward terminal channel, Champlain Canal. White Plains, NY. 271 pp.

Malcolm Pirnie, Inc. 1977b. Engineering report, investigation of conditions associated with the removal of the Ft. Edward dam, Ft. Edward, NY: Review of 1975 report. Malcolm Pirnie, Inc., White Plains, NY. 141 pp.

Malcolm Pirnie, Inc. 1977c. Evaluation of the feasibility of dredging PCB contaminated sediments in the upper Hudson River: Revised interim report of base data. Malcolm Pirnie, Inc., White Plains, NY. 324 pp.

Malcolm Pirnie, Inc. 1978a. Phase I engineering report, dredging of PCB contaminated hot spot, upper Hudson River, New York. Malcolm Pirnie, Inc., White Plains, NY.

Malcolm Pirnie, Inc. 1978b. Feasibility report, dredging of PCB-contaminated river bed materials from the upper Hudson River, New York. Vol. 1: Summary. Vol. 2: Engineering studies. Vol. 3: Environmental assessment and data base. Malcolm Pirnie, Inc., White Plains, NY.

Malcolm Pirnie, Inc. 1978c. Environmental assessment of remedial measures at the remnant deposits of the former Ft. Edward pool, Ft. Edward, NY. Malcolm Pirnie, Inc., White Plains, NY. 173 pp.

Malcolm Pirnie, Inc. 1979. Removal and encapsulation of PCB-contaminated Hudson River bed materials. Prepared for N.Y.S. Dept. of Environmental Conservation and Dept. of Transportation, Albany, NY. Malcolm Pirnie, Inc., White Plains, NY.

Malcolm Pirnie, Inc. 1980a. Engineering report, PCB hot spot dredging program, upper Hudson River. Containment site investigations: Program Report No. 1. Prepared for N.Y.S. Dept. of Environmental Conservation, Albany, NY.

Malcolm Pirnie, Inc. 1980b. Dredging alternatives report. Prepared for N.Y.S. Dept. of Environmental Conservation, Albany, NY. (Draft.)

Malcolm Pirnie, Inc. 1980c. PCB hot spot dredging program, upper Hudson River. Dredging system report: Program Report No. 2. Prepared for N.Y.S. Dept. of Environmental Conservation, Albany, NY.

Malcolm Pirnie, Inc. 1980d. Design report, PCB hot spot dredging program containment site. Prepared for N.Y.S. Dept. of Environmental Conservation, Albany, NY.

Malcolm Pirnie, Inc. 1980e. Draft environmental impact statement, N.Y.S. environmental quality review, PCB hot spot dredging program, upper Hudson River, New York. Prepared for N.Y.S. Dept. of Environmental Conservation, Albany, NY.

Malcolm Pirnie, Inc. 1981. Draft PCB hot spot dredging program, upper Hudson River, New York: Rescoping report. Malcolm Pirnie, Inc., White Plains, NY.

Matusik, J.J. 1978. Unpublished data on heavy metals in Hudson River sediments. N.Y.S. Dept. of Health, Radiological Science Laboratory, Albany, NY.

Maugh, T.H. 1975. Chemical pollutants: polychlorinated biphenyls still a threat. Science 190:1189.

Mehrle, P.M., T.A. Haines, S. Hamilton, J.L. Ludke, F.L. Mayer, and M.A. Ribick. 1982. Relationship between body contaminants and bone development in east-coast striped bass. Trans. Am. Fish. Soc. 111:231–241.

Miner, W. and T.J. Tofflemire. 1976. Ft. Edward maintenance dredging monitoring report. N.Y.S. Dept. of Environmental Conservation, Albany, NY.

Moskowitz, P., W. Hang, J. Silberman, D. Ross, and J. Highland. 1977. Troubled waters: toxic chemicals in the Hudson River. Environmental Defense Fund and New York Public Interest Group, Inc., New York, NY. 205 pp.

Mueller, J.A., LT.A. Gerrish, and M.C. Casey. 1982. Contaminant inputs to the Hudson-Raritan estuary. NOAA Technical Memorandum OMPA-21. NOAA, Office of Marine Pollution Assessment, Boulder, CO.

Nadeau, R.J. and R. Davis. 1974. Investigation of polychlorinated biphenyls in the Hudson River: Hudson Falls-Ft. Edward area. EPA Region II Report.

Nadeau, R.J. and R.A. Davis. 1976. Polychlorinated biphenyls in the Hudson River (Hudson Falls-Ft. Edward. NY). Bull. Environ. Contamin. Toxicol. 16(4):436–444.

National Research Council. 1979. Polychlorinated biphenyls. A report prepared by the Committee on the Assessment of Polychlorinated Biphenyls in the Environment. National Academy of Sciences, Washington, DC. 182 pp.

Natural Resources Defense Council, Inc. 1975. PCBs in fish: the fruits of inaction. NRDC Newsletter 4(3):1–2.

Nau-Ritter, G.M., C.F. Wurster, and R.G. Rowland. 1982. Polychlorinated biphenyls (PCB) desorbed from clay particles inhibit photosynthesis by natural phytoplankton communities. Environ. Poll. Ser. A. 28(3):177–182.

New York State Dept. of Environmental Conservation. 1975a. Monitoring PCBs in fish taken from the Hudson River, Albany, NY. Bureau of Environmental Protection, Div. of Fish and Wildlife, N.Y.S. Dept. of Environmental Conservation, Albany, NY.

New York State Dept. of Environmental Conservation. 1975b. PCB monitoring in the upper Hudson River basin. Division of Pure Waters, N.Y.S. Dept. of Environmental Conservation, Albany, NY. 110 pp.

New York State Dept. of Environmental Conservation. 1975c. File no. 2833 in the matter of alleged violations of Sections 17-0501, 17-0511, and 11-0503 of the Environmental Conservation Law of the State of New York by General Electric Co. 1. Complaint. 2. Exhibit I. Proposed abatement order. 3. Agreement. 4. Final order (Sept. 8, 1976.) *In* Technical Paper No. 58. N.Y.S. Dept. of Environmental Conservation, Albany, NY.

New York State Dept. of Environmental Conservation. 1976a. PCB data in Hudson River fish, sediments, water and wastewater. N.Y.S. Dept. of Environmental Conservation, Albany, NY.

New York State Dept. of Environmental Conservation. 1976b. Hudson River PCB monitoring: Data summary—past, present, proposed. Division of Pure Waters and Division of Fish and Wildlife, N.Y.S. Dept. of Environmental Conservation, Albany, NY. 106 pp.

New York State Dept. of Environmental Conservation. 1976c. Summary of data collected relative to Hudson River dredging. N.Y.S. Dept. of Environmental Conservation, Albany, NY.

New York State Dept. of Environmental Conservation. 1977. Hudson River PCB study description and detailed work plan. N.Y.S. Dept. of Environmental Conservation, Albany, NY.

New York State Dept. of Environmental Conservation. 1978a. The Hudson River: a reclamation plan. N.Y.S. Dept. of Environmental Conservation, Albany, NY. (Brochure ES-P55.) (20M-6/78)

New York State Dept. of Environmental Conservation. 1978b. Toxic substances

Bibliography B: Open Literature

impacting on fish and wildlife. Monthly Report 12. Division of Fish and Wildlife, N.Y.S. Dept. of Environmental Conservation, Albany, NY. Mar. 20, 1978.

New York State Dept. of Environmental Conservation. 1979a. Hudson River PCB study description and detailed work plan: Implementation of PCB settlement. Technical Paper No. 58. N.Y.S. Dept. of Environmental Conservation, Albany, NY. (Revised January 1979.)

New York State Dept. of Environmental Conservation. 1979b. Toxic substances in New York's environment. Vol. 1. An interim report. Vol. 2. Appendices. A joint report of N.Y.S. Dept. of Environmental Conservation and Dept. of Health, Albany, NY.

New York State Dept. of Environmental Conservation. 1979c. Hudson River basin water and related land resources. Level B study report and environmental impact statement. Hudson River Basin Study Group. N.Y.S. Dept. of Environmental Conservation, Albany, NY.

New York State Dept. of Environmental Conservation. 1980a. PCBs in the Hudson River: a reclamation program. N.Y.S. Dept. of Environmental Conservation, Albany, NY. (Brochure SW-P23, 9/80.)

New York State Dept. of Environmental Conservation. 1980b. PCBs in the land: a remedial program. N.Y.S. Dept. of Environmental Conservation, Albany, NY. (Brochure SW-P25, 11/80.)

New York State Dept. of Environmental Conservation. 1980c. (Unpublished data.) Monitoring PCB in pumpkinseed sunfish *Leopomis gibbosus* in HR in two areas: Newburgh and Albany. Robert Brant, N.Y.S. Dept. of Environmental Conservation, Fisheries, New Paltz, NY; Ron Sloan, N.Y.S. Dept. of Environmental Conservation, Albany, NY.

New York State Dept. of Environmental Conservation. 1981a. Public participation work plan for the Hudson River PCB cleanup project. Submitted to: U.S. Environmental Protection Agency, Region II, 26 Federal Plaza, New York, NY.

New York State Dept. of Environmental Conservation. 1981b. Hudson River PCB reclamation project: Step 1 grant application. Submitted to: Regional Administrator, U.S. Environmental Protection Agency, Region II, 26 Federal Plaza, New York, NY.

New York State Dept. of Environmental Conservation. 1981c. Hudson River PCB reclamation project: Response to executive summary of draft environmental impact statement. N.Y.S. Dept. of Environmental Conservation, Albany, NY.

New York State Dept. of Environmental Conservation. 1981d. Quality assurance plan for the Hudson River PCB reclamation project. N.Y.S. Dept. of Environmental Conservation, Albany, NY.

New York State Dept. of Environmental Conservation. 1981e. Toxic substances in fish and wildlife. 1979 and 1980 annual reports. Vol. 4, No. 1. Technical Report 81-1 (BEP). Division of Fish and Wildlife, N.Y.S. Dept. of Environmental Conservation, Albany, NY.

New York State Dept. of Environmental Conservation. 1981f. Toxic substances in fish and wildlife: May 1 to Nov. 1, 1981. Vol. 4, No. 2. Technical Report 82-1 (BEP). Division of Fish and Wildlife, N.Y.S. Dept. of Environmental Conservation, Albany, NY.

New York State Dept. of Environmental Conservation. 1982a. Hudson estuary fisheries development program progress report for fiscal year 1981-1982. N.Y.S. Dept. of Environmental Conservation, Albany, NY.

New York State Dept. of Environmental Conservation. 1982b. Environmental monitoring program: Hudson River PCB reclamation demonstration project. Bureau of Water Research, N.Y.S. Dept. of Environmental Conservation, Albany, NY.

New York State Dept. of Environmental Conservation. 1982c. Newsletter from the Hudson River PCB reclamation project. Vol. 1., Feb. 11, 1981 through Vol. II., Dec. 27, 1982. N.Y.S. Dept. of Environmental Conservation, Albany, NY.

New York State Dept. of Environmental Conservation. 1982d. The Hudson River PCB reclamation project: An update. N.Y.S. Dept. of Environmental Conservation, Albany, NY. 16 pp.

New York State Dept. of Environmental Conservation. 1982e. Decision-time for PCB removal. N.Y.S. Dept. of Environmental Conservation, Water Bull. (Nov.)

New York State Dept. of Health. (Undated.) PCBs: polychlorinated biphenyls. Health concerns, environmental effects, advisories. N.Y.S. Dept. of Health, Albany, NY. (Brochure.)

New York State Dept. of Health. 1975. Occurrence of polychlorinated biphenyls in New York State water supplies: a status report. N.Y.S. Dept. of Health, Albany, NY.

New York State Industrial Hazardous Waste Facility Siting Board. 1982. Final environmental impact statement in the matter of the application by the N.Y.S. Dept. of Environmental Conservation, Albany, NY.

New York State Industrial Hazardous Waste Facility Siting Board. 1982. Decision in the matter of the application by the N.Y.S. Dept. of Environmental Conservation, Albany, NY.

Normandeau Associates, Inc. 1977. Hudson River survey 1976-1977 with crosssection and planimetric maps. Normandeau Associates, Inc., Bedford, NH. 351 pp.

Nottis, G.N. 1979. Reconnaissance-level information on the geology and seismology of candidate PCB encapsulation sites nos. 10, 12, 23, 48, 20, and 16. N.Y.S. Dept. of Environmental Conservation, Albany, NY.

NUS. 1983. Feasibility study: Hudson River PCBs site, New York. EPA Work assignment. No. 01-2V84.0. Contract No. 68-01-6699. NUS Corp., Pittsburgh, PA.

O'Brien and Gere, Engineers, Inc. 1978. PCB analysis of Hudson River samples: Final report. O'Brien and Gere, Engineers, Inc., Syracuse, NY. 150 pp.

O'Brien and Gere, Engineers, Inc. 1982. Hudson River water PCB treatability study. Prepared for N.Y.S. Dept. of Environmental Conservation, Albany, NY.

O'Connor, J. 1984. PCBs: dietary dose and burdens in striped bass from the Hudson River. Northeastern Environ. Sci. 3(3/4):153-159.

Olsen, C.R. 1979. Radionuclides sedimentation and the accumulation of pollutants in the Hudson River estuary. Ph.D. Dissertation. Columbia University, New York, NY.

Olsen, C.R., H.J. Simpson, R.F. Bopp, S.C. Williams, T.H. Peng, and B.L. Deck. 1978. A geochemical analysis of the sediments and sedimentation in the Hudson estuary. J. Sedimentary Petrology 48:401-418.

Pastel, M., B. Bush, and J.S. Kim. 1980. Accumulation of polychlorinated biphenyls in American shad during their migration in the Hudson River, Spring 1977. Pesticides Monitoring J. 14(1):11–22.

Pavlou, S.P. and W. Hom. 1979. PCB removal from the Duwamish River estuary: Implications to the management alternative for the Hudson River PCB cleanup. Ann. N.Y. Acad. Sci. 320:651–671.

Peters, L.S. and J.M. O'Connor. 1982. Factors affecting short-term PCB and DDT accumulation by zooplankton and fish from the Hudson estuary, pp. 451–465 In Environmental Stress and the New York Bight: Science and Management (G. Mayer, ed.). Estuarine Research Federation, Columbia, SC.

Pizza, J.C. and J.M. O'Connor. 1983. PCB dynamics in Hudson River striped bass. II. Accumulation from dietary sources. Aquatic Toxicol. 3(4):313–328.

Powers, C.D., G.M. Nau-Ritter, R.G. Rowland, and C.F. Wurster. 1982. Field and laboratory studies of the toxicity to phytoplankton of polychlorinated biphenyls (PCBs) desorbed from fine clays and natural suspended particulates. J. Great Lakes Res. 8(2):350–357.

Pratt, M. 1979. Hudson PCB cleanup in limbo. New York Environmental News VI(13):4–6.

Reed, J.S. and J.C. Henningson. 1984. The management of contaminated dredged material from the Hudson River. Northeastern Environ. Sci. 3(3/4):217–221.

Roscigno, P. and L. Punto. (Undated.) Preliminary findings: Physical and biological parameters relative to PCB's in the Hudson River ecosystem. The Louis Calder Conservation and Ecology Study Center of Fordham University, Armonk, NY.

Sanders, J.E. 1982. The PCB pollution problem of the upper Hudson River from the perspective of the Hudson River PCB Settlement Advisory Committee. Northeastern Environ. Sci. 1(1):7–18.

Schroeder, R.A. and C.R. Barnes. 1983a. Polychlorinated biphenyl concentrations in Hudson River water and treated drinking water at Waterford, NY. U.S. Geological Survey, Water Resources Investigations Report 83-4188, Albany, NY. 13 pp.

Schroeder, R.A. and C.R. Barnes. 1983b. Trends in polychlorinated biphenyl concentrations in Hudson River water five years after elimination of point sources. U.S. Geological Survey, Water Resources Investigations Report 83-4206, Albany, NY. 28 pp.

Severo, R. 1980. "Hudson: Portrait of a river under attack." The New York Times, Sept. 9, 1980.

Shen, T.T. and T.J. Tofflemire. 1979. Air pollution aspects of land disposal of toxic waste. Technical Paper No. 59. N.Y.S. Dept. of Environmental Conservation, Albany, NY.

Shen, T.T. 1982. Air quality assessment for land disposal of industrial wastes. Environ. Management 6(4):297–305.

Sheppard, J.D. 1976. Valuation of the Hudson River fishery resources: Past, present and future. N.Y.S. Dept. of Environmental Conservation, Albany, NY. 69 pp.

Simpson, H.J., R.F. Bopp, B.L. Deck, S. Warren, and N. Kostyk. 1984. Polychlorinated biphenyls in the Hudson River: the value of individual packed-column peak analysis. Northeastern Environ. Sci. 3(3/4):160-166.

Simpson, K.W. 1982. PCBs in multiplate residues and caddisfly larvae from the

Hudson River, 1977–1980. Technical Paper (in prep.). N.Y.S. Dept. of Environmental Conservation, Albany, NY.

Simpson, K.W., R.C. Mt. Pleasant, and B. Bush. 1977. The use of artificial substrates for monitoring toxic organic compounds: A preliminary evaluation involving the PCB problem in the Hudson River. N.Y.S. Dept. of Health, Albany, NY. 18 pp.

Skea, J.C., H.A. Simonin, H.J. Dean, J.R. Colquhoun, J.J. Spagnoli, and G.D. Veith. 1979. Bioaccumulation of Aroclor 1016 in Hudson River fish. Bull. Environ. Contam. Toxicol. 22:332–336.

Skinner, L.C. 1976. PCB's in the Hudson River. Paper presented at Northeastern Division meeting of Am. Fish. Soc., Hershey, PA.

Sloan, R.J. 1978. PCB's in Hudson River fish. Technical Report to Hudson River PCB Advisory Committee. Sponsored by N.Y.S. Dept. of Environmental Conservation, Jan. 19–20, 1978. Albany, NY.

Sloan, R.J. 1979. Monitoring PCB contamination levels in Hudson River fish. Presented at "PCB's in the Hudson River: 2 1/2 years of research." Technical Report to Hudson River PCB Advisory Committee, June 11–12, Albany, NY.

Sloan, R.J. and J.D. Sheppard. 1978. Spatial and temporal distributions of PCB concentrations in Hudson River fish: A preliminary assessment. Fish. Wildl. Abstr., 34th Northeast Fish and Wildl. Conf., White Sulfur Springs, WV. 32 pp.

Sloan, R., R. Armstrong, and E.G. Horn. 1980. Declining levels of PCB in Hudson River fish: Prospects for recovery of the fishery, pp. 34–35 in 36th Ann. Northeast Fish and Wildl. Conf. Fish. Abstrs., Apr. 27–30. Ellenville, NY.

Sloan, R.J. and R.W. Armstrong. 1981. Fighting for cleaner fish. The Conservationist 35(5):36–41.

Sloan, R.J. and R.W. Armstrong. 1986. PCB patterns in Hudson River fish. II. Migrant/marine species. *In* Fisheries Research in the Hudson River (C.L. Smith, ed.). Proceedings of a workshop sponsored by the Hudson River Environmental Society, September, 1981. State University of N.Y. Press, Albany, NY. (in press.)

Sloan, R.J., K.W. Simpson, R.A. Schroeder, and C.R. Barnes. 1983. Temporal trends toward stability of Hudson River PCB contamination. Bull. Environ. Contam. Toxicol. 31(4):377–385.

Sloan, R.J., M. Brown, R. Brandt, and C. Barnes. 1984. Hudson River PCB relationships between fish, water and sediment. Northeastern Environ. Sci. 3(3/4):138-152.

Sofaer, A.D. 1976. Interim opinion and order, unpublished, *In* The Matter of Violations of ECL by GE Co. File No. 2833, N.Y.S. Dept. of Environmental Conservation, Albany, NY. 77 pp.

Sofaer, A.D. 1976. Recommendation of settlement, unpublished opinion in the matter of violations of ECL by GE Co., NY. File No. 2833. N.Y.S. Dept. of Environmental Conservation, Albany, NY. 14 pp.

Spagnoli, J.J. and L.C. Skinner. 1975. PCBs in fish from selected waters in New York State. Technical Paper No. 75-14. Bureau of Environmental Protection, Division of Fish and Wildlife. N.Y.S. Dept. of Environmental Conservation, Albany, NY. 47 pp.

Spagnoli, J.J. and L.C. Skinner. 1977. PCBs in fish from selected waters of New York State. Pesticides Monitoring J. 11(2):69–87.

Stalling, D.L. and F.L. Mayer, Jr. 1972. Toxicities of PCBs to fish and environmental residues. Environmental Health Perspectives 1:159–164.

Stone, W. 1978. How we have polluted our planet. The Conservationist 33(2):EQN III–VI.

Stone, W.B., E. Kiviat, and S.A. Butkas. 1980. Toxicants in snapping turtles. N.Y. Fish Game J. 27(1):39–50.

Stone, W.B. and J.C. Okoniewski. 1983. Organochlorine toxicants in great horned owls *(Bubo virginianus)* from New York, 1981–1982. Northeast Environ. Sci. 2(1):1–7.

Syrotynski, S. 1978. Current problems in New York State. B. Detection frequency of heavy metals, trace elements, pesticides, and other organic chemicals in water, *In* Toxic Materials in the Environment: Collected Papers from a Workshop Held in Albany, NY on Feb. 8–10, 1978. Technical Paper No. 52. N.Y.S. Dept. of Environmental Conservation, Albany, NY.

Thomann, R.V. and J.P. St. John. 1979. The fate of PCB's in the Hudson River ecosystem. Ann. N.Y. Acad. Sci. 320:610–629.

Thomann, R.V., P.C. Pacquin, and J.J. Fitzpatrick. 1980. Modeling of PCB fate in the Hudson River system, pp. 130–137 *In* Proceedings of Natl. Conf. Environ. Eng. (W. Saukin, ed.). ASCE Environ. Eng. Div. Spec. Conf.

Thomas, R.F., R.C. Mt. Pleasant, and S.P. Maslansky. 1979. Removal and disposal of PCB-contaminated river bed materials, pp. 167–172 *In* Proceedings of the 1979 National Conf. on Hazardous Material Risk Assessment, Disposal and Management, Apr. 25–27, Miami Beach, FL.

Tofflemire, T.J. 1976. Preliminary report on sediment characteristics and water column interactions relative to dredging the upper Hudson River for PCB removal. Prepared for N.Y.S. Dept. of Environmental Conservation, Albany, NY. 82 pp.

Tofflemire, T.J. 1978. Dredging mass removal experiment. 1978. Bureau of Water Research, N.Y.S. Dept. of Environmental Conservation, Albany, NY.

Tofflemire, T.J. 1979. Improving the efficiency of dredging several feet of contaminated sediment off the top of an uncontaminated sediment. Technical Paper No. 61. N.Y.S. Dept. of Environmental Conservation, Albany, NY.

Tofflemire, T.J. 1984. PCB transport in the Ft. Edward area. Northeastern Environ. Sci. 3(3/4):203-208.

Tofflemire, T.J. and T.F. Zimmie. 1977. Hudson River sediment distributions and water interactions relative to PCB: Preliminary indications. Kepone Symposium II, U.S. EPA, Region III. Easton, MD. 26 pp.

Tofflemire, T.J., L.J. Hetling and S.O. Quinn. 1979. PCB in the upper Hudson River: sediment distributions, water interactions and dredging. Technical Paper No. 55. Bureau of Water Research, N.Y.S. Dept. of Environmental Conservation, Albany, NY. 68 pp.

Tofflemire, T.J. and S.O. Quinn. 1979. PCB in the upper Hudson River: Mapping and sediment relationships. Technical Paper No. 56. N.Y.S. Dept. of Environmental Conservation, Albany, NY.

Tofflemire, T.J. and T.T. Shen. 1979. Volatilization of PCB from sediment and water: Experimental and field data, *In* Proceedings of Mid-Atlantic Ind. Waste Conf. 11:100–109.

Tofflemire, T.J., S.O. Quinn, and P.R. Hague. 1979. PCB in the Hudson River:

Mapping, sediment sampling and data analysis. Technical Paper No. 57. N.Y.S. Dept. of Environmental Conservation, Albany, NY.

Tofflemire, T.J., S.O. Quinn, and I.G. Carcich. 1980. Sediment and water sampling and analysis for toxics relative to PCB in the Hudson River. Technical Paper No. 64. N.Y.S. Dept. of Environmental Conservation, (WM-P23), Albany, NY.

Tofflemire, T.J., T.T. Shen, and E.H. Buckley. 1981. Volatilization of PCB from sediment and water: Experimental and field data. Technical Paper No. 63. N.Y.S. Dept. of Environmental Conservation, Albany, NY. Also published in 1983, pp. 411–422. *In* Physical Behavior of PCBs in the Great Lakes (D. Mackay *et al.*, eds.). Ann Arbor Science Publishers, Woburn, MA.

Turk, J.T. 1980. Applications of Hudson River basin PCB-transport studies, pp. 171–183 *In* Contaminants and Sediments, Vol. I. (R.A. Baker, ed.). Ann Arbor Science Pub., Inc., Ann Arbor, MI.

Turk, J.T. and D.E. Troutman. 1981. Relationship of water quality of Hudson River, New York, during peak discharges to geologic characteristics of contributing subbasins. Water Resources Investigations 80-108. U.S. Geological Survey, Albany, NY. 15 pp.

Turk, J.T. and D.E. Troutman. 1981. Polychlorinated biphenyl transport in the Hudson River, New York. Water Resources Investigations 81-9. U.S. Geological Survey, Albany, NY. 11 pp.

U.S. Dept. of the Army. 1981. Public Notice No. 10663. Application No. 81-068 for PCB hot spot dredging and disposal program. N.Y. District Corps of Engineers, 26 Federal Plaza, New York, NY.

U.S. Dept. of the Army. 1981. Public Notice No. 10728. Application No. 81-068 for PCB hot spot dredging and disposal program. N.Y. District Corps of Engineers, 26 Federal Plaza, New York, NY.

U.S. Geological Survey. 1975. Chemical quality of water in community systems in New York: May 1974 to May 1975. Open File Report 77-731. U.S. Geological Survey, Albany, NY.

U.S. Geological Survey. (Issued annually.) Water resources data for New York water years 1975–1982. Vol. 1: Eastern New York excluding Long Island. U.S. Geological Survey, Albany, NY.

U.S. Environmental Protection Agency, Region II. 1977. PCBs in lower Hudson River sediments: a preliminary survey. EPA 12/11/76 - 12/15/76. U.S. EPA Surveillance and Analysis Division, Edison, NJ. 40 pp.

U.S. Environmental Protection Agency, Region II. 1981. Draft environmental impact statement on the Hudson River PCB reclamation demonstration project. Prepared by WAPORA Inc. under the direction of EPA Region II personnel. U.S. EPA Region II, 26 Federal Plaza, New York, NY.

U.S. Environmental Protection Agency, Region II. 1981. Supplemental draft environmental impact statement on the Hudson River PCB reclamation demonstration project. U.S. EPA Region II, 26 Federal Plaza, New York, NY.

U.S. Environmental Protection Agency, Region II. 1982. Final environmental impact statement on the Hudson River PCB reclamation demonstration project. U.S. EPA Region II, 26 Federal Plaza, New York, NY.

United States Environmental Protection Agency, Region II. 1982. Record of decision for the environmental impact statement on the Hudson River PCB recla-

mation demonstration project. U.S. EPA Region II, 26 Federal Plaza, New York, NY.
U.S. 96th Congress. 1979. Bill H.R. 5849 to amend Title I of the Federal Water Pollution Control Act to authorize a demonstration project for the reclamation of the Hudson River. Nov. 26, 1979.
Werner, M.B. 1981. The use of a freshwater mollusc (*Elliptio complanatus*) in biological monitoring programs. B. The freshwater mussel as a biological monitor of PCB concentrations in the Hudson River. Toxic Substances Control Unit, N.Y.S. Dept. of Environmental Conservation, Albany, NY.
West, R.H. and P.G. Hatcher. 1980. Polychlorinated biphenyls in sewage sludge and sediments of the New York Bight. Mar. Poll. Bull. 11(5):126–129.
West, R.H., P.G. Hatcher, and D.K. Atwood. 1976. Polychlorinated biphenyls and DDTs in sediments and sewage sludge of the New York Bight. National Oceanographic and Atmospheric Administration, Boulder, CO. 42 pp.
Weston Environmental Consultants-Designers. 1978. Migration of PCBs from landfills and dredge spoil sites in the Hudson River valley, New York: Final report. Prepared for N.Y.S. Dept. of Environmental Conservation, Albany, NY and Weston Environmental Consultants, West Chester, PA.
Wurster, C.F. and H.B. O'Connors. 1980. PCBs: Research accomplishments and proposal for future research. Memorandum to Italo Carcich, Director, Bureau of Water Research, N.Y.S. Dept. of Environmental Conservation, Albany, NY.
Zimmie, T.F. 1981. Determining rates of cohesive sediment erosion for the Hudson River. Final report to the N.Y.S. Dept. of Environmental Conservation, Albany, NY.
Zimmie, T.F. 1984. Transport of PCB in Hudson River bedload sediments. Northeastern Environ. Sci. 3(3/4):191-196.
Zimmie, T.F. and T.J. Tofflemire. 1977. Disposal of PCB contaminated sediment in the Hudson River, pp. 850–864 *In* American Society of Civil Engineers Spec. Conf. of Geotech. Eng. Div., June 1977, Ann Arbor, MI.

C. Radionuclides

Anderson, K. (Date unknown.) Tritium in New York State waters. Bureau of Radiological Services, N.Y.S. Department of Health. Radioactivity Bull. 66–6.
Bopp, R.F., H.J. Simpson, and C.R. Olsen. 1977. PCBs and Cs-137 in sediment of the Hudson estuary. Eos: Trans. Am. Geophys. Union 58(6):407.
Bopp, R.F., H.J. Simpson, C.R. Olsen, R.M. Trier, and N. Kostyk. 1982. Chlorinated hydrocarbons and radionuclide chronologies in sediments of the Hudson River and estuary, New York. Environ. Sci. Technol. 16:666–676.
Cohen, L.K. and T.J. Kneip. 1973. Environmental tritium studies at a PWR (pressurized water reactor) power plant, pp. 623-639 *In* Tritium. Messenger Graphics Publ.
*Davies, S., F. Cosolito, and M. Eisenbud. 1966. Radioactivity in the Hudson River, pp. 289–311 *In* Hudson River Ecology, 1st Symp. Hudson River Valley Comm., Tuxedo, NY.
Eisenbud, M., A. Perlmutter, and A.W. McCrone. 1965. Radioecological survey of the Hudson River. Prog. Rep. No. 1. Div. Radiological Health. U.S. Public Health Service Contract PH-86-64-Neg. 141, Washington, D.C.
Eisenbud, M., A. Perlmutter, and A.W. McCrone. 1966. Radioecological survey

of the Hudson River. Prog. Rep. No. 2. Div. Radiological Health. U.S. Publ. Health Service Contract PH-86-64-Neg. 141, Washington, D.C.

Hairr, L.M. 1974. Investigation of factors influencing radiocesium cycling in estuarine sediments of the Hudson River. Ph.D. Dissertation. New York University. University Microfilms Order No. 76-19, 6-5 NY.

Hammond, D.E., H.J. Simpson, and G. Mathieu. 1975. Methane and Radon-222 as tracers for mechanisms of exchange across the sediment-water interface in the Hudson River estuary, pp. 119–132 *In* Marine Chemistry in the Coastal Environment (T.M. Church, ed.). Am. Chemical Soc., Washington, D.C. 710 pp.

Hammond, D.E., H.J. Simpson, and G. Mathieu. 1976. Distribution of Radon-222 in the Delaware and Hudson estuaries as an indicator of migration rates of dissolved species across the sediment-water interface. Eos: Am. Geophys. Union Trans. 57(3):151.

Hammond, D.E., H.J. Simpson, and G. Mathieu. 1977. Radon-222 distribution and transport across the sediment-water interface in the Hudson River estuary. J. Geophys. Res. 82(27):3913–3920.

Jinks, S.M. 1975. An investigation of the factors influencing radiocesium concentrations of fish inhabiting natural aquatic ecosystems. Ph.D. Dissertation. New York University, New York, NY.

Jinks, S.M., M.E. Wrenn, B.J. Friedman, and L.M. Hairr. 1974. A "critical pathway" evaluation of the radiological impact resulting from liquid waste discharges at the Indian Point 1 Nuclear Power Station. Paper 18 *In* Hudson River Ecology, 3rd Symp. Hudson River Environ. Soc., Bronx, NY. 35 pp.

Jinks, S.M. and M.E. Wrenn. 1976. Radiocesium transport in the Hudson River estuary, pp. 207–227 *In* Environmental Toxicity of Aquatic Radionuclides: Models and Mechanisms (M.W. Miller and J.N. Stannard, eds.). 8th Rochester Intern. Conf. Rochester, NY, June 2–4, 1975. Ann Arbor Science Publ., Ann Arbor, MI.

Kahn, B., B. Shleien, and C. Weaver. 1971. Environmental experience with radioactive effluents from operating nuclear power plants. Environ. Prot. Agency, Cincinnati, OH.

Kneip, T.J., G.P. Howells, and M.E. Wrenn. 1970. Trace elements, radionuclides, and pesticides in the Hudson River. FAO Tech. Conf. on Marine Pollution and Its Effects on Living Resources and Fishing, Rome, Italy.

Lentsch, J. 1975. The fate of gamma-emitting radionuclides released into the Hudson River estuary and an evaluation of their environmental significance. Ph.D. Dissertation. New York University, New York, NY.

Lentsch, J.W., M.E. Wrenn, T.J. Kneip, and M. Eisenbud. 1970. Manmade radionuclides in the Hudson River estuary, pp. 499–528 *In* Conf. 701106, Proc. 5th Ann. Health Physics Soc. Midyear Topical Symp., Health Physics Aspects of Nuclear Facility Siting, Vol. 2. Nov. 5–6, 1970.

Lentsch, J., T.J. Kneip, M.E. Wrenn, G.P. Howells, and M. Eisenbud. 1971. Stable manganese and manganese-54 distribution in the physical and biological components of the Hudson River estuary, pp. 752–768 *In* Radionuclides in Ecosystems. 3rd Nat. Symp. on Radioecology, Oak Ridge National Laboratory, Oak Ridge, TN.

Linsalata, P. 1979. Determination of ^{239,249}PU, and ^{238}PU in Hudson River estuary sediments. Health Phys. 37(6):804–805.

Mauro, J. 1973. The pathway by which *Fundulus heteroclitus* obtains Cs-137 from the Hudson River in the vicinity of Indian Point. Ph.D. Dissertation. New York University, New York, NY.

Mauro, J. and M.E. Wrenn. 1974. An investigation into the reasons for lack of a trophic level effect for Cs-137 in *Fundulus heteroclitus* and its food in the Hudson River estuary, pp. 279–284 *In* Proc. 3rd Internat. Congress of the Internat. Radiation Protection Association, CONF-730907-P1.

*McCrone, A.W. 1966. The Hudson River estuary, hydrology, sediments, and pollution. Geogr. Rev. 56:175–189.

Olsen, C.R. 1979. Radionuclides sedimentation and the accumulation of pollutants in the Hudson River estuary. Ph.D. Dissertation, Columbia University, New York, NY.

Olsen, C.R., H.T. Simpson, and R.M. Trier. 1977. Anthropogenic radionuclides as tracers for recent sediment deposition in the Hudson estuary. Eos: Am. Geophys. Union Trans. 58(6):406.

Paschoa, A.S. 1975. The natural radiation dose from radium to *Gammarus* in the Hudson River. Ph.D. Dissertation. New York University, New York, NY. 207 pp.

*Paschoa, A., M.E. Wrenn, and M. Eisenbud. 1976. Natural radiation dose to *Gammarus* from Hudson River. Paper 11 *In* Hudson River Ecology, 4th Symp. Hudson River Environ. Soc., Bronx, NY.

Paschoa, A.S., M.E. Wrenn, and M. Eisenbud. 1979. Natural radiation dose to *Gammarus* from Hudson River. Radioprotection 14(2):99–115.

Simpson, H.J. and S.C. Williams. 1976. Plutonium and cesium radionuclides in the Hudson River estuary. Ann. Tech. Prog. Rep., Dec. 1, 1975 to Nov. 30, 1976. Lamont-Doherty Geol. Observ., Palisades, NY. NTIS-AN-77-003-701.

*Simpson, H.J., C.R. Olsen, R.M. Trier, and S.C. Williams. 1976. Man-made radionuclides and sedimentation in the Hudson River estuary. Science 194:179–183.

Williams, S.C. 1976. Plutonium and cesium radionuclides in the Hudson River estuary. Nuclear Sci. Abstr. 33(3):576.

Wrenn, M.E., J.. Lentsch, M. Eisenbud, G.J. Lauer, and G.P. Howells. 1971. Radiocesium distribution in water, sediment, and biota in the Hudson River estuary from 1964 through 1970, pp. 334–343 *In* 3rd National Symp. on Radioecology, May 10–12, 1971. Oak Ridge, TN.

Wrenn, M.E., S.M. Jinks, N. Cohen, and L.M. Hairr. 1974. Cs-137 and Cs-134 distribution in sediment, water, and biota of the lower Hudson River and their dosimetric implications for man, pp. 279–284 *In* Proc. 3rd Internat. Congress of the Internat. Radiation Protection Association, CONF-730907-P1.

Wrenn, M.E., S.M. Jinks, L.M. Hairr, A.S. Paschoa, and J.W. Lentsch. 1975. Natural activity in Hudson River estuary samples and their influence on the detection limits for gamma emitting radionuclides using NaI gamma spectrometry, pp. 897–916 *In* Proc. 2nd Internat. Symp. on the Natural Radiation Environment, CONF-720805-P2.

*Wrenn, M.E., S.M. Jinks, G.R. Pack, and J. Lentsch. 1976. A mathematical model describing radiocesium transport in the Hudson River estuary. Paper 10 *In* Hudson River Ecology, 4th Symp. Hudson River Environ. Soc., Bronx, NY.

D. Other Toxic Substances

Andelman, P.A. 1972. "60,000 gallon oil spill fouls 16 miles of the Hudson." New York Times, Jan. 22, 1972, p. M 46.

Armstrong, R. and R.J. Sloan. 1980. Trends in levels of several known chemical contaminants in fish from New York State waters. Technical Report 80-2. Bur. Environ. Protection, Division of Fish and Wildl. N.Y.S. Dept. of Environmental Conservation, Albany, NY.

Bondietti, E.A., F.H. Sweeton, T. Tamura, R.M. Perhac, L.D. Hulett, and T.J. Kneip. 1973. Characterization of cadmium and nickel contaminated sediments from Foundry Cove, NY, pp. 211–224 *In* Proc. First Ann. NSF Trace Contaminants Conf., Oak Ridge National Laboratory, Oak Ridge, TN. Aug. 8–10.

Bopp, L.H., A.M. Chakrabarty, and H.L. Ehrlich. 1983. Chromate resistance plasmid in *Pseudomonas fluorescens*. J. Bacteriol. 155(3):1105–1109.

Bower, P.M. 1976. Burdens of industrial cadmium and nickel in the sediments of Foundry Cove, Cold Spring, NY. M.A. Thesis. Queens College, City University of NY, Flushing, NY.

Boyce Thompson Institute. 1972. Croton Point ecology; assessment of environmental impact of waste disposal area. Westchester Dept. Public Works. (unpublished.)

Boyle, R.H. 1974. The Croton: Troubled river of history. The Conservationist 28(3):6–11 (Dec. to Jan., 1973–1974).

Boyle, R.H. 1979. The Croton River: Mini-estuary, pp. 27–34 *In* Fish Stories (D.W. Bennett, ed.). Am. Littoral Soc., Sandy Hook. Highlands, NJ.

Buchler, K. and H.I. Hirschfield. 1974. Cadmium in an aquatic ecosystem; effects on phytoplankton organisms. Presented at 2nd Ann. Nat. Sci. Foundation, RANN (Research Applied to the National Needs) Trace Contaminants Conf., Asilomair, CA.

Carlson, G., G. Anders, and R.L. Collin (eds.). 1978. Toxic materials in the environment. Collected Papers from a Workshop Held in Albany, NY, Feb. 1978. N.Y.S. Dept. of Environmental Conservation, Albany, NY.

Carmody, D.J. 1972. The distribution of five heavy metals in the sediments of the New York Bight. Ph.D. Thesis. Columbia University, New York, NY. 121 pp.

Carmody, D.J., J.B. Pearce, and W.E. Yasso. 1973. Trace metals in sediments of New York Bight. Mar. Pollution Bull. 4:132–135.

Catanzaro, E.J. 1976. Some relationship between exchangeable copper and lead and particulate matter in a sample of Hudson River water. Environ. Sci. Technol. 10(4):386–388.

Catanzaro, E.J. and J.O. Nriagu (eds.). 1976. Mass spectrometric-isotope dilution determinations of copper and lead in Hudson River water, pp. 513–517 *In* Environmental Biogeochemistry. Vol. 2. Metals Transfer and Ecological Mass Balances. Ann Arbor Science Publishers, Inc., Ann Arbor, MI.

Greeley, J. 1959. Battle report on water chestnut. The Conservationist 13(4):32–33.

Greeley, J. 1965. Eradication of water chestnut. The Conservationist 19(4):13–15.

Gross, M.G. 1970. Analysis of dredged wastes, fly ash, and waste chemicals—New York metropolitan region. Mar. Sci. Center, Technical Report 7. S.U.N.Y. at Stony Brook, NY.

*Hazen, R. and T. Kneip. 1976. The distribution of cadmium in the sediments of Foundry Cove. Paper 8 *In* Hudson River Ecology, 4th Symp. Hudson River Environ. Soc., Bronx, NY.

Helsinger, M.H. and G.M. Friedman. 1982. Distribution and incorporation of trace elements in the bottom sediments of the Hudson River and its tributaries. Northeastern Environ. Sci. 1(1):33–47.

Henderson, C., A. Inglis, and W.L. Johnson. 1972. Mercury residues in fish, 1969–1970. National Pesticide Monitoring Program. Pesticide Monitoring J. 6(3):144–159.

Hetling, L.E. and J.P. Lawler. 1980. Organics and their impact, *In* Hudson River Ecology, 5th Symp. Hudson River Environ. Soc., Vassar College, Poughkeepsie, NY.

Howells, G.P. and D. Bath. 1969. Trace elements in the Hudson River. Development of a biological monitoring system and a survey of trace metals, radionuclides and pesticide residues in the Hudson River. New York University Medical Center, Institute of Environmental Medicine, New York, NY.

*Howells, G.P., T.J. Kneip, and M. Eisenbud. 1970. Water quality in industrial areas: Profile of a river. Environ. Sci. Tech. 4(1):26–35.

Howells, G.P. and D. Bath. 1971. Ecological survey of the Hudson River. Trace elements in the Hudson River, *In* Hudson River Ecology, 2nd Symp., 1969. N.Y.S. Dept. of Environmental Conservation, Albany, NY.

*Intorre, B. and P. DeRienzo. 1974. The estuary and industrial wastes: Power plants, pp. 169–177 *In* Hudson River Colloquium (O.A. Roels, ed.). Ann. N.Y. Acad. Sci. 250.

Kennikoh, K.B., H.I. Hirschfield, and T.J. Kneip. 1978. Cadmium toxicity in planktonic organisms of a freshwater food web. Environ. Res. 15(3):357–367.

*Ketchum, B.H. 1974. Population, resources, and pollution, and their impact on the Hudson estuary, pp. 144–156 *In* Hudson River Colloquium (O.A. Roels, ed.). Ann. N.Y. Acad. Sci. 250.

Klein, L.A., M. Lang, N. Nash, and S.L. Kirschner. 1974. Sources of metals in New York City wastewater. J. Poll. Control Fed. 46(12):2653–2662.

Klinkhammer, G.P. 1977. Measuring metals in the Hudson River gives clue to water quality. Maritimes 21(4):7–9.

Klinkhammer, G.P. 1977. The distribution and partitioning of some trace metals in the Hudson River estuary. M.S. Thesis. University of Rhode Island, Providence, RI. 125 pp.

Klinkhammer, G.P., M. Bender, and J. Simpson. 1975. A trace metal model of the Hudson estuary. American Chemical Society, 169th National Meeting. Special Symposium on Marine Chemistry in the Coastal Environment 14. (abstract only.)

Klinkhammer, G.P., M. Bender, and J. Simpson. 1976. The partitioning of some trace metals in the Hudson River estuary. Eos: Trans. Am. Geophys. Union 57(4):255.

*Klinkhammer, G.P. and M.L. Bender. 1981. Trace metal distributions in the Hudson River estuary. Est. Coastal and Shelf Sci. 12:629–643.

Kneip, T.J. 1968. Trace contaminants in the Hudson River, pp. 356–373 *In* National Estuarine Pollution Study Proceedings. New York, NY.

*Kneip, T.J., G.P. Howells, and M.E. Wrenn. 1969. Trace elements, radionuclides, and pesticide residues in the Hudson River, pp. 162–199 In Hudson River Ecology, 2nd Symp. N.Y.S. Dept. of Environmental Conservation, Albany, NY.

Kneip, T.J., G.P. Howells, D.W. Bath, and A. Perlmutter. 1969. Biological monitoring experiments, pesticides. Development of a biological monitoring system and a survey of trace metals, radionuclides and pesticide residues in the Hudson River. New York University Medical Center, Institute of Environmental Medicine, New York, NY.

Kneip, T.J., G.P. Howells, and M.E. Wrenn. 1970. Trace elements, radionuclides, and pesticide residues in the Hudson River. FAO Tech. Conf. on Marine Pollution and Its Effects on Living Resources and Fishing. Rome, Italy.

Kneip, T.J. and H.I. Hirschfield. 1975. Cadmium in an aquatic ecosystem: Distribution and effects, In 2nd Ann. Progress Report. New York University Medical Center, Institute of Environmental Medicine, New York, NY. 102 pp.

Kneip, T.J. and R.E. Hazen. 1979. Deposit and mobility of cadmium in a marsh-cove ecosystem and the relation of cadmium concentration in biota. Environ. Heath Perspectives 28:67–73.

Kneip, T.J., J. O'Connor, A. Wiedow, and R. Hazen. 1979. Final report. Cadmium in Foundry Cove crabs: Health hazard assessment. Report to N.Y.S. Health Research Council. 35 pp.

Koepp, S.J., E.D. Santoro, and G. DiNardo. 1986. Heavy metals in finfish and selected macroinvertebrates of the lower Hudson estuary. In Fisheries Research in the Hudson River (C.L. Smith, ed.). Proceedings of a workshop sponsored by the Hudson River Environmental Society, September, 1981. State University of N.Y. Press, Albany, NY. (in press.)

*Litchfield, C.D., M.A. Devanas, J. Zindulis, M. Meskill, J. Freedman, and C. McClean. 1982. Influence of cadmium on the microbial populations and processes in sediments from the New York Bight Apex, pp. 587–604 In Ecological Stress and the New York Bight: Science and Management (G.F. Mayer, ed.). Estuarine Research Federation, Columbia, SC.

*McCrone, A.W. 1966. The Hudson River estuary: hydrology, sediments, and pollution. Geogr. Rev. 56(2):175–189.

*Mehrle, P.M., T.A. Haines, S. Hamilton, J.L. Ludke, F.L. Mayer, and M.A. Ribick. 1982. Relationship between body contaminants and bone development in east-coast striped bass. Trans. Am. Fish. Soc. 111:231–241.

Moskowitz, P., W. Hang, J. Silberman, D. Ross, and J. Highland. 1977. Troubled waters. Toxic chemicals in the Hudson River. Environ. Defense Fund and the New York Public Interest Res. Group, Inc., New York, NY.

*Mt. Pleasant, R.C. 1966. Industrial and domestic waste discharges into the Hudson River from the Troy Lock to the George Washington Bridge, pp. 258–267 In Hudson River Ecology, 1st Symp. Hudson River Valley Comm., Tuxedo, NY.

*Occhiogrosso, T., W. Waller, and G. Lauer. 1976. Effects of heavy metals on benthic macroinvertebrate densities in Foundry Cove on the Hudson River. Paper 9 In Hudson River Ecology, 4th Symp. Hudson River Environ. Soc., Bronx, NY.

Olsen, C.R., H.J. Simpson, R.F. Bopp, R.M. Trier, and S.C. Williams. 1978.

Pollution history and sediment accumulation patterns in the Hudson estuary. Lamont-Doherty Geological Observatory, Palisades, NY. 26 pp.

*Pakkala, I.S., W.H. Gutenmann, D.J. Lisk, G.E. Burdick, and E.J. Harris. 1972. A survey of the selenium content of fish from 49 New York State waters. Pesticide Monitoring J. 6(2):107–114.

*Pakkala, I.S., M.N. White, D.J. Lisk, G.E. Burdick, and E.J. Harris. 1972. Arsenic content of fish from New York State Waters. N.Y. Fish and Game J. 19(1):12–31.

Panuzio, F.L. 1966. Solid refuse, pp. 312–325 In Hudson River Ecology, 1st Symp. Hudson River Valley Comm.

*Pressman, W. and S.C. Kirschner. 1976. Heavy metal contributions to the lower harbor from sewage outfalls. Paper 7 In Hudson River Ecology, 4th Symp. Hudson River Environ. Soc., Bronx, NY.

Quinn, T. 1979. Toxic substances in New York's environment, an interim report and appendices. Technical Report. N.Y.S. Environmental Conservation and N.Y.S. Dept. of Health, Albany, NY. 48+ pp.

*Rehwoldt, R., G. Bida, and B. Nerrie. 1971. Acute toxicity of copper, nickel and zinc ions to some Hudson River fish species. Bull. Environ. Contam. Toxicol. 6(5):445–448.

Rehwoldt, R., L.W. Menapace, B. Nerrie, and D. Alessandrello. 1972. The effect of increased temperature upon the acute toxicity of some heavy metal ions. Bull. Environ. Contam. Toxicol. 8(2):91–96.

Rehwoldt, R., L. Lasko, C. Shaw, and E. Withowski. 1972. The acute toxicity of some heavy metal ions toward benthic organisms. Bull. Environ. Contam. Toxicol. 50(5):291–294.

Rehwoldt, R., L. Lasko, C. Shaw, and E. Wirhowski. 1974. Toxicity study of two oil spill reagents toward Hudson River fish species. Bull. Environ. Contam. Toxicol. 11(2):159–162.

Rehwoldt, R., E. Kelley, and M. Mahoney. 1977. Investigations into the acute toxicity and some chronic effects of selected herbicides and pesticides on several fresh water fish species. Bull. Environ. Contam. Toxicol. 18(3):361–365.

Rehwoldt, R., W. Mastrianni, E. Kelley, and J. Stall. 1978. Historical and current heavy metal residues in Hudson River fish. Bull. Environ. Contam. Toxicol. 19(3):335–339.

Schofield, C.D., Jr. 1973. The ecological significances of air pollution-induced changes in water quality of dilute-lake districts in the Northeast. Trans. Northeast Fish Wildl. Conf. 1972, pp. 98–112.

Schroeder, H. 1973. The trace elements and man. Devin-Adair, Old Greenwich, CT. 171 pp.

Simpson, H.J., P.M. Bower, S.C. Williams, and Y.H. Li. 1978. Heavy metals in the sediments of Foundry Cove, Cold Spring, NY. Environ. Sci. Technol. 12(6):683–687.

Sloan, R. and R. Karcher. 1984. On the origin of high cadmium concentrations in Hudson River blue crab (*Callinectes sapidus* Rathbun). Northeastern Environ. Sci. 3(3/4):222-232.

Stevens, D.B. 1974. Hudson River pollution control—New York State. Paper 5 In Hudson River Ecology, 3rd Symp. Hudson River Environ. Soc., Bear Mountain, NY.

Stevens, D.B. and W.A. Bruce. 1969. Hudson River oil pollutions, pp. 457–473 *In* Hudson River Ecology, 2nd Symp. N.Y.S. Dept. of Environmental Conservation, Albany, NY.

Stone, W.B., E. Kiviat, and S.A. Butkas. 1980. Toxicants in snapping turtles. N.Y. Fish and Game J. 27(1):39–50.

Thomann, R.V. 1981. Equilibrium model of fate of microcontaminants in diverse aquatic food chains. Can. J. Fish. Aquat. Sci. 38(3):280–296.

*Tofflemire, T.J. and L. Hetling. 1969. Pollution sources and loads in the lower Hudson River, pp. 78–146 *In* Hudson River Ecology, 2nd Symp. N.Y.S. Dept. of Environmental Conservation, Albany, NY.

*Tong, S.C., W.H. Gutenmann, D.J. Lisk, G.E. Burdick, and E.J. Harris. 1972. Trace metals in New York State fish. N.Y. Fish and Game J. 19(2):123–131.

U.S. Army Engineers. 1971. Aquatic plant control program, Hudson and Mohawk Rivers, New York (Final EIS). U.S. Army Corps of Engineers. NTIS PB-200-003-F. 17 pp.

U.S. Army Engineers. 1971. Control of water chestnut with 2,4-D amine (Final EIS). U.S. Army Corps of Engineers, New York District. NTIS. 27 pp.

U.S. Dept. of Health, Education, and Welfare. 1965. Proceedings, Volume 1. Conference in the matter of pollution of the interstate waters of the Hudson River and its tributaries—New York and New Jersey. Sept. 28–30, 1965. U.S. Dept. of Health, Education, and Welfare, New York, NY. 360 pp.

U.S. Environmental Protection Agency. 1975. Preliminary assessment of suspected carcinogens in drinking water: An interim report to Congress. Office of Toxic Substances, EPA, Washington, D.C. 33 pp. + 214 pp. (2 vols.)

Wall Street Journal. 1970. "U.S. says G.M. pollutes Hudson River; company promises quick action." Wall Street Journal, Dec. 16, 1970, p. 12.

*Whipple, J.A. 1982. The impact of estuarine degradation and chronic pollution on populations of anadromous striped bass (*Morone saxatilis*) in the San Francisco Bay-Delta, California. Report to NOAA. Office of Marine Pollution Assessment, Seattle, WA.

Wich, K.F. 1968. Water chestnut eradication program in New York State. Presented at 22nd Annual Northeastern Weed Control Conference, NY. Jan. 1968. 6 pp. (Mimeo.)

Williams, S.C., H.J. Simpson, C.R. Olsen, and R.F. Bopp. 1977. Sources of heavy metals in sediments of the Hudson River. Mar. Chemistry 6:195–213.

*Wills, J.H. 1966. Pollution of the Hudson River by agricultural chemicals, pp. 268–288 *In* Hudson River Ecology, 1st Symp. Hudson River Valley Comm., Tuxedo, NY.

Wolfe, D.A., D.F. Boesch, A. Calabrese, J.J. Lee, C.D. Litchfield, R.J. Livingston, A.D. Michael, J.M. O'Connor, M. Pilson, and L.V. Sick. 1982. Effects of toxic substances on communities and ecosystems, pp. 67–68 *In* Ecological Stress and the New York Bight: Science and Management (G.F. Mayer, ed.). Estuarine Research Federation, Columbia, SC. 715 pp.

Zubarik, L.S., J.M. O'Connor, J.H. Thorpe, and J.W. Gibbons (eds.). 1978. Radioisotope study of mercury uptake by Hudson River biota, pp. 273–289 *In* Ecological Symposia on Energy and Environmental Stress in Aquatic Systems. Augusta, GA. Nov. 2, 1977.

IV. General Ecological Studies and Surveys

Boyce Thompson Institute. 1973. Maintenance of a functional environment in the Hudson River estuary. 2nd Annual Report to the Rockefeller Foundation. (Unpublished.)

Buckley, E.H. 1971. Maintenance of a functional environment in the lower portion of the Hudson River estuary: An appraisal of the estuary. Contrib. Boyce Thompson Inst. 24(14):387-396, or Reprint 2233.

Chamberlain, J.L. 1974. An ecological interpretation: Mohawk River flood plain. Oneida Co. Dept. of Planning, Utica, NY. 62 pp.

Coastal Zone Management Study Program. 1977. Final report on significant coastal related fish and wildlife habitats of New York State (Task 7.3). N.Y.S. Dept. of Environmental Conservation, Division of Land Resources and Forest Management, Coastal Zone Management Study Program, Albany, NY.

Eisenbud, M. and G.P. Howells (eds.). 1968. Ecological survey of the Hudson River: Progress Report No. 3. New York University, Dept. of Environmental Medicine, New York, NY. 261 pp.

Hall, C.A.S. 1977. Models and the decision making process: the Hudson River power plant case, pp. 346-364 *In* Ecosystem Modeling in Theory and Practice: An Introduction with Case Histories (C.A.S. Hall and J.W. Day, Jr., eds.).

Howells, G.P., M. Eisenbud, and T.J. Kneip. 1971. Ecology of the estuary of the lower Hudson River. FAO Fish. Rep. 99:132-133.

Kiviat, E. 1974. A fresh-water tidal marsh on the Hudson, Tivoli North Bay. Paper 14 *In* Hudson River Ecology, 3rd Symp. Hudson River Environ. Soc., Bear Mountain, NY.

Kiviat, E. 1979. Hudson estuary shore zone: Ecology and management. M.S. Thesis. S.U.N.Y. at New Paltz, NY. 159 pp.

Lane, P.A. and J.A. Wright. 1977. Response of the structure and stability of four aquatic communities in the Hudson River to thermal effluents. Energy and Environmental Stress in Aquatic Systems, Nov. 2-4. Savannah River Ecology Laboratory, Savannah, GA.

Lauer, G.H., D. Pimentel, A. MacBeth, B. Salwen, and J. Seddon. 1976. Biological communities. Hudson Basin Project Task Group Report 8. The Rockefeller Foundation and Mid-Hudson Pattern, Poughkeepsie, NY. 155 pp.

McLaughlin, D.B., J.A. Elder, G.T. Orlob, D.F. Kibler, and D.E. Evenson. 1975. A conceptual representation of the New York Bight ecosystem. NOAA, Boulder, CO. NTIS-PB-252 543. 373 pp.

McLaughlin, D.B. and J.A. Elder. 1976. A conceptual representation of the New York Bight ecosystem, pp. 249-259 *In* Middle Atlantic Continental Shelf and The New York Bight. American Society of Limnology and Oceanography, Inc. Special Symposia. Vol. 2 (M.G. Gross, ed.). Allen Press, Inc., Lawrence, KA.

Moore, E. (ed.) 1937. A biological survey of the lower Hudson watershed. Supplement to 26th Annual Report, 1936. N.Y.S. Conservation Dept., Albany, NY. 373 pp.

O'Reilly, J.E., J.P. Thomas, and C. Evens. 1976. Annual primary production in the lower New York bay. Paper 19 *In* Hudson River Ecology, 4th Symp. Hudson River Environ. Soc., Bronx, NY.

Pierce, M. 1971. A comparative study of the aquatic life in the Hudson River at Danskammer Point and Howland. Marist College, Poughkeepsie, NY.

Prescott, G.W. 1966. A synopsis of some results obtained from a biological chemical analysis of the Hudson River MP 115 to MP 11. Michigan State University, East Lansing, MI.

Sirois, D.L. 1974. Community metabolism and water quality in the lower Hudson River estuary. Paper 15 *In* Hudson River Ecology, 3rd Symp. Hudson River Environ. Soc., Bear Mountain, NY.

Storm, P.C. and R.L. Heffner. 1976. A comparison of phytoplankton abundance, chlorophyll "a" and water quality factors in the Hudson River and its tributaries. Paper 17 *In* Hudson River Ecology, 4th Symp. Hudson River Environ. Soc., Bronx, NY.

Weinstein, L.H. (ed.) 1977. An Atlas of Biologic Resources of the Hudson Estuary. Boyce Thompson Institute for Plant Research. Yonkers, NY. 104 pp.

V. Regulations and Impact Statements

Alpine Geophysical Associates, Inc. 1973. Evaluation of environmental impact of the proposed interstate Route 518, West Side Highway from the Battery to 52nd Street, Borough of Manhattan. Submitted to Parsons, Brinckerhoff, Quade, and Douglas, Inc., New York, NY.

Atomic Energy Commission. 1971. Report on allegations, Indian Point No. 1 Plant of Consolidated Edison Co. of NY, Inc. (for periods of Aug. 1962 to June 1976).

Atomic Energy Commission. 1972. Final environmental statement—Indian Point No. 2. U.S. Atomic Energy Commission.

Atomic Energy Commission. 1972. Hudson River, part of commercial inland fisheries. FW-P35. Pamphlet of N.Y.S. Dept. of Environmental Conservation, Albany, NY.

Carlson, F.T. and J.S. McCann. 1968. Hudson river fisheries investigations, 1965–1968. Evaluations of proposed pumped storage project at Cornwall, NY in relation to the fish in the Hudson River. HRPC Field Investigations by N.E. Biol. Inc.

Carlson, F.T. and J.A. McCann. 1969. Evaluation of a proposed pumped storage project at Cornwall, NY in relation to the fish in the Hudson River, Hudson River Fish Investigations, 1965–1968. Hudson River Policy Comm., N.Y.S. Conservation Dept. 50 pp.

Environmental Protection Agency. 1980. Hudson river settlement agreement. Dec. 19, 1980.

Ginter, J.J.C. 1972. New Jersey striped bass fishing laws. Part of New Jersey Fish and Game Laws. Sections 23:5-5.1 to 23:5-5.8, 23:9-5, 23:9-8; 23:9-9, 23:9-16; 23:9-23, 23:9-26; 23:9-30, 23:9-34, 23:9-39, 23:9-40, 23:9-44, and 23:9-50.

Ginter, J.J.C. 1974. Marine fishery management law in New York. SS-403-II, Marine fisheries conservation in New York State: Policy and practice of marine fisheries management. Vol. 2, Appendix 1. 21 pp.

Hudson River Study Group. 1978. Present legal institutional structures for water resources management—Technical Paper No. 2. Prepared by the Cornell Law School, Ithaca, NY for N.Y.S. Dept. of Environmental Conservation, Albany, NY.

Hudson River Study Group. 1978. An assessment of possible social and community impacts of alternative plans for the development of the Hudson River

Basin. Prepared by N.Y.S. College of Human Ecology, Cornell University, for N.Y.S. Dept. of Environmental Conservation, Albany, NY.

Hudson River Study Group. 1979. Hudson River basin water and related land resources. Level B study report and environmental impact statement. N.Y.S. Dept. of Environmental Conservation, Albany, NY.

Hudson River Study Group. 1979. Selected background information. Technical Paper 3, Vol. I and Vol. II. Prepared by the Hudson River Basin, Level B, Water and Related Land Resources Study. N.Y.S. Dept. of Environmental Conservation, Albany, NY.

Krause, C. 1979. Hudson River power plant case: Oak Ridge scientists play a role. Oak Ridge National Laboratory Review 12(1):26–34.

Lawler, Matusky, and Skelly, Engineers. 1974. West Side Highway Project. Technical report on water quality. Part I. Water quality and hydrodynamic effects. Prepared for N.Y. State.

Lawler, Matusky, and Skelly, Engineers. 1980. Report and photographic documentation for the Battery Park City underwater colonization study. Prepared for N.Y.S. Dept. of Transportation and Parsons, Brinckerhoff, Quade, and Douglas, Inc., NY.

Lawler, Matusky, and Skelly, Engineers. 1980. Biological and water quality data collected in the Hudson River near the proposed Westway Project during 1979–1980. Prepared for the N.Y.S. Dept. of Transportation and System Design Concepts, Inc.

Luce, C.F. 1970. Power for tomorrow: The siting dilemma. Environmental Law 1(1):60–71.

Medeiros, W.H. 1974. Legal mechanisms to rehabilitate the Hudson River shad fishery. N.Y.S. Assembly Scientific Staff and N.Y.S. Sea Grant Program, Albany, NY. 65 pp + Appendix.

New York State Dept. of Environmental Conservation. 1981. SPDES discharge permit for facility ID No. NY-000 6262. Central Hudson Gas and Electric Corp., Danskammer Point Generating Station. (Draft.)

New York State Dept. of Environmental Conservation. 1981. SPDES discharge permit for facility ID No. NY 000 0231. Central Hudson Gas and Electric Corp., Roseton Generating Station. (Draft.)

New York State Dept. of Environmental Conservation. 1981. SPDES discharge permit for facility ID No. NY-000 4472. Consolidated Edison Co. of NY, Inc., and PASNY. Indian Point Generating Station Units 2 and 3. (Draft.)

New York State Dept. of Environmental Conservation. 1981. SPDES discharge permit for facility ID No. NY-000 8010. Orange and Rockland Utilities, Inc., Bowline Point Generating Station. (Draft.)

Rockland County Planning Board. 1974. Piermont community development goals and Erie Pier assessment. Piermont Planning Commission, Village Hall. Piermont, NY. 22 pp.

Sandler, R. and D. Schoenbrod. (eds.). 1981. The Hudson River Power Plant Settlement. Materials Prepared for a Conference. New York University School of Law, New York, NY.

Sasaki Associates. 1975. Environmental impact analysis for the proposed state part at Rockwood Hall, Mt. Pleasant, NY. N.Y.S. Office of Parks and Recreation, Taconic State Park, and Recreation Comm.

Schmidt, R.E. and N. Zeizing. 1983. Compliance by Hudson River principal surface dischargers with state pollution discharge elimination system, New York, effluent limits. Northeastern Environ. Sci. 2(1):49–54.

Talbot, A.R. 1983. Settling things. Six case studies in environmental mediation. Conservation Foundation and Ford Foundation, Washington, D.C. 101 pp.

Trubeck, D.M. 1977. Allocating the burden of environmental uncertainty: The NRC interprets NEPA's substantive mandate. Wisconsin Law Review (1977):747–776.

U.S. Army Corps of Engineers. 1971. Aquatic plant control program, Hudson and Mohawk Rivers, New York. (Final environmental impact statement.) U.S. Dept. of the Army, Corps of Engineers, New York District, NY. NTIS-PB-200-003-F. 17 pp.

U.S. Army Corps of Engineers. 1972. Maintenance of the Hudson River channel, New York, navigation project. (Final environmental impact statement.) U.S. Dept. of the Army, Corps of Engineers, New York District, NY. NTIS-EIS-NY-73-007-F. 12 pp.

U.S. Army Corps of Engineers. 1977. Northeastern U.S. water supply study. New York metropolitan area. Hudson River Project draft environmental impact statement. U.S. Army Corps of Engineers, North Atlantic Division.

U.S. Army Engineering District. 1977. Draft environmental statement. Bowline Point Generating Station. Haverstraw, NY.

U.S. Atomic Energy Commission. 1972. Final environmental statement related to operation of Indian Point Nuclear Generating Plant Unit 2. Consolidated Edison Co. of NY, Inc., Docket No. 50-247. Vol. I (Oct. 30, 1972), Vol. II (Sept. 1972).

U.S. Atomic Energy Commission. 1973. Draft environmental statement by the Directorate of Licensing related to operation of Indian Point Nuclear Generating Plant Unit 3. Consolidated Edison Co. of NY, Inc., Docket No. 50-286.

U.S. Dept. of Commerce and State of New York. 1982. Hudson River Estuarine sanctuary, draft environmental impact statement. National Oceanic and Atmospheric Administration and N.Y.S. Dept. of Environmental Conservation, Albany, NY.

U.S. Environmental Protection Agency. 1973. Water quality criteria, 1972. A report of the Committee on Water Quality Criteria prepared by Environmental Studies Board, National Academy of Sciences/National Academy of Engineering, Washington, D.C.

U.S. Environmental Protection Agency. 1974. Development document for effluent limitations guidelines and new source performance standards for the steam electric power generating point source category. PB-240 853. National Technical Information Service, Springfield, VA.

U.S. Environmental Protection Agency. 1974. Methods for chemical analysis of water and wastes. Environmental Monitoring and Support Laboratory, Cincinnati, OH. EPA 625/6-74-003.

U.S. Environmental Protection Agency. 1974. 316(a) Technical guidance thermal discharges. Water Planning Division, Office of Water and Hazardous Materials, Washington, D.C. 187 pp.

U.S. Environmental Protection Agency. 1977. Draft interagency 316(a) technical guidance manual and guide for thermal effects sections of nuclear facilities environmental impact statements, Washington, D.C. 79 pp.

U.S. Environmental Protection Agency. 1980. Hudson River Settlement Agreement. Dec. 19, 1980.
U.S. House of Representatives. 1974. Storm King Mountain (pump storage plant) hearings. Subcommittee on Fisheries, Committee on Merchant Marine and Fisheries. U.S. House of Representatives, 93 long: 2d sess., Feb. 19, 20, 1974. 395 pp.
U.S. Nuclear Regulatory Commission. 1976. Final environmental statement for selection of the preferred closed cycle cooling system at Indian Point Unit 2. Washington, D.C.
U.S. Nuclear Regulatory Commission. 1976. Final environmental statement for facility amendment for extension of operation with once through cooling, Indian Point Unit 2. Washington, D.C.
U.S. Office of Federal Register. 1982. Polychlorinated biphenyls (PCB's); manufacturing, processing, distribution in commerce, and use prohibitions; used in closed and controlled waste manufacturing processes. Federal Register 47(204):46980–46996.

VI. Bibliographies

Atomic Industrial Forum. 1976. Part II. A bibliography of 1974–1976 literature on the Hudson River (south of Troy, NY). Arlington, VA.
Coutant, C.C., S.S. Talmaye, B.N. Collier, R.F. Carrier, and N.S. Dailey. 1976. Thermal effects on aquatic organisms annotated bibliography of the 1975 literature. Oak Ridge National Laboratory, Environmental Sciences Division No. 93. ORNL/EIS-88. 235 pp.
Embree, W.N. and D.A. Wiltshire. 1978. Estuarine research: An annotated bibliography of selected literature, with emphasis on the Hudson River estuary, New York and New Jersey. Geological Survey, Water Resources Division. Open File Report 78-782. Albany, NY. 58 pp. (Also, New York State Dept. of Environmental Conservation Technical Paper No. 5.)
Gordon, R. and M. Spaulding. 1974. A bibliography of numerical models for tidal rivers, estuaries, and coastal waters. University of Rhode Island Mar. Tech. Rep. 32. Kingston, RI. 55 pp.
Horn, E.G. 1978. Annotated bibliography of reports and publications related to PCBs in the Hudson River. N.Y.S. Dept. of Environmental Conservation, Albany, NY. 27 pp.
Horseman, L.O. and R.J. Kernehan. 1976. An indexed bibliography of the striped bass, *Morone saxatilis*, 1970–1976. Ichthyological Associates, Inc. Bull. No. 13. 118 pp.
Horvath, R.F., G.E. Carroll, S.A. Covell, A.J. Evjen, and J. Schoof. 1984. Annotated bibliography of New York Bight, Hudson-Raritan estuarine system and contiguous coastal waters: 1973–1981. (2 vols.)
Hudson River Basin Study Group. 1978. Estuarine research: An annotated bibliography of selected literature with emphasis on the Hudson River estuary, New York and New Jersey. Prepared by W.N. Embree and D.A. Wiltshire. Technical Paper No. 5. N.Y.S. Dept. of Environmental Conservation, Albany, NY.
Kiviat, E. 1981. Hudson River estuary shore zone. Annotated natural history bibliography. Scenic Hudson, Poughkeepsie, NY. 76 pp.

Marine Sciences Research Center. 1973. A keyword-indexed bibliography of the marine environment in the New York Bight and adjacent estuaries. S.U.N.Y at Stony Brook, NY. 271 pp.

Massman, W.H. 1967. A bibliography of the striped bass. Sport Fish. Inst. Bull. 25 pp.

Mayer, R., J. Jefferin, E. Enjor, and R.S. Davidson. 1974. Bioenvironmental effects associated with nuclear power plants: A selected bibliography (first addendum). Battelle Memorial Institute, Columbus Laboratories, Columbus, OH. BRI-E-654.

Middlebrooks, E.J., R.J. Gaspar, R.D. Gaspar, J.H. Reynolds, and D.B. Porcella. 1973. Effects of temperature on the toxicity to the aquatic biota of wastewater discharges: A compilation of the literature. Utah State University, Logan, UT. PRWG-105-1.

NOAA. 1974. Bibliography of the New York Bight. Part 1—List of citations. National Oceanographic and Atmospheric Administration. Rockville, MD. NTIS-COM-74-50357. 194 pp.

NOAA. 1974. Bibliography of the New York Bight. Part 2—Indexes. National Oceanographic and Atmospheric Administration. Rockville, MD. NTIS-COMM-74-50357. 495 pp.

The Oceanic Society. 1980. Hudson River fishery management program study. A Report to the N.Y.S. Dept. of Environmental Conservation, Albany, NY.

Pearce, J.B., E. Waldhauer, and M. Trafford. 1971. Bibliography to studies in the New York Bight: Including references to effects of sewage sludge. National Oceanographic and Atmospheric Administration, National Marine Fisheries Service, Highlands, NJ. 51 pp.

Pfuderer, H.A., S.S. Talmage, B.N. Collier, W. Van Winkle, Jr., and C.P. Goodyear. 1975. Striped Bass—a selected annotated bibliography. Oak Ridge National Laboratory, Oak Ridge, TN. ORNL-EIS-75-73. 158 pp.

Raney, E.C., B.W. Menzel, and E.C. Weller. 1972. Heated effluents and effects on aquatic life with emphasis on fisheries—a bibliography. Ichthyological Associates Bulletin No. 9. U.S. Atomic Energy Commission. Office of Information Services, Technical Information Office TID-3918.

Rogers, B.A. and D.T. Westin. 1975. A bibliography on the biology of the striped bass, *Morone saxatilis* (Walbaum). Mar. Tech. Rept. No. 37, University of Rhode Island. 134 pp.

U.S. Atomic Energy Commission. 1973. Toxicity of power plant chemicals to aquatic life (C.D. Becker and T.O. Thatcher, eds.). Battelle Pacific Northwest Laboratories, Washington, D.C. WASH-1249-UC-11.

U.S. Dept. of Commerce. 1976. Environmental impacts of electric power generating plants on the Hudson River. Computerized literature research conducted by the National Technical Information Service, Springfield, VA.

VII. General Papers

Anderson, R.E. 1976. The site selection and evaluation process in water-oriented recreation planning on the Hudson River of Albany and Rensselaer Counties, New York. Masters Thesis. S.U.N.Y. at Albany, NY. 155 pp.

Atomic Industrial Forum. 1976. Part I: Identification of power plants on the Hudson River in New York State. Arlington, VA.

Barnthouse, L.W., J. Boreman, S.W. Christensen, C.P. Goodyear, W. Van Winkle, and D.S. Vaughan. 1984. Population biology in the courtroom: The lesson of the Hudson River controversy. Bioscience 34(1):14–19.

Birnie, R.W. and E.S. Posmentier. 1980. Identification of lateral spectral contrasts in the lower Hudson River estuary using Landsat digital data, pp. 1663–1672 *In* Proceedings of the Fourteenth International Symposium on Remote Sensing of Environment, Apr. 23–30. San Jose, Costa Rica. Environmental Research Institute of Michigan, Ann Arbor, MI.

Bowers, B.T., G.P. Larson, A. Michaels, and W.M. Phillips. 1968. Waste management. A report of the Second Regional Plan. Regional Plan Assoc. Bull. 107. 107 pp.

Boyce Thompson Institute of Plant Research. 1974. Biological studies in the Hudson River.

Boyle, R.H. 1969. The Hudson River. A Natural and Unnatural History. W.W. Norton and Co., Inc. 304 pp.

Boyle, R.H. 1971. Hudson River lives. Audubon 73(2):14–58.

Burmeister, W.F. 1974. Appalachian Waters 2: The Hudson River and Its Tributaries. Appalachian Books, Oakland, VA. 488 pp.

Carmer, C.L. 1939. The Hudson. Farrar and Rinehart, Inc., New York. 434 pp.

Christensen, S.W., W. Van Winkle, L.W. Barnthouse, and D.S. Vaughan. 1981. Science and the law: Confluence and conflict on the Hudson River. Environ. Impact Assessment Review 2(1):63–88.

Clark, J. and W. Brownell. 1973. Electric power plants in the coastal zone: Environmental issues. Am. Littoral Soc. Special Publ. No. 7.

Cronin, L.E. 1964. The role of man in estuarine processes, *In* Estuaries (G.H. Lauff, ed.). Am. Assoc. for Adv. Sci., Washington, D.C. Publ. No. 83.

Environmental Action Council of Manhattan College and the College of Mount Saint Vincent. 1973. A remaining resource: A survey of the Bronx shoreline of the Hudson River. Manhattan College and College of Mount Saint Vincent, Riverdale, NY. 29 pp.

Funk, R.E. 1976. Recent contributions to Hudson Valley prehistory. State Education Dept., Memoir. N.Y.S. Museum, Albany, NY. 325 pp.

Goldstein, M. 1968. Study of the impact of nuclear power plants on the Hudson River and adjacent lands. Report to the N.Y.S. Hudson River Valley Comm.

Gould, J.D. 1969. Conservation interests in the Hudson River, pp. 30–39 *In* Hudson River Ecology, 2nd Symp. N.Y.S. Dept. of Environmental Conservation, Albany, NY.

Hall, A.J. 1978. The Hudson: "That River's Alive." National Geographic 153(1):62–68.

Hope, J. 1975. A River for the Living: The Hudson and Its People. Barre Publishers, Barre, MA. 224 pp.

Howells, G.P. 1972. The estuary of the Hudson River, USA Proc. Royal Soc. London, B. 180:521–534.

Hudson River Basin Study Group. 1978. Hudson River basin level B water and related land resources study: Guidelines for identifying and evaluating scenic resources. N.Y.S. Dept. of Environmental Conservation, Technical Paper No. 4. Albany, NY. 106 pp.

Hudson River Valley Commission. 1966. Industrial trends in the Hudson Valley.

Prepared by Arthur D. Little, Inc. for N.Y.S. Hudson River Valley Comm. Bear Mountain, NY.

Hudson River Valley Commission. 1966. The Hudson River as a resource. N.Y.S. Dept. of Environmental Conservation, Division of Water Resources. 75 pp. + Appendix.

Hudson River Valley Commission. 1966. The Hudson: Population. N.Y.S. Office of Regional Development.

Hudson River Valley Commission. 1966. The Hudson: Riverfront commerce. N.Y.S. Office of Regional Development.

Ketchum, B.H. 1974. Population, resources and pollution and their impact on the Hudson estuary, In Hudson River Colloquium (O.A. Roels, ed.). Ann. N.Y. Acad. Sci. 250:144–156.

Martin, R.M. 1973. Hudson River, pp. 149–174 In Our Environment: The Outlook for 1980. Part 1 (A.J. Van Tassel, ed.). Lexington Books, MA.

Mitchell, J.G. 1972. The restoration of a river. Sat. Rev. 55:35–37.

New York (City) Dept. of City Planning. 1974. The New York City waterfront: Comprehensive planning workshop B. Probst. N.Y.C. Planning Commission. 159 pp.

New York State Dept. of Environmental Conservation. 1976. Hudson River basin water and related land resources study. Amended Plan of Study. 52 pp.

New York State Museum and Science Service. 1966. The Hudson: Biological resources. A report on areas of biological significance in the Hudson River valley. Prepared by S.J. Smith, J.A. Wilcox, and E.M. Reilly for N.Y.S. Hudson River Valley Commission. Iona Island. Bear Mountain, NY. 43 pp.

New York State Office of Parks and Recreation. 1978. Hudson River basin recreation study, 1978.

Raymond, Parish, Pine, and Weiner, Inc. and A. Gussow. 1979. Hudson River valley study. N.Y.S. Dept. of Environmental Conservation, New Paltz, NY. 180 pp.

Reetz, G.R. 1975. Water resources development and wilderness values: A study of the upper Hudson River. Sciences Center. Cornell University, Ithaca, NY. NTIS-PB-243 736.

Richardson, R.W., Jr. and G. Tauber (eds.). 1979. The Hudson River Basin: Environmental Problems and Institutional Response. Academic Press, New York, (2 vols.)

Squires, D.F. 1981. The bight of the big apple. The New York Sea Grant Institute of the State University of New York and Cornell University.

U.S. Dept. of Interior, Bureau of Outdoor Recreation. 1966. Focus on the Hudson. USDI, BOR. Government Printing Office, Washington, D.C. 50 pp.

VIII. Non-Hudson

Austin, H.M. 1976. The need for standardization in pre- and post-operational ecological surveys. N.Y. Fish and Game J. 23(2):180–182.

Brook, A.J. and A.L. Baker. 1972. Clorination at power plants: Impact on phytoplankton productivity. Science 176:1414–1415.

California State Water Resources Control Board. 1982. The striped bass decline in the San Francisco Bay-Delta estuary. Striped Bass Working Group. 58 pp.

Clark, J.R. 1977. Coastal Ecosystem Management. John Wiley & Sons, New York, NY.
Dovel, W.L., J.A. Mihursky, and A.J. McErlean. 1969. Life history aspects of the hogchoker, *Trinectes maculatus*, in the Patuxent River estuary, Maryland. Chesapeake Science 10(2):104–119.
Hart, C.W. and S.L.H. Fuller (eds.). 1979. Pollution Ecology of Estuarine Invertebrates. Academic Press, New York, NY.
Heinle, D.R., D.A. Flemer, and J.F. Ustach. 1977. Contributions of tidal marshlands to mid-Atlantic estuarine food chains, *In* Estuarine Processes, Vol. II. Academic Press, New York, NY.
Hughes, J.S. 1973. Acute toxicity of thirty chemicals to striped bass (*Morone saxatilis*). Western Assoc. State Game and Fish Comm., Salt Lake City, UT. 15 pp.
Mauck, W.L., P.M. Mehrle, and F.L. Mayer. 1978. Effects of polychlorinated biphenyl Aroclor 1254 on growth, survival, and bone development in brook trout (*Salvelinus fontinalis*). J. Fish. Res. Bd. Canada. 35:1084–1088.
Meyerhoff. 1975. Acute toxicity of benzene, a component of crude oil, to juvenile striped bass (*Morone saxatilis*). J. Fish. Res. Bd. Canada. 32:1864–1866.
Mihursky, J.A., W.R. Boynton, E.M. Setzler-Hamilton, K.V. Wood, and T.T. Polgar. 1981. Freshwater influences on striped bass population dynamics, pp. 149–167 *In* Proceedings of the National Symposium on Freshwater Inflow to Estuaries (R. Cross and D. Williams, eds.). U.S. Fish and Wildlife Service. Office of Biological Services. FWS/OBS-81/04. (2 vols.)
Mrozek, E., Jr., E.D. Seneca, and L.L. Hobbs. 1982. Polychlorinated biphenyl uptake and translation by *Spartina alterniflora* Loisel. Water, Air and Soil Pollution 17(1):3–15.
Sanders, H.D. and J.H. Chandler. 1972. Biological magnification of a polychlorinated biphenyl (Aroclor 1254) from water by aquatic invertebrates. Bull. Environ. Contam. Toxicol. 7(5):257–263.
Ulanowicz, R.E. and T.T. Polgar. 1980. Influences of anadromous spawning behavior and optimal environmental conditions upon striped bass (*Morone saxatilis*) year-class success. Can. J. Fish. Aquatic Sci. 37(2):143–154.
Van Winkle, W. (ed.) 1977. Proceedings of the Conference on Assessing the Effects of Power Plant Induced Mortality on Fish Populations. Pergamon Press, New York, NY. 380 pp.

Index

Adaptive environmental assessment and management, 168–169
Alewives, 175–177
Algae, 22
   Hudson River open literature highlights on, 204–206
Atlantic tomcod,
   effects of PCBs on, 101–102
   far field data on, 178–179
   literature on, 211
   near field data on, 175–177

Benthic invertebrates, 23–25
   at Bowline-Lovett, 190–191
   at Roseton-Danskammer, 196
Blueback herring, 175–177

Central Hudson Gas and Electric Corp., 46, 173
Clean Water Act, 3–4
Consolidated Edison of New York, Co., 44, 46, 173
Cooling systems
   closed-cycle, 43–44
   once-through, 45–48
Cumulative impact assessment, 169–170

Density-dependent compensation arguments, 70, 71
Dissolved oxygen, 13

Ecological Analysists, Inc., 51, 58, 174
Ecosystem processes, 162–165
Ecosystem approach, 2
Energy budget, 163, 164
Environmental Impact Assessment (EIA), problems with, 4
Environmental Impact Statement (EIS), 4
   of estuarine ecosystems, 161–162
   of PCB Reclamation Project, 92, 118
   of Westway Project, 135–138
Environmental policy acts, national, 3–4
Environmental policy development, early, 2–3
Estuarine ecosystems
   biological models of, 166
   community metabolic studies of, 163
   EIS assessments of, 161–162
   energy or materials budget models of, 164–166
   environmental assessment of, 160–168
   institutional changes for environmental management of, 167–168

Environmental Impact Statement (EIS) (*cont.*)
   mathematical models of, 166
   population level vs community ecosystem level approach to, 160–161
   Representative Important Species study of, 163–164
   research needs and useful approaches to, 162–167
   systems approach to, 161
   trophic analyses in, 163

Federal Water Pollution Control Act, 3–4
Fish, 25–27
   far-field data sets on, 178–179
   Hudson River open literature highlights on, 220–222
   near-field data sets on, 175–177
   PCBs and, 100–102

Geology and hydrology, Hudson River open literature highlights on, 223–225

Hudson River ecosystem, 6–39
   adequacy of data collection on, 156–157
   algae in, 22
   benthic invertebrates in, 23–25
   computerization of data base on, 159
   depth of, 8
   dissolved oxygen in, 13
   estuary of, 6, 7
   expanding data for impact estimation on, 157–158
   fish in, 25–27
   flushing time of, 9–10
   food web of, 20–27
   human influences on, 30–34
   macrophytes in, 22
   major pollutants of, 14–16
   microzooplankton in, 23
   morphometric and hydrologic characteristics of, 6–10
   nitrogen in, 17–20
   nutrients of, 16–20
   organic carbon inputs in, 27–30
   pH, alkalinity, hardness of, 13–14
   phosphorus in, 16–17
   physical and chemical aspects of, 6–16
   phytoplankton in, 20–22
   regulatory clout of EIA work on, 158–159
   review of EIA work on, 158
   salinity of, 10–12
   temperature of, 12–13
   tidal influence on, 8
   zooplankton in, 23
Hudson River Fishermen's Assoc. (HRFA), 45–46
Hudson River Foundation, 168
Hudson River open literature highlights, 204–238
   algae and macrophytes, 204–206
   fish, 220–222
   geology and hydrology, 223–225
   invertebrates, 207–209
   PCBs, 234–237
   shad, 216–218
   striped bass, 210–212
   sturgeon, 217–219
   temperature and salinity, 226–227
   tomcod, 213
   toxics, radionuclides, metals, 232–235
   water quality, 228–231
Hydroscience, 108–111

Indian Point Unit 2
   once-through cooling systems and, 45–46
   phase 1 impact assessment of, 68–69
Invertebrates, Hudson River open literature highlights on, 207–209

Lawler, Matusky, and Skelly Engineers, Inc. (LMS)
   and PCBs, 93, 105, 107–108
   and power plants, 58, 66
   and Westway, 137, 142–144

Macrophytes, 22
   Hudson River open literature highlights on, 204–206

Macrozooplankton, at Bowline-Lovett, 188
Malcolm Pirnie, Inc. (MPI), 93, 94, 144, 147
Mathematical models,
  and environmental impact assessment, 165
  limits to predictability, 165
  PCB impacts, prediction of, 104–113
  power plant impacts, prediction of, 66–74
  strengths and weaknesses of, 165
Metals, Hudson River open literature highlights on, 232–234
Microzooplankton, 23
  at Bowline-Lovett, 187
  at Roseton-Danskammer, 194–195

National Environmental Policy Act (NEPA), 4
National Pollutant Discharge Elimination System, 46–47
Nuclear Regulatory Comm. (NRC), 56, 58
NYS Dept. of Environmental Conservation (DEC), 15
  and PCBs, 95, 103, 105, 114
  and power plants, 45
NYS Dept. of Health (DOH), 92, 93

Oak Ridge National Lab., 56, 58, 66, 69
Orange and Rockland Utilities Corp., 46, 73
Organic carbon inputs, 27–30
Oxygen, *see* Dissolved oxygen

PCBs, 83–130
  in air and landfill, 99–100
  in biota, 100–104
  consensus issues on, 122–123
  description and usage history of, 84–85
  discharges into Hudson of, 86
  discovery of and hearings on GE's responsibility for, 86–90
  distribution and toxicity of, 85–86

downriver vs upriver reactions to, 120–121
ecosystem fate model of, 108–110
federal involvement in cleanup of, 117–120
fish and, 100–102
Hudson River open literature highlights on, 236–239
integration: role of models in assessment of, 104–105
lower trophic relationships and, 102–104
physical sediment and transport model of, 105–108
proposed dredging project for, 114–117
remedial actions concerning, 113–120
scientific assessment of Hudson River problems with, 91–94
in sediments, 94–97
settlement agreement for GE on, 90–91
strengthening scientific studies on, 121
successes and failures of models for, 110–113
terrestrial organisms and, 104
in water column, 97–99
Phytoplankton, 20–22
  at Bowline-Lovett, 184–185
  at Roseton-Danskammer, 192–193
Power plant operation, 40–82
  basic ecosystems studies and, 48–51
  closed-cycle cooling in, 43–44
  descriptions of, 40–44
  entrainment and impingement studies and, 57–64
  Indian Point Unit Two, 45–46
  litigation and hearings pertaining to, 44–46
  locations of, 40–41
  once-through cooling systems and, 45–48
  organism loss due to, 41
  Phase 1 models for impact assessment on, 66–71
  Phase 2 models for impact assessment on, 71–73
  Phase 3 models for impact assessment on, 73–74

Power plant operation (*cont.*)
  Scenic Hudson Preservation Conference v. Federal Power Commission and, 45
  scientific research and, 48
  Storm King pumped storage project, 44–45
  water for cooling of, 41–44

Quirk, Lawler, Matusky, Inc., 66, 136, 137, 174

Radionuclides, Hudson River open literature highlights on, 232–235
Representative Important Species study, 163–164
Rivers and Harbors Act, 3

Salinity, Hudson River open literature highlights on, 226–227
Scenic Hudson Preservation Conference v. Federal Power Commission, 45
Shad, Hudson River open literature highlights on, 212–214
Sierra Club, 138–141
Storm King pumped storage project, 44–45
  phase 1 impact assessment of, 66–69
Striped bass, 27
  basic ecosystems studies on, 48–51
  ecology and natural history of, 51–57
  entrainment and impingement of, 57–64
  Hudson River open literature highlights on, 210–212
  mortality rates of larvae and young-of-year juveniles of, 52–55
  older, seasonal movements of, 55–56
  PCBs and, 101–102
  probable origins of, 56–57
  temperature effects on, 54–55
Sturgeon, Hudson River open literature highlights on, 217–219

Temperature, Hudson River open literature highlights on, 224–225

Texas Instruments, 52, 58, 71, 74, 174
Tomcod, Hudson River open literature on, 211
Toxics, Hudson River open literature highlights on, 230–233

U.S. Army Corps of Engineers, 8, 30, 135, 138–141, 148
U.S. Geological Survey, 92, 97
Utility-sponsored research, 173–199
  fisheries data sets, 174–180
    far-field, 178–179
    near-field, 175–177
  non-fisheries data sets, 180–197
    benthic sampling at Roseton-Danskammer, 196
    benthic sampling at Bowline-Lovett, 190–191
    Bowline-Lovett entrainment studies, 186
    Bowline-Lovett near-field studies, 183–186
    Indian Point ecological studies, 183
    Indian Point near-field studies, 181–183
    Kingston-Ossining aquatic studies, 189
    microzooplankton at Bowline-Lovett, 187–188
    microzooplankton at Roseton-Danskammer, 194–195
    phytoplankton at Bowline Point and Lovett, 184–185
    phytoplankton at Roseton-Danskammer, 192–193
    Roseton-Danskammer habitat selection, 189
    Roseton-Danskammer near-field studies, 186–189
  water quality data sets, 197–199

Water quality, Hudson River open literature highlights on, 226–229
Weston Environmental Consultants, 100
Westway, 131–154
  air pollution issues over, 139
  final phases of, 148–153
  fisheries issues over, 139–141

history of, 132–135
impact assessment process for, 135–138
  estuarine habitat in, 136
  fish populations and, 137–138
legal controversies over, 139–141
scientific controversies over, 141–148
  estimating and interpreting uncertainty and, 146–148
  objectivity and, 142–145
  sampling technique and, 141–142
  tokenism and, 145–146
White perch
  entrainment and impingement studies on, 61–64
  year-class strength reduction of, 64

Zooplankton, 23

Springer Series on Environmental Management
Robert S. DeSanto, Series Editor

**Gradient Modeling:
Resource and Fire Management**
Stephen R. Kessell

1979/432 pp./175 illus./27 tables/cloth
ISBN 0-387-90379-8

**Disaster Planning:
The Preservation of Life and Property**
Harold D. Foster

1980/275 pp./48 illus./cloth
ISBN 0-387-90498-0

**Air Pollution and Forests:
Interactions between Air Contaminants
and Forest Ecosystems**
William H. Smith

1981/379 pp./60 illus./cloth
ISBN 0-387-90501-4

**Environmental Effects
of Off-Road Vehicles:
Impacts and Management
in Arid Regions**
*R.H. Webb
H.G. Wilshire (Editors)*

1983/560 pp./149 illus./cloth
ISBN 0-387-90737-8

**Natural Hazard Risk Assessment
and Public Policy:
Anticipating the Unexpected**
*William J. Petak
Arthur A. Atkisson*

1982/489 pp./89 illus./cloth
ISBN 0-387-90645-2

**Global Fisheries:
Perspectives for the '80s**
B.J. Rothschild (Editor)

1983/approx. 224 pp./11 illus./cloth
ISBN 0-387-90772-6

**Heavy Metals in Natural Waters:
Applied Monitoring and Impact
Assessment**
*James W. Moore
S. Ramamoorthy*

1984/256 pp./48 illus./cloth
ISBN 0-387-90885-4

**Landscape Ecology:
Theory and Applications**
*Zev Naveh
Arthur S. Lieberman*

1984/376 pp./78 illus./cloth
ISBN 0-387-90849-8

**Organic Chemicals in Natural Waters:
Applied Monitoring and Impact
Assessment**
*James W. Moore
S. Ramamoorthy*

1984/282 pp./81 illus./cloth
ISBN 0-387-96034-1

The Hudson River Ecosystem
*Karin E. Limburg
Mary Ann Moran
William H. McDowell*

1986/344 pp./44 illus./cloth
ISBN 0-387-96220-4